## DATE DUE

| FE 10 00 | | | |
|---|---|---|---|
| DE 2 1 0 | | | |
| | | | |
| | | | |
| | | | |
| | | | |
| | | | |
| | | | |
| | | | |
| | | | |
| | | | |
| | | | |
| | | | |
| | | | |
| | | | |
| | | | |
| | | | |

DEMCO 38-296

# TO LIGHT SUCH A CANDLE

# TO LIGHT SUCH A CANDLE

## *Chapters in the History of Science and Technology*

by

## KEITH J. LAIDLER

*Professor Emeritus of Chemistry, The University of Ottawa*

Oxford    New York    Melbourne

Oxford University Press

1998

*t Clarendon Street, Oxford OX2 6DP*

*d   New York*

*k   Bogota   Bombay   Buenos Aires*

*Salaam   Delhi   Florence   Hong Kong*

*mpur   Madras   Madrid   Melbourne*

*Mexico City   Nairobi   Paris   Singapore   Taipei   Tokyo   Toronto   Warsaw*

*and associated companies in*
*Berlin   Ibadan*

*Oxford is a trade mark of Oxford University Press*

*Published in the United States*
*by Oxford University Press Inc., New York*

*© Keith Laidler, 1998*

*A catalogue record for this book is available from the British Library*

*Library of Congress Cataloguing in Publication Data*
*Laidler, Keith James, 1916–*
*To light such a candle : chapters in the history of science and*
*technology / by Keith J. Laidler.*
*Includes bibliographical references and index.*
*1. Science–History.   2. Technology–History.   3. Discoveries in*
*science–History.   4. Scientists–Biography.   5. Engineers–*
*Biography.   I. Title.*
*Q125.L25   1997      609–dc21      97–24690*
*ISBN 0 19 850056 4*

*Typeset by EXPO Holdings, Malaysia*

*Printed in Great Britain by*
*Bookcraft (Berth) Ltd*
*Midsomer Norton, Avon*

'Be of good comfort, Master Ridley, and play the man.
  We shall this day light such a candle…as shall never be put out.'
  Words spoken by Bishop Hugh Latimer to Bishop Nicholas
  Ridley on the day of their martyrdom, 16th October, 1555.


'Science is the soul of prosperity of nations, and the living source
of all progress.'
'There is no such thing as applied science, only applications of
science.'
  Louis Pasteur, address at Lyon on 11 September 1872,
  reported in *Comptes Rendus*, September, 1872.

# *Preface*

Most scientists feel that there is a need for better communication between themselves and the general public. Politicians make many decisions relating to science and technology, but scientists are often convinced that the wrong decisions were made because of ignorance of the true issues. Popular movements frequently take up causes without a correct assessment of the science involved—or so it appears to many scientists. How can these misunderstandings be avoided?

In this book I have tried to deal with what I think may be the heart of the difficulty, the confusion that exists in people's minds between science and technology. The relationship between the two is a complex and changing one. I have taken seven themes in science and technology, and have tried to explain in straightforward language how they have developed. The seven themes, in Chapters 2 to 8, can be briefly summarized as: steam engines, photography, electric power, radio transmission, electronics, large molecules, and nuclear power. All of these themes belong to what is called hard science, which is the kind of science that can be formulated mathematically, and can be tested by experiment.

The first two themes are different from the others, in that technology, not based on science at all, came first, and led to great advances both in science and technology. In the other five, which came later, pure science was followed by technology that would have been impossible without the science.

What is striking when we compare these seven themes is that there is a great diversity in the way they developed. There is obviously no simple relationship between science and technology: a brief account of the matter is sure to get it wrong. One lesson we learn is that when pure science is being done, and sometimes long after it is done, it is usually impossible to predict what the consequences will be. This shows how important it is to support pure science even if it has no obvious consequences. The important criterion should be the quality of the work, not its possible practical applications.

It may be useful to summarize here the conclusions that seem to arise inevitably from the discussions in the present book:

1.  Pure research should be judged entirely on the basis of its quality, and not in terms of possible practical applications.

2.  Technology and engineering must be based on pure science; the time for empirical invention is long past.

3.  Decisions about science and technology must be based on a careful consideration of all the factors involved.

Although this book is primarily intended to be a 'popular' exposition of some aspects of science and technology, I have made every effort to be accurate, both historically and scientifically. I have kept in mind Lord Macaulay's admonition about the writing of history: 'He who is deficient in the art of selection may, by showing nothing but the truth, produce all the effects of the greatest falsehood'. I have done much selecting, but hope that the result is a fair picture of the way scientists and technicians go about their work in the hard sciences, and of what they have achieved.

Above all, I have tried to write as clearly as possible. I am puzzled, and also a little concerned, by the fact that there are some books about science, which I think are written very obscurely and do not give a good account of the subject, which have nevertheless sold well. Could it be that the public thinks that an opaquely written book must be a good one? Do members of the public say to themselves 'I can't understand a word of this book; it must be a good one, and the author very clever'? I remember in my early days of teaching, being told by a student that the students in my class understood my lectures very well. Then she spoilt everything by adding, 'But none of us can understand Professor X at all; but then, *he* is very brilliant'. The truth is, I am afraid, that there is no inverse correlation between obscurity and brilliance. Being convinced of this I have followed a precept for writing that was stated by Sir Peter Medawar, and which I slightly modify as follows:

'Correctness, cogency, and clarity, these three; but the greatest of these is clarity.'

The title of this book is based on the idea that over the past centuries a few gifted individuals, some of them scientists and some technologists, have been led to 'light such a candle' that the material lives of all of us have been transformed.

*Ottawa*                                                                   Keith J Laidler
August 1997

# Acknowledgements

I am greatly indebted to many people who have helped enormously while I have been writing this book. Over the years Dr John Shorter, Reader Emeritus at the University of Hull, and Mr A. V. Simcock of Oxford's Museum for the History of Science, have provided me with many ideas and much valuable information. Their constructive and critical comments on what I have written have contributed greatly to its accuracy and clarity. Correspondence with Professor Brian Gowenlock, formerly of the Heriot–Watt University in Edinburgh, has also been extremely helpful. The same is true of the many letters I have received from Professor Charles Tanford, formerly of Duke University.

My son Jim Laidler has read the whole of the book with great care, and has kept me up to date on a number of technical matters, particularly involving computers, of which I had been innocent. He has also helped greatly with correcting the proofs.

Several of my scientific colleagues have read much of the book, and have made helpful comments. I am particularly grateful to Drs John Holmes and Tony Durst of the University of Ottawa, and to Dr John Morton, formerly of the National Research Council of Canada. I began this book almost completely ignorant of molecular biology and the details of the theory of evolution, and had to seek expert advice. Dr Walter Gratzer, of the Randall Institute of King's College, London, has been most generous with his time in suggesting books that I should read, and in commenting on what I have written on that subject in Chapter 7. Correspondence with Dr Jack Morrell has also greatly helped with that chapter.

Since this book is primarily intended for readers who are not scientists, I needed the help of friends who would be willing to judge the book from that point of view, and give me appropriate advice. I am particularly grateful to Heather Hoy, and to Jim, Elizabeth, and Alexander Reicker. Their suggestions led to a number of changes in the early versions of the book.

# *Figures*

For the photograph of the Newcomen engine that belonged to James Watt (Fig. 2.5) I am indebted to the Hunterian museum of the University of Glasgow, and especially to Dr Lawrence Keppie.

The Kirk Session of the Parish Church of Galston, Ayrshire, Scotland, was most helpful in sending me photographs of the Revd Dr Stirling, including the one reproduced as Figure 2.9. I thank in particular Mr Robert Murray, Session Clerk.

For the caricatures of Count Rumford (Fig. 2.13) and W. R. Grove (Fig. 2.19) I am indebted to the Royal Institution, particularly Mrs I. M. McCabe.

I thank Professor Michael Kasha for sending me the photograph he took of G. N. Lewis (Fig. 2.28) and for allowing me to reproduce it.

For the portrait of Sir Oliver Lodge (Fig. 5.13) I am grateful to Mr Adrian Allan, Archivist, and Dr David Edwards, University of Liverpool, and to Sir Oliver's grandson Mr Oliver Lodge.

In a few cases (e.g., Figs 4.20, 5.15) copies have been made of original diagrams, for clarification and sometimes simplification. Thanks are due to Eva Szabo of the University of Ottawa for making the copies, in which she carefully preserved the style of the original diagrams.

Every effort has been made to get in touch with persons and organizations who might hold the copyright on diagrams or portraits used. If notified the author and publisher will be glad to rectify any omissions.

# Contents

# ONE

# Science and technology

Suppose we ask a well-informed non-scientist to name the most important scientific advances of the last couple of centuries. The reply is apt to include such topics as the invention of radio, colour television, the invention of the laser, and the computer. A scientist asked the same question will probably mention Maxwell's theory of electromagnetic radiation, the quantum theory, its extension into quantum mechanics, and the theory of relativity. Why is there such a remarkable difference between the two sets of answers?

A scientist might say that the non-scientist has failed to distinguish correctly between science and technology, and has listed advances in technology rather than advances in science. Evidently there is a serious discrepancy between the public's and the scientist's idea of science. In particular, the distinction between science and technology is understood differently by the public and the scientist. In this book I try to illustrate, by means of many examples, the relationship between science and technology.

If we try to discuss briefly the distinction between science and technology, we are almost sure to go wrong. It helps a little if we say that science is done with the main object of discovering the truth, with little regard for practical applications, whereas with technology it is the applications that determine the way the work is carried out. However, we must recognize that science done with the 'purest' motives—with no regard at all to applications—often leads to practical applications of the greatest importance. As we shall see in Chapter 4, Faraday's pure research on electricity opened the door to the vast modern electrical industry. James Clerk Maxwell's theory of electromagnetic radiation (Chapter 5) led directly to the discovery of radio transmission, something of which Maxwell had no conception. We will meet many more examples in this book.

We will also see that empirical technology, not based on science at all, may lead directly to scientific work of fundamental importance. The early steam engines were by no means science-based, but they led directly to the science of thermodynamics, one of the most fundamental branches of pure science. This example at once refutes the suggestion, sometimes made, that the pure science comes first, and that the applied scientist or engineer then applies the pure science to practical ends. That is certainly the case in recent years; it is unlikely that any important practical innovation could today be made by anyone not thoroughly versed in the underlying pure science.

## EMPIRICAL INVENTION

In the past, of course, many innovations of the greatest importance were made without any help from science, and we call them *empirical* inventions. The wheel, the canoe, the spear, and the spade were invented in prehistoric times. In the early Middle Ages, a number of empirical inventions were made which were ultimately to have an important impact on science and society. The weight-driven clock, with a pendulum and controlled by an escapement, is regarded as one of the most ingenious inventions ever made; we do not know who the inventor was, but such clocks were installed in the late thirteenth century. When this invention was made, nothing was known of the science that lay behind it. To understand the action of the escapement in a weight-driven clock we must know something of gravitational acceleration and of momentum and inertia; until the work of Galileo and Newton in the sixteenth and seventeenth centuries these concepts had not been thought of.

The blast furnace was invented in the twelfth century. The printing press was invented by Johan Gutenberg (*c*.1400–1468) in the fifteenth century, his Bible being produced in 1455. The watch, and the first clock that did not have a pendulum and could therefore be used at sea, were invented in the seventeenth century. The spinning jenny and the steam engine were both invented in the eighteenth century.

Two of the most recent purely empirical inventions are photography in the early nineteenth century, and the zipper (zip fastener) in the late nineteenth century. These inventions owed almost nothing to science. Today it seems unlikely that there will be any more purely empirical inventions.

Considerable attention to inventions was paid by the statesman and philosopher Francis Bacon (1561–1626)—the same man who did not write the plays of Shakespeare. Throughout his career as a lawyer and statesman he gave much thought to philosophy and science, and his ideas were set out in his *Novum Organum*, published in 1620, and in his *New Atlantis*, published posthumously in 1627. One of his important contributions was to make a distinction between science-based inventions and empirical inventions. He asked whether the invention would have been comprehensible to one of the ancient Greek scientists, such as Archimedes (287–212 BC). Confronted with a spinning jenny, an early steam engine, a treadle-operated sewing machine, or a simple lawn mower, Archimedes would have had no difficulty understanding how it worked. The same is true of the water closet installed in 1589 by Sir John Harington (1561–1612) at his home in Kelston, Somerset. Even that much more modern invention, the zipper (zip fastener), would have intrigued Archimedes but would not have baffled him.

Another criterion for deciding whether an invention is empirical or not, one that leads to the same answer, is that an invention made by someone completely ignorant of science must be empirical. James Hargreaves (*c*.1720–1778), who invented the spinning jenny, was an illiterate weaver and carpenter. The watch, and the first clock that did not have a pendulum, were invented by Thomas Tompion (1639–1713), who began his career as a furrier and was then apprenticed to a clockmaker. Two great scientists, Christiaan Huygens (1629–1695) and Robert Hooke (1635–1703), may also have played some part in the invention, by introducing the idea of the balance spring, but the invention was nevertheless almost entirely empirical. Louis Jacques Mandé Daguerre (1787–1851), who first announced a photographic technique in 1839, was a painter of theatrical scenery and a showman, with no background in science.

Early empirical inventions were important in leading to what has been called the Scientific Revolution of the seventeenth century. It was in that century that scientific academies such as the Royal Society were founded. Much of the work they did was to try to understand some of the mechanical devices that had been invented. Later, in the nineteenth century, scientists devoted themselves to understanding steam engines, and were led to the science of thermodynamics (Chapter 2). A little later the technique of photography was invented, and it contributed greatly to transforming the techniques by which pure science is carried out. As we shall see in Chapter 3 it led directly to many advances in science, particularly in astronomy, spectroscopy, and the discovery of new elements.

By contrast, the zipper has made our lives slightly more convenient by relieving us from fiddling with buttons, and may have spared us some embarrassing moments, but it has not transformed our lives, and has created hardly a ripple in the river of science.

## SCIENCE-BASED TECHNOLOGY

Of necessity, all modern technology of any importance is strongly science-based. Even by Francis Bacon's time there had been a few science-based inventions, such as the cannon and the mariner's compass. Bacon discussed some of these, and astutely pointed out that these inventions had been made outside the relevant technologies. It is unlikely that the early military engineers would have had the technical expertise to invent a gun; the invention was instead made by someone who experimented with the burning of explosive substances in a confined space. A sailor, however skilled in his own techniques, is unlikely to have devised a mariner's compass. That invention was made by investigators who had the curiosity to experiment with magnetic materials.

Bacon also pointed out that many of the empirical inventions he knew of, such as the mariner's compass, would have been ridiculed if they had been forecast before the basic science had been carried out. As Mark Twain said, it is difficult to make predictions, particularly about the future. Scientists, even the greatest of them, have been notoriously unreliable in perceiving the consequences of their work. Michael Faraday's discovery of electromagnetic induction led to transformers and to the wide distribution of electric power (Chapter 4), but there is nothing in his writings that suggests that Faraday ever thought of such a result. James Clerk Maxwell's electromagnetic theory (Chapter 5) led directly to radio transmission and all that followed from it, including television, but Maxwell himself never wrote of such a possibility. Even after Heinrich Rudolph Hertz first achieved radio transmission in 1885, he explicitly denied that it could ever be used in telegraphy. He said that impossibly large reflectors would be needed, and that waves travel in straight lines, whereas the earth is round! These arguments seemed irrefutable, but by 1901 Marconi had sent a radio signal across the Atlantic. Lord Rutherford, who did so much important work on nuclear processes (Chapter 8), ridiculed the idea that they could be harnessed to provide large amounts of energy.

Since the middle of the nineteenth century almost all inventions have been science-based. The diesel engine was based very explicitly on thermodynamic principles expounded in *Theorie und Konstruction eines rationallen Wärme-Motors*, published by Rudolf Diesel (1858–1913) in 1893. Other important inventions that could only have been made by persons with a sound scientific background are computers, television, plastics, antibiotics, and body scanners. An interesting exception is the zip fastener, a purely empirical invention. It was originally patented in 1893, and not produced in its modern form until 1913.

## SCIENCE AND TECHNOLOGY COMBINED

In some cases it is hard to say whether work is science or technology. Good examples are provided by many of the investigations described in Chapter 7 on the structures of large molecules. Most pure scientists would, without any doubt, claim Dorothy Hodgkin as a pure scientist and not a technologist. Her great work, however, was on the structures of penicillin, vitamin $B_{12}$, and insulin, all of great practical importance in medicine. Was her work science or technology? The question, expressed in that simple form, is unanswerable. Good science and good technology can be—and superb science and superb technology will always be—one and the same.

It is a common belief today that there are, and always have been, two classes of individuals, namely scientists and technologists (or engineers), and that the scientist arrives at basic information that is then used by the technologist. This misconception may have arisen because modern educational institutions, for purely administrative convenience, draw a distinction between science and engineering. Universities, for example, usually have separate faculties of science and engineering. However, the work of individuals in the two faculties overlaps greatly. Members of engineering faculties are often doing basic scientific research; indeed most of the pure research in some fields, such as surface tension and the properties of metals, is now done by engineers. Conversely, much of the research done in science faculties is very applied; the basic research of chemists and biochemists is often closely related to practical medical problems (Chapter 7), and the work of physicists may relate to nuclear energy production (Chapter 8).

Before the latter part of the nineteenth century the distinction between science and technology would not have been explicitly considered. Engineering faculties were not established in many universities until after the turn of the century, but much technical work was done by people classed as scientists. Robert Boyle (1627–1691) and Robert Hooke (1635–1703) are now thought of as pure scientists, but did a lot of technical work. In the nineteenth century William Thomson (Lord Kelvin, 1824–1907) was for the most part a pure scientist but, literally with his own arms, he helped to lay the transatlantic cable. James Clerk Maxwell (1831–1879), chiefly famous as a theoretical physicist, did much practical work and designed many scientific instruments. Many other examples will be found in this book.

Society has been changed in many respects as a result of the work of the inventors and scientific investigators of the past three hundred years. General living conditions have been transformed by the new technologies. However, fundamental difficulties remain: for the most part they lie outside the field of science. Political and economic systems simply have not kept up with the great changes brought about by advances in science and technology. As a result there are vast numbers of people who are either no better off or are even worse off as a result of the changes. The peoples of this world have indeed become somewhat polarized into a small group that has entered the technological age, and a much larger group that has been left outside. These problems will be overcome only if those in a position to influence political and economic systems gain more understanding of the way in which science advances, and of how it affects technology and people's lives. One of my reasons for writing this book has been to help them to do so.

## 'HARD' AND 'SOFT' SCIENCE

Almost all of the science covered in this book is of the kind sometimes referred to by scientists as *hard* science. The word hard used in this context is perhaps unfortunate, as it is easily misunderstood. *The New Shorter Oxford Dictionary* gives twenty distinct meanings for the adjective hard, divided into four categories. Hard is often used with a meaning the opposite of which is easy, and I should say at once that this is not the meaning I use here. A hard science is not necessarily harder for a student to grasp than one that is not so hard, although people often believe that it is. First-year university students who are required to take one of the sciences often choose biology rather than chemistry or physics, because they think that biology, which is softer than physics and chemistry, is easier; quite often they decide later they have made a mistake. Of course, we differ from one another in how difficult we find a subject to be. I myself always found it easier to study the hard sciences than biology, which is why all of my research has been in physics and chemistry. I have no difficulty with understanding the concepts of biology, but I do not have a good memory for the kinds of facts that are important in biology. To other people the situation is quite different.

The meaning that we need here for hard, with reference to hard science, is instead a meaning the opposite of which is soft. Helpful meanings given by the *Oxford Dictionary* (Nos. 8b and 8c) are 'factual, reliable, substantiated; unable to be denied or explained away'. That is the right idea, but unfortunately it puts the matter rather too strongly, and we will have to discuss it further. For the adjective 'soft' the *Oxford Dictionary* gives twenty-three meanings, of which the most relevant for us (No. 21) is 'of a science as to method: not amenable to precise mathematical treatment, or to experimental verification or refutation'. Again, that is putting the matter a little too strongly.

What complicates our discussion of these matters, and perhaps makes it confusing, is that all of the scientific disciplines have hard areas and soft areas. By a scientific discipline I mean the main divisions of science with which we are all familiar—physics (including applied mathematics, which is a mathematical formulation of physical problems), chemistry, astronomy, biology, and geology. Of these disciplines, physics is the hardest; in fact, nearly all of it is hard. This is certainly true of the laws of mechanics, of thermodynamics, of electricity, and of the fundamental structures of atoms and molecules. All of these topics can be formulated mathematically, and the conclusions about them have been amply confirmed by experimental test. Here and there, however, we find areas of physics that are less hard. For example, not all of us are convinced by some of the physicists' suggestions about fundamental particles, so that there is some softness. But as time goes on, and more experiments are done, these uncertainties become resolved.

As to chemistry, some of it is very hard, and some of it soft. Until the latter part of the nineteenth century, most of chemistry was soft, as it consisted of a mass of descriptive material, with little theory to connect it all together. As time passed, however, and with advances in physics, more and more chemistry is becoming hard. There is one important branch of chemistry, physical chemistry (or chemical physics, which is almost the same thing), which is very hard, because it is really physics applied to chemistry; to put it another way, the methods of physics are used, but they are applied to chemical problems, which roughly speaking means that they are applied to situations in which there is a change from one chemical substance to another. The difference between physics and physical chemistry is by no means a sharp one. Two people who were trained as physicists, and who were generally regarded as physicists, won their Nobel Prizes not for physics but for chemistry; they were Ernest Rutherford, for his work on atomic nuclei, and Gerhard Herzberg, for his work on the spectra of atoms and molecules. Rutherford, incidentally, was perhaps not entirely pleased at being called a chemist; in his speech at the Nobel banquet he remarked that 'he had dealt with many different trans-formations with various time periods, but the quickest he had met was his own transformation from a physicist to a chemist'.

As time goes on, physics enters into chemistry more and more. Until fairly recently, organic chemistry (the chemistry of substances containing the element carbon) and inorganic chemistry (the chemistry of those that do not) were largely descriptive subjects. Now physical methods are being applied to the structures of compounds and their behaviour, with the result that many aspects of these subjects are very hard. Today a publication by a person regarded as a chemist often reads very much like a paper in physics.

Astronomy is strongly based on physics, to the extent that a person who wishes to be an astronomer usually takes a degree in physics first; most universities, in fact, do not give a bachelor's degree in astronomy, but require a degree in physics before a student can study astronomy. It might thus appear that astronomy is a hard science. However, the amount of experimentation one can do in astronomy is strictly limited. One can, and indeed must, apply mathematics to astronomical problems, but since the theories usually cannot be tested experimentally, the conclusions remain to some degree speculative. A good example is provided by Saturn's rings, which we will discuss in a little detail in Chapter 5. For a long time the nature of the rings remained a mystery. In the middle of the nineteenth century Maxwell applied a rigorous mathematical treatment to the rings, and considered a number of possibilities. He finally eliminated all but one theory. His conclusion was necessarily tentative, so that we are in the domain of soft science. Much more recently the *Voyager* spacecraft has made some observations that agree with Maxwell's conclusions, and that aspect of the science has therefore hardened.

Often astronomers enter the field of cosmology, and develop theories about the origin of the universe. This is obviously very soft science, as one cannot possibly do experiments to test the theories. In spite of this, cosmologists often put their theories forward with great assurance, as if they are quite sure they are right. Since a number of mutually exclusive theories have been put forward, they cannot all be right! The Soviet physicist Lev Landau (1908–1968), who won the 1968 Nobel Prize for physics, perhaps had the last word on this with his pithy comment that cosmologists 'are often wrong but never in doubt'.

Geology and biology are both soft sciences to a considerable extent. Much of these disciplines is descriptive, incapable of mathematical treatment. In some areas of biology it is often difficult or impossible to do controlled experiments, and even when one can do experiments the results are not as highly reproducible as they are in physics and chemistry. This, of course, is no criticism of biologists; it is just that biological systems show many variations.

It follows from our discussion of the hard and soft sciences that there will be an important difference between books written about different aspects of science. Books on the hard sciences—which the present one is for the most part—do not cover much material that is controversial. When, for example, we talk about the laws of mechanics, or the quantum theory, there is not much room for argument. In biology, on the other hand, the situation is rather different. Richard Dawkins has written a number of very lucid books in which he emphasizes the soft science that relates to the theory of evolution. In a number of places he mentions other distinguished biologists, explains their ideas, and then gives reasons why he cannot agree with them. There is one unfortunate consequence that may arise from this; people who are prejudiced against the theory of evolution on non-scientific grounds can misrepresent these disagreements, and argue that the whole theory is wrong. In fact the disagreements are only on points of detail.

As time goes on, physical methods are entering more and more into biology. Before World War I there was little of this, and the field was very soft, although of course some splendid work was done, particularly the theory of natural selection put forward simultaneously and independently by Charles Darwin and Alfred Russel Wallace. A certain amount of hardness, however, began to be introduced into biology in the early years of the twentieth century. Karl Pearson (1857–1936) did important mathematical work on statistics and probability theory, and applied it to genetics and other biological problems. Soon after the First World War the biochemist John Burdon Sanderson Haldane (1892–1964) began pioneering work on the physical chemistry of the behaviour of enzymes, which are the biological catalysts, and also on the mathematics of population genetics and natural selection. Later, in 1937, Cyril Norman Hinshelwood (1897–1967) began to apply the methods of chemical kinetics to the growth of bacteria. There are many more examples.

Physicists skilled in the determination of structures by X-rays began to apply their techniques to the large molecules that occur in biological systems, a matter that is considered in Chapter 7. Certain aspects of biology therefore became much harder. Two events were of special importance in this connection. In 1951 Linus Pauling suggested a basic structure which occurs in many protein molecules. Of even greater importance was the helical structure suggested in 1953 for deoxyribonucleic acid (DNA) by James Watson and Francis Crick. Since a gene is a section of a DNA molecule, this work quickly led to the detailed understanding of the functioning of the genes. As a result, modern work in this new field of molecular biology, or molecular genetics, is the application of physics to biological problems, and is therefore hard science.

This development has had particularly profound implications for the theory of natural selection. Because that theory completely transforms our understanding of living systems it undoubtedly ranks in importance with the other great accomplishments of science—Newton's laws of motion, Maxwell's theory of electromagnetic radiation, Planck's quantum theory, and Einstein's theory of relativity. But for many years there was a profound difference in status between Darwinian theory and the other theories. The others were in physics and were hard science. All of them could be put into mathematical form and could be tested; all survived the tests.

Darwin's theory, on the other hand, was in soft science. It could not be put into mathematical form, and the amount of testing that could be done was limited. Nevertheless, most scientists who looked into the matter, and certainly most biologists who studied the matter in detail, found themselves convinced by the arguments put forward by Darwin and others. Opposition to the theory was mainly based on religious ideas, which scientists do not consider to be relevant (this matter is discussed in much more detail at the end of Chapter 7). In any case, today the traditional churches have abandoned their opposition in the face of overwhelmingly strong scientific arguments.

The fact that Darwinian ideas were originally in the area of soft science was a weakness. Now that certain important advances in molecular biology have been made, however, all that has changed. Some of the results of molecular biology are mentioned briefly in Chapter 7. What we emphasize here is that all that has been done in molecular biology—hard science—has given support to the theory of natural selection in its broad outline (a few details had to be modified). It is now known that the information that tells how a cell is to function is coded in certain specific chemical units along a DNA molecule. The order in which these units occur can now be determined. The genetic code, by means of which the information in the genes is used in the synthesis of proteins, is now known precisely. As a practical result, and this is of great importance, a whole new field of genetic engineering has sprung up. Molecular biologists are now able to improve the quality of plants and animals by the transfer of genes, with predictable results. This is certainly hard science, and it is entirely consistent with the theory of natural selection, thus giving it strong support.

The achievements of Darwin and Wallace were perhaps even more remarkable than those of the other great scientists, because they were working in a soft science where it is more difficult to avoid error than in a hard science. The fact that their basic ideas have been so completely vindicated by hard science makes their accomplishment quite extraordinary. Scientists now feel as confident of the truth of Darwinian theory as they are of the theory of relativity. It is surprising to them that there still remains on the part of non-scientists some disbelief in the theory.

## 'HARD' AND 'SOFT' TECHNOLOGY

*The New Shorter Oxford Dictionary* gives as its first definition of technology 'The branch of knowledge that deals with the mechanical arts or applied sciences'. *The American Heritage Dictionary* has, quite consistently, 'The application of science, especially to commercial objectives'. A technology that is based on, or is closely related to, a hard science is conveniently called a hard technology. This is obviously true of engines of all kinds, of the distribution of electricity, and of radio and television. It is also true of the manufacture of plastics, which is based on hard areas of chemistry.

This distinction between hard and soft technology is not often made, but I think it is useful as it helps in our assessment of the reliability of technology. Products of a hard technology tend to be more reliable than those of a soft technology. When we buy a car, or a television set or a video cassette recorder (VCR), we usually find that it works reasonably well; if it does not, a competent technician can usually overcome the problem fairly easily. At any rate, after being repaired a car is unlikely to behave in a completely unexpected way; it will not fly into the air or transmit a television signal.

A product of a soft technology, on the other hand, may not do what it is expected to do. A drug given to a patient to cure a particular disease may instead aggravate it, or even produce another condition. When a drug is based on a soft science there is an inherent lack of reproducibility in its action. Similarly, techniques like winemaking and the manufacture of perfumes cannot be done with the same sort of reliability that is possible with the making of a car or an electronic device.

An example may help to make the distinction between hard and soft technologies clearer. If we have a television set that is not functioning, we know from experience that if we take it to a highly competent technician, it will be put right. We may be told that we need a new picture tube, but that everything else functions perfectly; we agree to have the work done, and everything is then satisfactory (if not, we know that the technician is incompetent—we do not doubt that someone else could have done the work well). By contrast, if we see the most competent medical experts in the world and are told that we need a heart transplant, we would be rather naive if we assumed that after the operation all would be well, without any doubt. We know, from our experience of what has happened to others in the same situation, that no such assurance can ever be given in a matter of that kind. Medicine is a soft technology, simply because of the enormous complexity of the human body. It will probably be a long time before a body can be repaired with the same assurance as a television set.

As time goes on, medical technologies are steadily hardening. The reason is that the basic science of how biological systems function has become much harder. A number of diseases are now understood at a molecular level. Sometimes, for example, a medical condition can be directly related to the behaviour of one particular enzyme. It is now possible to find out the structures of many enzymes by X-ray and other methods. If this has been done with an enzyme that is causing a medical problem, it may be a simple matter to design a molecule that will attach itself to the enzyme, and prevent it from acting. The resulting drug, being a product of hard science, is less likely to fail to act properly. This is the situation in the modern treatment of gout, that painful disease of the joints. For many years sufferers from gout—like Martin Luther—had to suffer great pain for many years, and little could be done to alleviate the pain. Now (always provided there are no complications) a few pills will cure the condition in a day or two. Of course, even when hard science is used in the design of a drug, disastrous mistakes can still occur, as soft science is used in almost every medical application.

A similar condition arises with those pharmaceutical drugs that are designed on the basis of knowledge of molecular biology. There are a number of diseases that can now be related to gene defects; examples are muscular dystrophy and cystic fibrosis. When this link can be made, the search for a remedy becomes much more straightforward, and is based on hard science. There can be no doubt that in the future pharmaceutical drugs will be more and more designed on the basis of hard basic science, and will be more effective and reliable.

Progress in pharmaceutical technology provides us with yet another important reason why basic research should be strongly supported by governments, in both the hard and the soft sciences.

## Good and bad science

It will be obvious that the distinction between hard and soft science has nothing to do with the distinction between good and bad science. Soft science can be excellent (for example, the theory of natural selection, as originally formulated by Darwin and Wallace), and hard science is occasionally very bad. There are several reasons for bad science, of which we will meet a few examples in this book. The most obvious one is sheer dishonesty, for personal gain or personal prestige. We will meet a blatant case of dishonest science for monetary gain in Chapter 2, involving the American swindler Robert Keeley in the nineteenth century. He made a lot of money by convincing investors that from a small amount of water he could create vast amounts of what he actually called atomic energy; he used trickery to appear to demonstrate that he could do so. This was one of the more spectacular and, for a time, successful of many claims to get something for nothing. Of course, scientific dishonesty is always found out, sooner rather than later. Swindlers should realize that they would do better to work in something other than science, especially hard science; one cannot get away with scientific dishonesty for very long, because experiments soon get repeated by others—and the more striking the reported results are, the sooner will the experiments be repeated.

Fraudulent science for personal prestige, in some cases just to gain a university degree, is by no means uncommon; most university teachers meet it during their careers.

Bad science sometimes results not from fraud but from a genuine mistake. Often the person responsible is so convinced as a result of the original mistake that self-delusion results; the mistake may be repeated many times even though there may be no conscious intention to deceive. An interesting example of this is the 'discovery' of 'n rays' by the Frenchman René Prosper Blondlot in 1903. These rays were alleged to increase the brightness of a spark, and produce other effects; they were supposed to be different from X rays and other known rays. Since Blondlot was a respected physicist some credence was given to his finding, especially in France, but other scientists failed to reproduce his work. Even when the deathblow to the existence of n rays was given, in a characteristically amusing way by the American physicist R W Wood, Blondlot never conceded that he was mistaken. From the beginning, self-delusion had consumed him.

A similar controversy arose more recently about 'cold fusion'. In the spring of 1989 Stanley Pons and Martin Fleischmann of the University of Utah announced that they could produce large amounts of heat by the expenditure of small amounts of electrical energy. The immediate reaction of much of the scientific community was that the results were unbelievable; such large amounts of energy could only be produced if nuclear processes were occurring, and that possibility seemed remote. Largely because of the somewhat sensational way in which the results were announced, a considerable stir was produced. What Pons and Fleischmann had done was surely hard science, as it was capable of being tested and verified. In spite of this there was some confusion about the claim for a few years; some scientists claimed to have confirmed the findings, others said that they could not do so. Complications arose from the fact that it was difficult for people to be sure that they were repeating the experiments under the original conditions; also, there is some evidence that something unexpected does occur in the reaction, but not what Pons and Fleischmann had asserted. In the end, conclusive experiments by many other scientists showed that the original claim could not be sustained. This example is instructive, as it shows that things are not necessarily cut and dried even in hard science. There can be uncertainty and controversy, but it will eventually be resolved.

# Two

# James Watt and the science of thermodynamics

One of the most important problems of science was to gain a proper under-standing of the nature of heat and of the relationship between heat and mechanical work. Today we all accept the fact that heat is a form of energy, and that heat can be converted into work, and work into heat. Until about the middle of the nineteenth century, however, this was by no means taken for granted. Another idea was popular in the minds of some investigators, and indeed had been strongly advocated by the great French chemist Antoine Lavoisier (1743–1794). Some of his experiments seemed to show that heat was a substance of a rather special kind; it had no weight and he referred to it as an 'imponderable fluid'. Lavoisier included 'calorique' in his list of the chemical elements, which are substances like oxygen and hydrogen which chemists could not break down into other substances. As Lavoisier was held in great esteem—and as most of his other ideas were certainly correct—his idea that heat is a substance was taken seriously by many eminent scientists.

The painstaking experiments of the Scottish chemist Joseph Black (1728–1799) also seemed to lead to the conclusion that heat is a substance. It was Black who first distinguished clearly between temperature and heat, and who showed how temperature measurements can be used to establish the quantity of heat. Black was struck by the fact that mercury, which is very dense, has a lower heat capacity than an equal volume of water. He thought, not unreasonably at the time, that more motion should be possible in a mat-erial of greater density; therefore, he concluded, heat cannot be motion. This is an instructive example of how the most careful experiments and intelligent reasoning can lead one astray if the time is not ripe.

The alternative—and as we now know correct—idea that heat is a form of motion was held by many distinguished people. Francis Bacon (1561–1626), whose interesting ideas about science and technology we met in the last chapter, produced strong arguments for believing that heat is not a substance but a form of energy. A similar view was taken by Robert Boyle (1627–1691) and by Isaac Newton (1642–1727). The idea was expressed very clearly by John Locke (1632–1704):

> Heat is a very brisk agitation of the insensible parts of the object [i.e., the atoms], which produces in us that sensation from which we denominate the object hot; so that what in our sensation is heat, in the object is nothing but motion.

Locke is now chiefly remembered as a philosopher, but he was active as a physician, and did a certain amount of work in experimental chemistry.

Reaching the correct conclusion about the nature of heat and its relationship to mechanical work was by no means a simple matter. Much help came from some purely technical and empirical work done earlier on the development of steam engines. The steam engine was invented, and indeed brought to a high degree of perfection, by men who had no training in science and little knowledge of it. Only after a particularly efficient steam engine had been built, by James Watt in the latter part of the eighteenth century, did the pure scientists begin to investigate how it worked. Their conclusions were embodied in the basic laws of thermodynamics. These laws, so important in all of science today, thus owe their origin not to people who were seeking the truth for its own sake, but to the empirical efforts of a few extremely ingenious engineers.

## EARLY STEAM ENGINES

The first steam engine of practical value was invented by Thomas Newcomen (1663–1729), who was born in Dartmouth, England, and began his career as a blacksmith. By 1712 he had constructed a steam engine for use mainly in pumping water out of coal mines (see Figs 2.1–2.3). Its principle was very simple. It had a single cylinder with a piston connected to a pivoted wooden beam. Steam from a boiler was admitted to the cylinder and the piston rose; then cold water was admitted to the cylinder, causing the steam to condense and the piston to fall. The pivoted beam performed about 12 strokes per minute, and a system of valves automatically controlled the admission of steam and of cold water to the cylinder.

If Archimedes had been shown a Newcomen engine he would have had no difficulty in understanding how it worked. On the basis of this criterion (which we noted in Chapter 1) the steam engine is an *empirical* invention rather than one based on scientific knowledge. In brief, in a steam engine (which was at first called a 'fire engine'), the steam admitted into the cylinder moves the piston and performs mechanical work. After the steam has cooled, so that it condenses to liquid water, much less work is required to move the piston back to its original position, and then the cycle can be continued. A net amount of work is therefore performed by the engine. We know now that this work is performed as a result of the heat provided to the engine, but this was by no means clear until well into the nineteenth century, as we will see later in this chapter.

Newcomen engines generated about five horsepower, the horsepower being a unit devised by James Watt to express the power of an engine in terms of the number of horses it replaced. We can also make an estimate of what is now called their *thermodynamic efficiency*; this is the ratio of work actually performed by an engine to the maximum amount of work that would

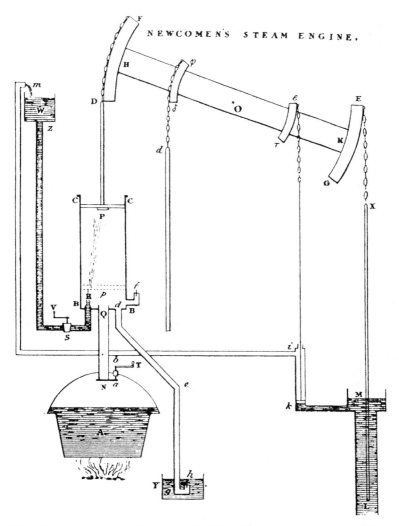

**Fig. 2.1** An engraving showing the principle of Newcomen's engine, from John Robison, *A system of natural philosophy*, Vol. 2, Murray, London, 1822. Steam from a boiler A passes into the cylinder CBBC, and raises the piston P. Cold water from the reservoir W then cools the cylinder, causing the piston to fall. Control is effected by a system of valves.

be performed if all of the heat supplied by the boiler could be completely converted into work. For a Newcomen engine this efficiency is estimated to be about 1%. In view of this low value it may seem surprising that a few Newcomen engines were still in practical use at least until the third decade of the twentieth century. They were at coal mines, where there is always a residue of poor quality coal which cannot be sold; it was better to burn it in an inefficient engine than to throw it away.

**Fig. 2.2**    An engraving showing an early steam engine at Griff, Warwickshire; from an engraving in J T Désagulier's *A course in experimental philosophy*, 1744 edition; the book was first published in London in 1734, and was a sequel to his *A system of experimental philosophy*, London, 1719. The Revd John Theophilus Désagulier (1683–1744) was born in France and was a Huguenot refugee in England. While serving in various church livings he carried out much experimental work in pure and applied science, some of it at Newton's suggestion.

## JAMES WATT'S INNOVATIONS

The great name in connection with steam engines is James Watt (1736–1819; Fig. 2.4). The circumstances under which Watt invented his much more efficient steam engine are especially interesting, and show us that good luck as well as mechanical genius is sometimes involved in a technical innovation. Watt was born in Greenock, Scotland, and in 1755, at the age of 19, he went

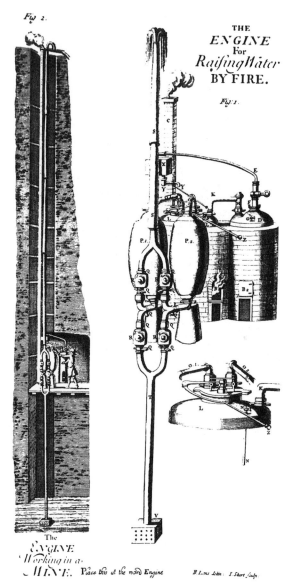

**Fig. 2.3** An engraving showing an early steam engine, from John Harris, *Lexicon technicum: or, an universal English dictionary of arts and sciences: explaining not only the terms of art, but the arts themselves*, London, 1704. In this book the Revd John Harris (1666–1719) obtained his information from the greatest authorities of his time, including Newton. The book, reprinted in 1966 by the Johnson Reprint Corporation, New York, is well worth consulting today.

**Fig. 2.4** James Watt (1736–1819), from a portrait in George Williamson, *Memorials of...James Watt*, 1856. Watt is famous for his invention of the separate condenser which led to his great improvements in the design of steam engines. As a unit of power (rate of doing work) he introduced the 'horsepower', but the modern unit is the watt (symbol, W), named in his honour (1 H.P. = 745.7 W).

Watt was a man of very fine character, and British prime minister Lord Liverpool said of him

'A more excellent and amiable man in all the relations of life I believe never existed'.

There is a large statue of Watt in Westminster Abbey, by the famous sculptor Sir Francis Legatt Chantrey (1771–1841).

to London to learn the trade of instrument maker. On his return to Glasgow a year later he wanted to set up a business as an instrument maker, but the Hammersmen's Guild put difficulties in his way because he had never served an apprenticeship. The University of Glasgow, however, allowed him to practise his trade on the University premises, and there he soon established a reputation for ingenuity and persistence. At the time the University was remarkably rich in talent, the economist and philosopher Adam Smith (1723–1790), and the chemist Joseph Black being two of its professors. During his stay in Glasgow, until 1774, Watt became friendly with some of the professors and students, notably Joseph Black who was doing work of great importance on the subject of heat. Watt also became friendly with John Robison (1739–1805), Black's student and successor as professor of chemistry. Undoubtedly these relationships helped Watt to attack his technical problems in a scientific way.

At the time the professor of natural philosophy (physics) at Glasgow was John Anderson (1726–1796), a rather remarkable man. He had previously been professor of oriental languages, and he had a strong social conscience.

He believed that his physics lectures should be available to artisans and others who were not members of the university, and threw his classes open to them. These 'anti-toga' classes did not receive the approval of the university authorities, and somewhat hard feelings developed. As a result Anderson left all of his considerable fortune to found a rival institution in Glasgow. It became known as Anderson College or the Andersonian Institution, and evolved into the present University of Strathclyde.

For demonstrations in his physics classes, Anderson had a model of a Newcomen engine (Fig. 2.5). It worked very badly, performing only a few strokes before stalling, and after finding that instrument makers in London were unable to help, Anderson asked Watt to overhaul it. Watt went about this task with great persistence, and carried out a number of investigations on the thermal effects of mixing steam and water. He concluded that the main trouble with the model was that, because the surface:volume ratio was much larger than in the full-scale engines, the loss of heat was relatively much greater. ('Scaling' effects of this kind had previously been recognized by

**Fig. 2.5**  The model of the Newcomen engine that belonged to John Anderson of the University of Glasgow, and was modified by James Watt; from George Williamson, *Memorials of...James Watt*, 1856). This model is still to be seen at the University's Hunterian Museum.

Newton). We now know that there were other factors beside the one mentioned by Watt. The cylinder of the model was made of brass, while in the engines themselves they were of iron and would be coated internally with iron oxide; this is another reason for greater heat loss in the model. In addition, the wall thickness in the model was disproportionately large, so that the cylinder had a larger heat capacity. Watt was able to get the model working (but only just!) by careful control of the amount of cooling water added, and by reducing leakage at the piston.

While working on the model of the Newcomen engine, Watt realized that it had a fundamental flaw. Heating up the cylinder with steam, and then cooling it with water, obviously involves much unnecessary wastage of heat and loss of efficiency. Watt then had the most important of his many innovative ideas, the *separate condenser*. We even know just when and where Watt had the idea: it was on Easter Sunday 1765, when he was walking past the Golf House on Glasgow Green. His suggestion was that there should be two cylinders connected together, one always kept hot, and the other, the condensing cylinder, always cold (see Fig. 2.6). At first the compression of the hot cylinder was brought about by atmospheric pressure, but in 1769 Watt realized that this produced unnecessary cooling, and that it was better to let steam do the compression; this was his second important innovation. In the same year he introduced his so-called *expansive principle*; instead of con-

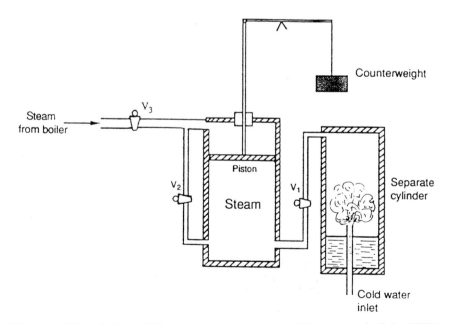

**Fig. 2.6**   The principle of Watt's separate condenser. A Watt engine built in 1779 is to be seen in the Birmingham (England) Museum, and on certain days the staff have it working—216 years after its construction, which must be something of a record!

tinuously admitting steam to bring about the compression, he cut off the supply and let the pressure fall. He estimated that greater efficiency resulted in this way.

In order to put his ideas into practice Watt had to get extensive financial backing. After some abortive enterprises he entered into a partnership with Matthew Boulton (1728–1809), a prosperous manufacturer at Soho, near Birmingham. Watt had obtained a patent for his engine in 1769, and manufacture was commenced at the Soho Engineering Works in 1774. Altogether about 500 engines were built during the Boulton–Watt partnership. The thermodynamic efficiencies of the first ones were about 8%, but by the end an efficiency of about 19% had been achieved. This was quite creditable since, as we will see later, the second law of thermodynamics imposes a maximum possible efficiency of about 25% under the conditions of the Watt engines.

One reason for the great success of the Boulton–Watt partnership was the very different and complementary temperaments of the two men. Watt, though a kindly man, was a somewhat dour Scot, who tended to take a pessimistic view of things. Boulton was a large, cheerful man who was always optimistic, and he gave Watt much encouragement. On one occasion when Watt was unduly depressed, Boulton wrote to him suggesting that he should say twice daily the 'Scotch [*sic*] prayer' (which reads 'The Lord grant us a guid conceit of airselves'; it should be explained that at the time the word 'conceit' did not have its present pejorative meaning, but meant rather a justifiable self-esteem. Watt was certainly entitled to much self-esteem, but it seems that he lacked it).

Another reason for success was that Watt and Boulton collaborated with John Wilkinson (1728–1808), a highly inventive ironmaster. Wilkinson's firm was able to bore cylinders much more accurately than had hitherto been achieved, and this greatly improved the efficiency of the steam engines. Another of Wilkinson's achievements was that he was the first to float an iron ship. Its launching in 1787 created a great sensation; it was thought that material denser than water would never float, and ships had always been made of wood. With some foresight Wilkinson constructed for himself an iron coffin, but his foresight had its limitations; by the time of his death his cross-sectional area had increased so much that he could not be squeezed into it.

Watt's engines found many applications. The first locomotive engine, illustrated in Fig. 2.7, was built in 1784 by William Murdock (1754–1839), who was Watt's assistant. At the time he was building stationary engines in Cornwall, and while living in Redruth built a model locomotive. Since there were no railway lines on which to try it he decided to make use of the steep banks of a lane leading to the local church. When he made his first trial, at night, the engine moved so fast that it out-distanced him and almost ran down the rector of the church who happened to be walking along the road. As he ran away in terror the rector assumed the fiery monster to be the embodiment of the devil.

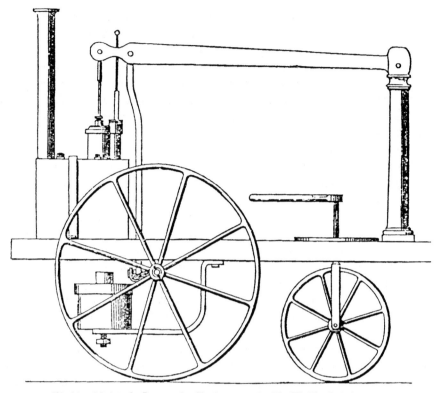

Working Model of a Locomotive Engine made by Mr. W. Murdock in 1784.

**Fig. 2.7**   William Murdoch's steam locomotive of 1784 (from J P Muirhead's *Life of James Watt*, 1858).

However, little use was made of the Watt engines in locomotives. The reason is that Watt was strongly opposed to the use of high-pressure steam, for reasons of safety. With improvements in engineering techniques, however, it became possible to construct high-pressure steam engines that were quite safe. The first successful high-pressure steam carriage was made by Richard Trevithick (1771–1833) in 1801. During the next few years Trevithick designed a number of stationary steam engines, steam carriages, and steam locomotives. He constructed the first steam locomotive that travelled on tracks in 1804, and showed that it provided considerable traction. Nothing that he did, however, was commercially successful, and he died in debt.

With high-pressure steam it was usual not to use a condenser, but simply to vent the steam to the surroundings. The non-condensing high-pressure engines were less efficient than the Watt engines (having a thermodynamic efficiency of about 10% rather than 19%), but the absence of a condenser made them much more compact. With a locomotive, as opposed to a ship, compactness was of great importance, and all the major developments of steam locomotives involved the use of non-condensing engines.

In 1807 the American engineer Robert Fulton (1765–1815) established the first regular steamship service with the paddle steamer *Clermont*, which carried a Watt engine and sailed on the Hudson River between Jersey City and Albany (Fig. 2.8). Fulton was a most remarkable man. He designed the world's first submarine, the *Nautilus*, and it was launched in 1800, but the US, British, and French governments all refused to support the project, and he abandoned it in 1806.

Without in any way detracting from the genius of Watt, we may reflect on the amount of luck that was involved in his work on the steam engine. Several circumstances fitted together very neatly. Watt had just established himself at the University of Glasgow, with the right skills, when John Anderson had trouble with his model Newcomen engine and asked for his help—and Watt was perhaps the only person in the world capable of giving it. Watt, needing financial support for his invention, began an association with Matthew Boulton, one of the few men in the world capable of leading it to industrial success. And just at the right time they gained the cooperation of John Wilkinson, who could bore a cylinder more precisely than anyone else in the world.

Perhaps the most important factor leading to the success of Watt's work on the steam engine was that besides being a great engineer he was also, without training but by instinct and through the influence of his associates, an excellent scientist. His technical developments were always supported by

**Fig. 2.8**  Engraving of the American steamship *Clermont*, which was equipped with a Watt engine.

his work in pure science. He carried out fundamental experiments on the condensation of steam and, quite independently of Joseph Black, discovered *latent heat*—the heat that is required, for example, to convert water into steam, and is released when the steam condenses. Watt was, incidentally, perhaps the first person to realize that water is a chemical compound and not an element.

## STIRLING ENGINES

Another type of engine, in which the cylinder is heated externally rather than being injected with steam, was developed in the early years of the nineteenth century by the Revd Robert Stirling (1790–1878; Fig. 2.9). Born in Cloag, Scotland, Stirling attended the University of Glasgow but probably did not learn any science there; he studied the classics, philosophy, theology, and mathematics. He was ordained a minister of the Church of Scotland in 1816, and most of his career was as pastor of the parish church of Galston in Ayrshire. The University of St Andrews awarded him the honorary degree of Doctor of Divinity in 1840.

Throughout his career Robert Stirling was greatly interested in engines of various kinds, as was his younger brother James who became an engineer by profession. In 1816, the year of his ordination, Robert patented a rather remarkable air engine, and in 1827 and 1840 he and his brother took out

**Fig. 2.9**  Photograph of a painting of the Revd Dr Robert Stirling (1790–1878). The painting is in the Parish Church of Galston, Ayrshire, Scotland.

further patents for improved air engines. The Stirling engines were mechanically somewhat complicated, and extremely ingenious—so much so that for many decades practically everyone misunderstood how they worked. Nearly 100 years later engines of a similar type were being 'invented', and some of them were failures because they lacked some essential feature which Robert Stirling had grasped intuitively. There is no doubt that the Stirling engine, like the weight-driven clock controlled by an escapement (Chapter 1), was one of the most ingenious of all inventions.

Figure 2.10 is a copy of the diagram given in the original patent specification of 1816, and is useful for explaining the general principles. The essential feature is that the gas, air in the early engines, remains in the cylinder, and is alternately heated and cooled; unlike the steam engine, the Stirling engine is of the closed-cycle type, in which the same gas is used over and over again. The iron cylinder contains a 'displacer' (9), through which the air can pass. There is also a completely separate piston (2), which confines the air within the cylinder.

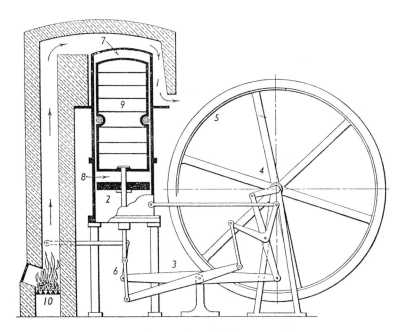

**Fig. 2.10** Copy of the diagram of the original Stirling engine, as shown in the 1816 patent. The diagram does not make it clear that the main piston (2) and the displacer (9) are moved independently. The plunger-displacer is porous, allowing the air to pass through it; it alternately absorbs and releases heat. Its purpose is to bring about less wastage of heat than necessarily occurs in a steam engine of the Watt type. The action of the engine is explained in the text. An engine of this type was used in 1818 for pumping water from a quarry. The closed end of the cylinder is kept hot by the products of combustion from the fire in the grate (10), while the region at the other end of the displacer is kept cold by exposure to the atmosphere, or by a water cooler.

By appropriately moving the displacer and the piston, the engine can perform mechanical work. The primary function of the displacer is to control the position of the gas in the cylinder, and therefore to control the temperature of the gas. The primary function of the piston is to compress the gas when it is cold, and to allow the gas to expand when it is hot. When the displacer is moved to come into contact with the piston the gas (7) is hot, and can perform work on the piston by expanding. When the displacer is moved to come into contact with the closed end of the cylinder the gas (8) is cold. The gas is then compressed by moving the piston, and the work required to compress the cold gas is less than the work performed by the engine when the hot gas expands. The engine therefore performs a net amount of work.

The Stirling brothers realized correctly that the function of the displacer was merely to act as a heat exchanger which would reduce the amount of heat wasted. They referred to their displacer as an 'economizer', and argued that there would be less loss of heat than in the usual type of steam engine. The function of the displacer was later often misunderstood, and it came to be called a 'regenerator', a term that is still used. The term is somewhat misleading, as it may be taken to mean that in some way it allows the same heat to be used over again, which is impossible. It is important to understand that the regenerator *merely conserves heat*.

There is no doubt that the idea behind the Stirling engines was a sound one, but unfortunately they did not meet with great technical success during the lifetimes of the two brothers. The reason for this was that steel technology was not sufficiently well developed for the engines to withstand the stresses involved; the engines were therefore ahead of their time. Also, it was difficult to construct suitable displacers that were sufficiently strong and would allow the gas to pass sufficently rapidly. Difficulty was also encountered with leakage of the air past the piston.

Since about 1937 there has been a considerable revival of interest in Stirling engines, particularly by the Philips Research Laboratories of N V Philips Gloelampenfabrieken, the large international company based at Eindhoven in The Netherlands and well known for its electrical and electronic products. Stirling engines have now been produced for use in heating and refrigeration devices, for use in automobiles, and even for artificial hearts. In modern engines the air is often replaced by hydrogen or helium; these have the advantage of a much higher thermal conductivity, which means that the heating and cooling of the gases occurs more rapidly.

Modern Stirling engines offer several advantages over steam engines and internal combustion engines. Since the working gas is confined within the cylinder, there is no exhaust emission. Since there are no intermittent explosions such as are found in an internal combustion engine, the noise level is much lower. The thermodynamic efficiencies are higher than in other engines, as the loss of heat has been minimized.

The disadvantages of modern Stirling engines include a high initial cost, because of the mechanical complexity of the heat-exchange system. It is also difficult to control the leakage of gas from the cylinder; this difficulty is accentuated by the use of hydrogen or helium, the light molecules of which move more rapidly than the molecules of the oxygen and nitrogen in the air. A disadvantage for automobile engines is that the start-up time is slow, since the heating and cooling systems take time to reach the required temperatures.

## SADI CARNOT AND THE MOTIVE POWER OF HEAT

Pure science, carried out for reasons of curiosity and without any regard for practical applications, played no part in the development of steam engines until early in the nineteenth century. Important steps were then taken in the field of thermodynamics, for the most part by investigators whose main concern was to understand the principles relating to the workings of nature. Paradoxically, however, the first great step in the pure science of thermodynamics was taken by one trained as an engineer, Sadi Carnot.

It is significant that in Carnot's only publication he lists a number of engineers by their last names only—with the exception of Watt, who is 'le célèbre Watt'. The book makes it evident that his theory of heat engines was greatly influenced by Watt's introduction of the separate condenser. Carnot made no mention of the Stirling engines, and there is no evidence that they ever came to his attention.

Nicolas Léonard Sadi Carnot (1796–1832; Fig. 2.11), was born in Paris and was a member of a distinguished family. His father, Lazare Nicolas Marguerite Carnot (1753–1823) is important in the history of mechanics as the author of *Essai sur les machines en général*, published in 1783, in which he discussed the principle of conservation of energy. He was also prominent in political and military spheres, being called the 'organisateur de la victoire' because of his activities under the First Republic of France. Sadi Carnot's nephew, Marie François Sadi Carnot (1837–1894) became President of the Third Republic of France in 1887.

Sadi Carnot was educated at the École Polytechnique as a military engineer, and saw active service in 1814. For a few years he held various routine military positions, but being frustrated by the work obtained a protracted leave of absence and took up residence in Paris where he undertook study and research in science and engineering. He was 28 years of age when he published his 118-page book, *Réflections sur la puissance motrice du feu et sur les machines propres à développer cette puissance*. This book, Carnot's only publication, was of great importance; Lord Kelvin described it as an 'epoch-making gift to science', and Sir Joseph Larmor as 'perhaps the most original in physical science'. It was clearly written but by no means a popular account; it presupposes some knowledge of steam engines, physics, and basic mathematics.

**Fig. 2.11**   Sadi Carnot (1796–1832), from a painting by Louis Léopold Boilly, done when Carnot was aged 17 and was in the uniform of the École Polytechnique in Paris.

In his book Carnot developed his highly original treatment of heat engines on the basis of his belief that heat is a substance, called caloric, and he thought that when, in an engine, heat flows from a higher to a lower temperature and work is done, the heat is actually conserved. Carnot discussed the analogy of a waterfall causing a wheel to turn (Fig. 2.12); there is no loss of water, the wheel being turned by the force of the water. In the same way he thought that the force of falling heat would cause the piston in an engine to move and perform work. We now know that when work is done as a result of a flow of heat, some of the heat is converted into an equivalent amount of work.

Carnot realized that a steam engine cannot function if every part of it is at the same temperature. His great contribution to science was to consider ideal types of engines operating between a higher temperature which we will call $T_h$ and a lower one, $T_c$. He imagined a gas to undergo a cycle of changes between the two temperatures, returning to its initial state but performing mechanical work.

One important contribution made by Carnot in his book was that he introduced for the first time the idea of *thermodynamic reversibility*. (He did not use the *word* 'reversibility', which was suggested much later). For a process to be thermodynamically reversible it must occur infinitely slowly and be reversible at every stage of the process; for example, if a substance is being cooled reversibly, the external temperature must at all times be infinitesimally lower than the temperature of the substance. A thermodynamically reversible process can only occur in our imagination; any process that really occurs must

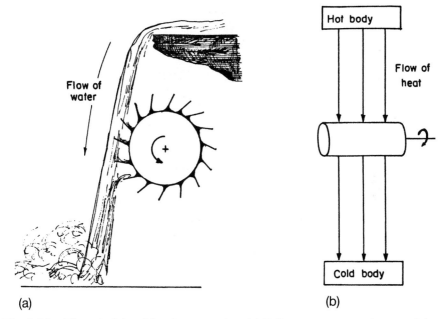

**Fig. 2.12** The principle of the Carnot engine. (a) Falling water can perform work by turning a water wheel. (b) Carnot thought that heat 'falling' from a higher to a lower temperature could produce work, without any loss of heat.

be irreversible in the thermodynamic sense, although if it is slow it may be close to reversible.

Carnot first imagined the gas to be undergoing the cycle of operations completely reversibly; then he imagined it to go round the cycle by some irreversible processes. He realized that the work done in the completely reversible process is the maximum possible, and that a consideration of the completely reversible engine is useful in giving the maximum amount of work that could be done when an engine consumes a given amount of fuel such as coal. Modern treatments of the Carnot cycle usually deal with the *efficiency* of the system, which is the fraction of the heat absorbed at the higher temperature that is converted into work; Carnot, believing in Lavoisier's caloric theory, did not consider that the work was done at the expense of heat that disappeared.

Instead, Carnot considered the maximum 'duty' of an engine, which is the amount of work it does for a given amount of fuel. He expressed the work done as the mass of water that could be lifted multiplied by how high it is lifted. One important conclusion he reached is that if heat falls from a higher temperature $T_h$ to a lower temperature $T_c$, the duty is larger the larger the difference between the two temperatures, just as in a waterfall (Fig. 2.12(a)) the greater the fall of the water the more work is done by the water wheel.

Carnot also found that for a given drop in temperature, $T_h - T_c$, the work is greater the smaller is $T_h$; thus a drop from 1°C to 0°C will produce more work than one from 100°C to 99°C.

Carnot discussed his conclusions with reference to some of the steam engines of his time. He was able to explain why a high-pressure steam engine is more efficient than a low-pressure one; in his own words (in translation):

> It is easy to see why the so-called high-pressure steam engines are better than the low-pressure ones; their advantage lies essentially in their ability to utilize a greater fall of caloric. Steam generated at a higher pressure is also at a higher temperature and as the temperature of the condenser is nearly always the same, the fall of caloric is evidently higher.

Another important conclusion reached by Carnot in his book is now referred to as *Carnot's theorem*. He considered two engines, both working between two particular temperatures. One of the engines worked reversibly, and the question he asked was whether the other could be designed in such a way as to produce more work from the same amount of fuel. He answered the question by first postulating that there could be such an engine, and he caused it to drive the first one backwards. He then showed that if this were to occur there would be a net flow of heat from the lower to the higher temperature. This, he pointed out, is contrary to experience; if it could happen, perpetual motion machines could be constructed. Today we express his conclusion by the statement that the efficiencies of all reversible engines working between two given temperatures must be the same, namely $(T_h - T_c)/T_h$. We will say a little more about these efficiencies later.

This theorem had great practical implications, which Carnot pointed out. Previously it had been thought that engines could be improved without changing the working temperature, by changing to different materials; perhaps a steam engine could be improved by changing from water to alcohol or oil, or some other material. Carnot had shown, however, that attention must be directed to the working temperature and not to the materials used.

In 1827 Carnot was required to return to active duty, with the rank of captain, but after less than a year's service he was able to return to Paris. He continued his studies on the theory of heat and the design of engines, but made no further publications.

Carnot's health was always fragile, and he died at the age of 36. There is uncertainty about the circumstances of his death. The official version, announced at the time, is that an attack of scarlet fever in June 1832, undermined his constitution, and that in August he fell victim to a cholera epidemic and died within a day. However, there is more recent evidence that he died in a hospital for the mentally disturbed near Paris. Those of his papers that survived showed that he was abandoning the view that heat is a substance, and was beginning to favour the idea that heat is a form of energy.

## THE NATURE OF HEAT

After Carnot's book appeared not much more could be done on the subject until the nature of heat had been clarified. As we discussed at the beginning of this chapter, ideas as to the nature of heat shifted considerably over the years. However, investigations carried out in the late eighteenth and early nineteenth centuries led slowly to the conviction that heat is a form of motion. Some of these were carried out by the remarkable administrator and investigator Benjamin Thompson (1753–1814; Fig. 2.13). Born in Woburn, Massachusetts, Thompson married a wealthy widow, Mrs Rolfe, and settled at Rumford, New Hampshire, which is now called Concord; he was commissioned a major in a New Hampshire regiment. On the outbreak of the War of Independence in 1776 Thompson favoured the British side, and fled to

**Fig. 2.13**  A caricature of Benjamin Thompson (1753–1814), who was born in Massachusetts, had much of his career in Europe, and was created Count Rumford. He founded the Royal Institution in London, and is shown warming regions of his anatomy, and apparently setting himself on fire, in front of a stove of his own invention. (Courtesy of the Royal Institution.)

England, abandoning his wife and baby daughter. In England he gave valuable information to the government on the situation in America and received an appointment in the Colonial Office. During his stay in England he carried out some experiments with gunpowder, and in 1779 was elected a Fellow of the Royal Society. In 1782 he returned to America with the rank of lieutenant-colonel, and after peace had been declared he was knighted, so becoming Sir Benjamin Thompson.

In 1784 Thompson entered the service of the Elector of Bavaria, and brought about a number of remarkable improvements in that state. He reformed the army, established a military academy, planned a poor-law system, disseminated the knowledge of nutrition and domestic economy, and improved the breeds of horses and cattle. For his services in Bavaria he was made head of the Bavarian war department and created a Count of the Holy Roman Empire; he chose for his title the town of Rumford where he had lived, and so became Count Rumford. He spent 1795–1796 in England, and endowed two Rumford medals of the Royal Society for research in light and heat. At about the same time, perhaps to show that there was no ill-feeling on his side, he endowed two medals for the American Academy of Arts and Sciences (this Academy, founded in Boston in 1780, is not to be confused with the National Academy of Sciences, based in Washington). On his return to Munich he found that the Elector, threatened by both France and Austria, had fled, and he became generalissimo and president of the Council of Regency. For a period he was concerned with the boring of cannon in Munich, and in experiments in which a horse operated a dull drill he was impressed by the fact that large amounts of heat, capable of boiling ice-cold water, were continuously produced for periods of up to $2\frac{1}{2}$ hours. He established that no weight change occurred during the process, and that the metal shavings had the same properties as the unbored metal. From these investigations he concluded that heat could not be a substance but that there was a conversion of mechanical work into heat, and he obtained a value—not a very accurate one—for the 'mechanical equivalent of heat'.

In 1799 Rumford left the Bavarian service and returned to England, where he became associated with the Society for Bettering the Condition and Increasing the Comforts of the Poor. He proposed the founding of a public institution for the diffusion of knowledge, for giving public lectures, and for applying science to practical ends. This suggestion led to the founding in London, on 7 March 1799, of the Royal Institution of Great Britain. In 1801 Rumford persuaded Humphry Davy (1778–1829), still only 22 years old, to join the Royal Institution as assistant lecturer in chemistry. This was a happy choice; a year later Davy had been promoted to become professor of chemistry and director of the laboratories, and he went on to do research of great distinction, being knighted in 1812.

Rumford's association with the Royal Institution was shortlived, as he soon quarrelled with its managers. In 1804 he moved to Paris where he married

Lavoisier's widow, and lived in her villa at Auteuil, where he died in 1814. By all accounts the marriage was far from a success. Rumford's scientific and organizational achievements were remarkable, but his complex character and behaviour were far from praiseworthy. He was admirably and pithily described by William H. Brock in an article in 1980 as

> a loyalist, traitor, spy, cryptographer, opportunist, womaniser, philanthopist, egotistical bore, soldier of fortune, military and technical adviser, inventor, plagiarist, expert on heat (especially fireplaces and ovens) and founder of the world's greatest showplace for the popularization of science, the Royal Institution.

Although Rumford's paper on the mechanical equivalent of heat was clearly and convincingly written there was much scepticism about its conclusions, many investigators continuing to regard heat as a substance. Evidence that was considered to be more convincing was obtained by the German physician Julius Robert Mayer (1814–1878; Fig. 2.14). In 1840 he sailed to Java on a Dutch vessel, and noticed that venous blood has a much redder colour in the tropics than in a colder climate. He concluded that this is due to a lower metabolic rate in warmer weather, so that there is smaller consumption of oxygen with the result that there is less contrast in colour between venous and arterial blood. He gave much thought to the relationship between food consumption, heat production, and work done. He also made a careful study of some of the heat studies on gases that had been made by various investigators, but did no experiments on the subject himself.

**Fig. 2.14** Julius Robert Mayer (1814–1878). A paper in 1842 in the *Annalen der Chemie und Physik* established his priority for providing evidence for the principle of conservation of energy, which is the first law of thermodynamics.

On the basis of evidence of this kind, Mayer arrived at the conclusion that heat and work are interconvertible. He considered both to be different forms of what he called force (by which he meant energy) and which is indestructible. On his return home Mayer prepared a paper on the subject and submitted it to the *Annalen der Physik*. His paper was couched in metaphysical terms, and since he was ignorant of mechanics he made many elementary errors; as a result his paper was rejected. Mayer was very angry and frustrated at this treatment, but he later became aware of the paper's limitations and revised it extensively, submitting it for publication in the *Annalen der Chemie und Physik*; it was then accepted and it appeared in May 1842. Mayer later claimed, with some justification, that this paper established his priority for the principle of conservation of energy—which is now referred to as the *first law of thermodynamics*.

Mayer published further books and articles, paying particular attention to the physiological aspects of the problem. For a time Mayer's ideas were either ignored or ridiculed, and this had a serious effect on his mental stability; he attempted suicide and was confined several times in mental institutions, sometimes in a straitjacket. In his later years he received due recognition, being elected a corresponding member of the French Académie des Sciences. He received the Copley Medal of the Royal Society in 1871.

These investigations convinced most people that heat is not a substance but a form of energy. It is interesting that although we now accept this view we still find it convenient to use language that implies that heat is a substance. We speak of heat 'flowing' from one body to another, and we refer to the 'quantity of heat' in a body. It is difficult to see how we could do differently. In any case, when we speak of the quantity of heat we mean the amount of energy involved.

Much more careful and extensive experiments on heat were carried out by Joule, and these deserve a section to themselves.

## JAMES PRESCOTT JOULE

James Prescott Joule (1818–1899; Fig. 2.15), a prosperous English amateur scientist, took a rather different approach to the problem of the interconversion of heat and work. He carried out a variety of careful experiments himself, and his investigations completely transformed the subject. He was born, and he died, near Manchester, and did all his research there; being a quiet and shy man he felt happier in a familiar environment. He and his brothers were privately educated at home, and received some of their lessons in mathematics and science from the great John Dalton (1766–1844), who had proposed his atomic theory in the first decade of the century. Dalton was about 70 years old when he taught the Joule boys.

The Joule family owned a brewery, but none of the brothers had much interest in running it, and it was sold on the death of their father; they then

**Fig. 2.15** James Prescott Joule (1818–1899). His experimental work, published at about the same time as Mayer's 1842 paper, provides much more convincing evidence for the first law of thermodynamics.

lived on dividends from the proceeds of the business. James Prescott Joule's experiments were carried out in laboratories he installed in the brewery, and later in his various homes. He obtained some subsidies from scientific bodies for some of the work, and later, after suffering financial losses, he received a government pension.

Joule's most important investigations were made when he was between the ages of 19 and 29. They were inspired in 1837 by the work of William Sturgeon, who also worked in Manchester, on electromagnets and electric motors, a matter which will be discussed in more detail in Chapter 4. He was also influenced by an idea that had become prevalent, but which Joule helped to disprove, that there was no limit to the power that could be obtained from a motor operated by an electric battery.

The invention of batteries for the generation of current electricity, which we will consider in Chapter 4, had led to this idea. Steam engines had to be supplied with fuel obtained from under the ground, and it was realized even in the early nineteenth century that the supply of fuel was limited. Electric batteries, on the other hand, led to what Professor Donald Caldwell has called an 'electrical euphoria'. One of the enthusiastic proponents of this point of view was Moritz Hermann von Jacobi (1801–1874), a rather remarkable man who had a high reputation in his time, but who is now largely forgotten. He was born in Potsdam, Prussia, but his career was spent further east, first at Dorpat which was then in Russia (it is now called Tartu and is in Estonia), and then in St Petersburg (later Leningrad and now again St Petersburg). In 1834 von Jacobi constructed what was perhaps the first electric motor (see Chapter 4), and he carried out interesting experiments with it.

In the following year von Jacobi published a paper that created something of a sensation at the time. He argued that if certain imperfections of the electric motor, such as friction and what is now called back-emf (electron motive force), could be eliminated, a motor would go on accelerating indefinitely, producing enormous amounts of power. These arguments, although fallacious, seemed compelling, and many electric motors were built, for a variety of purposes.

At first Joule too was carried away by this 'electrical euphoria'. He carried out careful experiments on the mechanical effect that could be obtained from a motor, and related it to the amount of metal used up in the battery operating the motor. He was disappointed to find that the consumption of a given amount of zinc in a battery would lead to the production of only about one-fifth of the mechanical work that would by produced by the same weight of coal in a steam engine. Since in addition zinc was much more expensive than coal, this means that an electric motor is far from being a competitor to a steam engine for the primary production of energy. Joule presented this pessimistic conclusion in 1841, in a lecture at the Victoria Museum in Manchester. For him the 'electrical euphoria' was over. As we shall see in Chapter 4, the great usefulness of electricity had to be realized not by operating motors but in many quite different ways.

Joule then decided to study the heating effect of an electric current. Using simple equipment, but working with great care, he established that a current passing through a wire of resistance $r$ generates heat in proportion to $r$, and in proportion to the square $i^2$ of the current passing; this is his well-known $i^2r$ law. He also concluded that the heat produced is equal to the energy released by the chemical action occurring in the battery. Although this conclusion had later to be modified slightly, it was important in establishing that, contrary to von Jacobi's prediction, energy could not be created from nothing. With regard to the applications of electricity to practical use, Joule wrote that 'electricity is a grand agent for carrying, arranging, and converting chemical heat'. This was indeed a shrewd prophecy.

In 1843 Joule carried out an investigation that is remarkable for its ingenuity. He enclosed the revolving part ('armature') of an electric generator (sometimes later called a 'dynamo') in a vessel containing water, and measured the heat generated when he rotated the armature for a fixed period of time. He also measured the heat produced by the current that was generated. In this way he established the equivalence between the heat produced as a result of rotating the armature, and the mechanical work required for the rotation.

In later experiments Joule produced heat in water by stirring it with large paddles (Fig. 2.16). In 1847 he lectured in a church hall in Manchester and argued that 'the hypothesis of heat being a substance must fall to the ground'. Joule's work was in marked contrast to that of Mayer, who had

**Fig. 2.16** A diagram made by Joule, showing the apparatus he used to measure the heat produced when a liquid is stirred. The rotation of handle f raises two weights e. Release of the handle allows the weights to descend, and large paddles stir the liquid in the container. The heat produced can then be related to the work done by the falling weights.

reached his conclusions on the basis of imaginative intuition and had based his arguments on the experimental evidence obtained by others. Joule's conclusions were careful inferences from skilfully designed and meticulously conducted experiments. For good measure, however, he added some metaphysical and religious arguments: 'Believing that the power to destroy belongs to the Creator alone...I affirm...that any theory that demands the annihilation of force [i.e. energy] is necessarily erroneous'. The trouble with this type of argument is that another person could apply it to heat rather than total energy.

Since he was neither a university graduate nor held a recognized scientific appointment, Joule at first met with difficulty in having his views accepted. In 1847, however, he attended the meeting in Oxford of the British Association for the Advancement of Science (BAAS, sometimes irreverently called the 'British Ass'; Fig. 2.17), and presented a paper at it. There he had the good fortune to meet William Thomson, who later after a most distinguished career became Lord Kelvin. Kelvin very soon gave Joule strong support.

William Thompson (1824–1907; Fig. 2.18), was of Scottish descent, but was born in Belfast, Ireland. He was educated at the University of Glasgow and at St Peter's College (now called Peterhouse), Cambridge. He graduated as Second Wrangler in the Mathematical Tripos in 1845, and was at once elected a Fellow of St Peter's. He then spent a year in Paris working on heat with the distinguished French chemist Victor Regnault (1810–1878). When he first met Joule in 1847 he had been appointed the year before, at the age of 22, professor of natural philosophy at the University of Glasgow; he was to hold that position for over half a century. He is best known today by his title

**Fig. 2.17**   A caricature which appeared in 1896 in the *Liverpool Echo*, an evening newpaper. The British Association for the Advancement of Science held its annual meeting in Liverpool in that year, and the caricature shows a member of the public expressing his astonishment over the event. A memorable feature of the meeting was that Professor Oliver Lodge, professor of physics, in whose department the meeting was held, gave on 25 February a lecture on X-rays, discovered by Röntgen less than a year previously. The lecture was so popular that the doors of the Arts Theatre had to be closed after a thousand people had been admitted.

of Lord Kelvin, and to avoid confusion it seems better for us to call him Kelvin from now on, even though he did not receive that title for many years.

Kelvin's scientific work covered a wide range, and he made many contributions of the greatest importance both to science and technology. In his earlier years he was intimately concerned with the new science of thermodynamics, and it was he who first used the word 'thermo-dynamic', in 1849.

Kelvin was impressed by Joule's paper at the BAAS meeting, and had some private discussions with him. By a curious chance, two weeks later Kelvin was on a walking tour in Switzerland and unexpectedly ran into Joule who was carrying a large thermometer; although on his honeymoon, with his bride waiting patiently in a carriage not far away, the enthusiastic Joule was making

**Fig. 2.18** William Thomson (1824–1907), who later became Lord Kelvin, and is now usually remembered by that name. He encouraged Joule in his belief that heat is a form of energy, and did important work leading to the second law of thermodynamics.

temperature measurements at the top and bottom of a large waterfall. These meetings, and Joule's papers, finally convinced Kelvin that Joule was right; heat is a mode of motion, not a substance, and in a heat engine such as had been considered by Carnot there is an actual conversion of heat into mechanical work.

By the middle of the nineteenth century this view of heat had become generally accepted. The theory was often referred to as the 'dynamical theory of heat' or the 'mechanical theory of heat'. What was implied by these expressions was that heat should be understood as relating to the science of mechanics or dynamics.

## The principle of conservation of energy: or the first law of thermodynamics

Deciding that heat is a form of energy and that heat and work can be interconverted is not quite the same as deciding that energy is conserved. It is possible to conceive of a universe in which heat and not energy is conserved; for a time that opinion was held by many investigators, including Kelvin, and such a universe would not be greatly different from our own. A number of earlier investigators had decided that energy is conserved in purely mechanical systems involving collisions between spheres. For example, the great Dutch physicist and astronomer Christiaan Huygens (1629–1695) had carried out many ingenious experiments on colliding spheres, some of which he demonstrated at meetings of the Royal Society, and these led him to the correct conclusion that energy is preserved in an elastic collision. Similar ideas had been expressed by Leibniz and by Émilie, Marquise du Châtelet (1706–1749), a woman of remarkable ability who is noteworthy for having been Voltaire's mistress, and for her excellent translation into French of Newton's *Principia*.

There are indications that the idea of the conservation of energy was becoming recognized in the latter part of the eighteenth century. In 1775, for example, the French Académie des Sciences passed a resolution that they would no longer consider any machine claiming to be a 'perpetuum mobile'. This is a machine that was claimed to go on operating for ever, without any supply of energy from outside (what the Academie resolution actually said was rather different, as the word energy was not used at that time). A little later, in 1783, Sadi Carnot's father, Lazare Nicolas Marguerite Carnot (1753–1823), in his *Essai sur les machines en général*, suggested explicitly that energy is conserved and cannot be created. Decades after that, however, it was still maintained by many competent investigators that ways might be found to create from nothing unlimited amounts of energy. We have already met one such idea, in the suggestion by von Jacobi in 1834 that a battery operating an electric motor can generate unlimited amounts of energy. A hypothetical (and non-functional) device of this sort, which violates the first law of thermodynamics, is referred to as a perpetual motion machine of the first kind.

It was Mayer and Joule's work on the interconversion of heat and work that provided the best circumstantial evidence that there is conservation of energy. In 1848 the two men became involved in a bitter priority dispute, carried out mainly through the French Académie des Sciences, with other scientists taking sides. The question of priority seems of little importance today and is impossible to resolve since there were no crucial experiments that led inevitably to the conclusion that energy is conserved; the conclusion was seen by both Mayer and Joule to be a probable one in view of the results of experiments relating to the interconversion of heat and work. Mayer's paper of 1842 does give him chronological priority. However, Joule's evidence, based on his numerous careful experiments carried out shortly afterwards, was more thorough and reliable than the less direct evidence adduced by Mayer from the experiments of others.

Perhaps the most convincing expositions of the principle of conservation of energy were those of Grove and of Helmholtz. William Robert Grove (1811–1896; Fig. 2.19) had an active legal career as well as a scientific one. He went to Brasenose College, Oxford, at a tender age and obtained his BA degree in 1830 at the age of 19. He then practised as a barrister for a period, at the same time carrying out some scientific work; as will be discussed in Chapter 4, he designed some interesting batteries, including in 1839 a fuel cell that was a century ahead of its time.

**Fig. 2.19** A caricature of William Robert Grove (1811–1896), who combined a scientific and a legal career, and was one of the first to state the fiirst law of thermo-dynamics. The caricature is by SPY, the name used by Sir Leslie Ward (1851–1920). It appeared in the famous political and social periodical *Vanity Fair* (not to be confused with the more recent women's fashion magazine of the same name). The caricature was entitled 'Galvanic Electricity', referring to Grove's important work on electric cells (see Chapter 4). (Courtesy of the Royal Institution.)

On suffering some health problems he accepted in 1841 the position of professor of experimental philosophy at the London Institution, which was a little distance from the heart of London, at Finsbury Circus (the building still stands but is now occupied by the School of Oriental Studies of the University of London; the London Institution no longer exists). Scientists find it puzzling that Grove, although he pursued his scientific career vigor-ously, apparently believed it to be less physically demanding than a legal one. In 1846 he published his book *On the correlation of physical forces* in which he made a clear statement of the principle of conservation of energy (it should be noted that at the time there was some confusion about the exact meanings of the words 'force' and 'energy', which today are regarded as quite distinct; when Grove talked about force he meant what we now call energy). He re-signed his professorship in 1846 and resumed his legal career, but always retained an interest in science. He became a Queen's Counsel in 1853, and in 1856 was one of the counsel who unsuccessfully defended the notorious William Palmer of Rugeley who was convicted and hanged for the murder by poisoning of an associate (he also undoubtedly killed several creditors, his wife, an uncle, four of his legitimate children, and several of his many illegitimate children, but was not charged with those murders). Palmer committed his murders by the administration of antimony, strychnine, or prussic acid, and sometimes to make sure by a combination of all three. Grove was presumably chosen as counsel because of his chemical knowledge; his cross examination related to the administration of poisons. Grove became a judge in 1871 and was knighted in 1872.

The great German physicist and physiologist Hermann Ludwig Ferdinand von Helmholtz (1821–1894), who later became a close friend of Kelvin, did outstanding work in both physiology and physics. In 1847 he published his book *Über die Enhaltung der Kraft* in which he dealt with the conservation of energy in a comprehensive way. (Note again the use of the word *Kraft*, force, to mean energy.) He discussed in detail the dynamical theory of heat, and showed that in inelastic collisions the energy that is apparently lost is converted into heat. This publication probably did more than any other to lead investigators to accept the modern position with regard to energy, heat, and work. Helmholtz's important work in thermodynamics is considered later in this chapter.

It should be mentioned that a qualification to the principle of conservation of energy is required in view of Einstein's general theory of relativity (Chapter 8). According to this theory, mass ($m$) and energy ($E$) are interconvertible, the relationship between the two being $E = mc^2$ where $c$ is the speed of light. When nuclear transformations occur this relationship assumes great importance, but in ordinary chemical reactions the mass changes are too small to detect. This is discussed further in Chapter 8.

## THE SECOND LAW OF THERMODYNAMICS: THE DISSIPATION OF ENERGY

Joule's suggestion that heat is a form of energy, and that it can be converted into other forms of energy such as mechanical work, was only slowly accepted by other scientists. Kelvin, for example, took a few years to accept the idea. Paradoxically, the reason for his initial reluctance was that he had studied heat conversion more deeply than Joule had; as we discuss further in Chapter 9, in scientific research there is sometimes an advantage in not knowing too much about a subject. Kelvin knew that if there was an interconversion of heat and work, there was something not entirely straightforward about it. Work could be converted into heat without any apparent complications, as in Rumford and Joule's experiments, but there were obviously some restrictions on the conversion of heat into work. Kelvin grappled with this problem for some time, and was led to a deeper understanding of the restrictions and to what is now known as the second law of thermodynamics.

Kelvin realized, from his study of Carnot's work, that when an engine operates, not all of the heat absorbed can be converted into mechanical work. Some heat must also simply pass from a higher temperature to a lower temperature, and Kelvin referred to this as the *dissipation of energy*. He saw that it follows that an engine cannot operate at a single temperature. For example, a ship cannot propel itself by abstracting heat from the surrounding water, and causing the water to cool; the heat must be obtained from something at a higher temperature, and there must be dissipation of heat, some heat passing from a higher to a lower temperature.

In 1849 Kelvin suggested that Carnot's ideas could be understood more clearly in terms of an absolute temperature, which is now defined as the Celsius temperature plus 273.15 degrees. The absolute temperature has been named in honour of Lord Kelvin, and the symbol used is K; 25°C, for example, is 25 + 273.15 = 298.15 K. For a reversible Carnot engine, operating between two absolute temperatures $T_h$ and $T_c$, there is a simple equation for the efficiency of the engine, defined as the ratio of the net work done in the cycle to the heat absorbed at $T_h$. The equation is

$$\text{efficiency} = \frac{T_h - T_c}{T_h}$$

We can interpret this equation as follows. The amount of heat absorbed at the higher temperature is proportional to $T_h$, and the amount of heat rejected at the lower temperature is proportional to $T_c$. The net heat absorbed is therefore proportional to $T_h - T_c$. The work done by the engine is equal to the net heat absorbed, so that the ratio of the work done to the heat absorbed at the higher temperature is given by the above expression.

In 1851 Kelvin stated that

> it is impossible...to derive mechanical effect from any portion of matter by cooling it below the temperature of the coldest of the surrounding objects.

This is one statement of what has come to be called the *second law of thermodynamics*.

It may help to give examples of efficiencies. Suppose that a Carnot engine is operating reversibly, which means that the efficiency is the maximum possible. Suppose first that the lower temperature is 27°C (300 K) and the higher is 100°C (373 K). The efficiency of the reversible engine is then (373 − 300)/373 = 0.2 = 20%. In other words, only one-fifth of the heat taken in at the higher temperature 300 K can be converted into work; four-fifths is dissipated, or wasted, by simply passing from the higher to the lower temperature. In practice, since an actual engine cannot be reversible (because then it would operate infinitely slowly) the efficiency will be less than this.

If instead the higher temperature is 400 K, because of the use of high pressure steam, the efficiency is raised to (400 − 300)/400 = 25%. Now one quarter of the heat absorbed at the higher temperature 400 K has been converted into work; the wastage or dissipation is three-quarters, which is less than four-fifths. It will be remembered that Carnot had made the important point that a higher efficiency may be achieved by the use of higher temperatures. Towards the end of the nineteenth century this conclusion was put into practice by Rudolf Diesel (1858–1913), who designed an engine, named after him, in which $T_h$ is much higher than in a steam engine because of the use of oil instead of water. In designing his engine Diesel gave full attention to thermodynamic principles, which he expounded in his *Theorie und Konstruktion eines rationallen Warme-Motors* (1893).

Hypothetical machines in which the second law is allegedly circumvented are known as perpetual motion machines of the second kind. Attempts to get around the second law have always ended in failure. One noteworthy attempt was made by the inventor John Ericsson (1803–1889). Born in Sweden, Ericsson spent some time in England before settling in the United States, becoming a citizen in 1848. In most respects he was a highly competent engineer, constructing locomotive engines and a screw propeller which greatly improved navigation. For the US Navy he designed the *Princeton*, built in 1844, the first warship with a screw propeller and with engines below the waterline. Later he designed for the Navy an armoured ship that had a revolving gun turret.

Ericsson was familiar with the Stirling engines, but he misunderstood their function. He applied the name 'regenerator' to the displacer or economizer that the Stirling brothers had regarded as functioning solely by minimizing the wastage of heat. He thought that the Stirling engine allowed the heat rejected at the lower temperature to be utilized again, contrary to what had been concluded by Carnot, and contrary to the second law of thermodynamics. He apparently thought that he could violate the first law also, for in 1855 he wrote that 'we will show practically that bundles of wire [i.e. the 'regenerator'] are capable of exerting more force than shiploads of coal'. Ironically, none of his engines actually had a satisfactory 'regenerator'.

Ericsson built a vessel, named the *Ericsson*, which was fitted with 'caloric engines', in which this rejected heat was supposed to be 'regenerated' and used again. It was hoped that the *Ericsson* would cross the Atlantic, but we are not surprised today to learn that it was a disaster. It completely failed its trials in 1853 and had to be refitted with steam engines. Even these did not improve the unfortunate vessel's luck, as it sank to the bottom of the sea in 1854. Ericsson does not deserve too much blame for his mistakes; when he designed the ship the second law was understood by hardly anyone.

There is no reason to doubt Ericsson's honesty, but the same cannot be said of a number of other individuals who at about the same time made proposals which clearly violated the two laws of thermodynamics. One of the most notorious of these was the American swindler Robert Keeley, who tried to involve prominent men like the financier Cornelius Vanderbilt

(1794–1877) in his schemes. He claimed to have made revolutionary discoveries about the forces of nature. One of his projects, by which he wheedled millions of dollars out of credulous investors, involved releasing vast amounts of what he called atomic energy (perhaps the first use of that term) from small amounts of water. He demonstrated the effect by running a motor, which actually operated by means of compressed air entering through hidden pipes. He had the effrontery to try to get the eminent physicist Joseph Henry, then director of the Smithsonian Institution in Washington, to support his nefarious schemes, but Henry was not deceived, and denounced them.

## CLAUSIUS'S ENTROPY: THE ARROW OF TIME

Ideas similar to those put forward by Kelvin, but expressed in a more precise form, were put forward by the German physicist Rudolf Julius Emmanuel Clausius (1822–1888; Fig. 2.20). Clausius carried out the most important of his thermodynamic work in Switzerland, at the University of Zürich, where he was professor of mathematical physics from 1855 to 1867. In 1850 he made a statement of the second law, and in 1854 he presented a detailed mathematical analysis of the Carnot cycle. He suggested a new physical property, which in 1865 he called *entropy*.

Clausius's mathematical treatment was presented rather obscurely, and even highly competent mathematicians like Kelvin were unable to understand it. One person who did understand the treatment was James Clerk Maxwell,

**Fig. 2.20** Julius Clausius (1822–1888), who perhaps did the most important work in connection with the second law of thermodynamics.

who explained it very clearly in later editions of his book *Theory of heat;* even he, however, got it mixed up the first time he explained it, and had to apologize in the next edition. The mathematical treatment is beyond the scope of the present book, and we will have to be content with giving some idea of what entropy is. We can say that

> entropy is a property which helps to provide a numerical measure of the extent to which the heat in a system is unavailable for conversion into mechanical work.

Often we are particularly interested in the heat produced in a process, and with how that heat is related to the work that can be done. An electric battery provides us with a good example. We have seen earlier in this chapter that Joule was interested in the interconversion of heat and work, and that he made some measurements on electric batteries. He showed that the work that was performed by a battery (for example in driving an electric motor) was done as a result of the chemical reaction that occurred in the cell. He went wrong, however, in concluding that all of the heat released by the chemical reaction would be converted by a motor into mechanical work. The truth turns out to be a little different. As in a steam engine, some of the heat may be unavailable for conversion into work. This is the case if the chemical system undergoes a decrease of entropy as the reaction in the battery occurs. If this is so, the thermodynamic result is that the work done by the chemical system in the battery is the heat evolved minus the product of the entropy decrease and the absolute temperature. In other words, the unavailable energy is

> entropy decrease in the system × absolute temperature.

(If instead there is an entropy increase in the chemical system, which is possible, the work done by the battery will be greater than the heat that would be evolved—the unavailable energy is negative. This sounds as if we are getting energy for nothing, but we are not; the point is that heat transferred from the environment ensures a true energy balance.)

It is easy to understand why Joule reached the wrong conclusion, since his experiments were done before the idea of entropy had been suggested. It is interesting that Kelvin made the same mistake even though he based his conclusion on what seemed to be excellent experimental evidence. Kelvin considered a Daniell cell (Chapter 4), a popular cell at the time, and from the heat evolved in the chemical reaction he calculated what the voltage of the cell should be. His result agreed so well with the experimental value that there seemed no doubt that heat in a battery was simply converted into work, with no complications. The trouble turned out to be that for the particular reaction occurring in that cell, the entropy change is practically zero, so that the unavailable energy is almost zero. This example is instructive in showing us how easy it is to go wrong in drawing a scientific conclusion, even using excellent experimental evidence and sound logic. A certain amount of bad luck may also be involved, in choosing an appropriate system on which to experiment. (Of course, luck can be on our side; we shall discuss an example of this in Chapter 5 in connection with Maxwell's photoimage of a Scottish ribbon).

Clausius concluded that when any spontaneous process occurs—such as a building falling down or an explosion occurring—there will be an increase in the entropy of the universe as a whole. The increase need not be in the system itself; there can be a loss of entropy in the system, but the process will only occur if there is a greater gain in the surroundings, resulting from the emission of heat. Clausius's statement of the second law was thus that

> The entropy of the universe tends towards a maximum.

He explained that his choice of the word entropy was partly based on a Greek word meaning a transformation, a turn-about, or a change of direction.

At first entropy was difficult to interpret and to understand. Later it became apparent that entropy is related to the probability that a system exists in a particular state. There is a natural tendency for systems to pass from orderly states to states of greater disorder. The analogy of a deck of cards is often used to explain entropy. A deck can be arranged in a particular way, or it can be shuffled, and the shuffled deck is more probable than the ordered one. Shuffling an ordered deck will almost certainly produce a disordered deck. It is highly unlikely that a shuffled deck will become ordered if it is further shuffled. The British philosopher-scientist Sir Arthur Eddington (1882–1944) referred to entropy as the 'arrow of time'. Time cannot go backwards but only forwards, and this is because at a later time the state of the universe has a greater probability than at an earlier time.

We are all familiar with processes that occur naturally as a result of this tendency for an ordered state to become a disordered one. When a lump of sugar is dissolved in coffee we know that the molecules of the sugar spread themselves throughout the liquid; however long we wait we do not find the cube reforming itself—although if we could wait a very long time (perhaps longer than the age of the universe!) it would do so and at once dissolve again. We know that if a bottle of perfume is left open, the perfume will spread around the room, and we do not expect that in our lifetimes the molecules of perfume will go back again into the bottle. Some of us have seen a demonstration experiment in which oxygen and hydrogen gases are brought together, and a flame is put to the mixture; the gases explode with the formation of water ($H_2O$). However long we wait, a glass of water will not suddenly decompose into hydrogen and oxygen. The basic reason is that there is a great increase of entropy when the gases are exploded together, largely because heat is given off and is dissipated into the surroundings. In principle heat given off could assemble in a glass of water and decompose it into hydrogen and oxygen, but the probability of this happening is extremely remote.

Entropy is so subtle a property that many scientists were unable to understand it when Clausius first suggested it. Kelvin, for example, never appreciated entropy, and maintained that the second law can be more easily understood in terms of the dissipation of heat, which is easily visualized. Kelvin's philosophy of science was that everything must be explained in terms of a mechanical model, and entropy cannot be explained in this way. Properties like volume, pressure, and temperature can be measured with simple instruments and can be appreciated even by people who do not know much about science. Entropy, on the other hand, is elusive; no instrument can directly measure an entropy change, which has to be calculated from data involving heat and temperature changes.

Credit for interpreting the second law on a statistical basis goes to the great Scottish scientist James Clerk Maxwell (1831–1879), about whom much more will be said in Chapter 5. Here we will just consider his 'demon', which was a great help in understanding the second law.

## MAXWELL'S DEMON: ENTROPY AND PROBABILITY

Scientific ideas usually come to light in formal scientific papers, but Maxwell's proposal was an exception. It took the form of an imaginary supernatural being, later called Maxwell's demon, which was born in a letter that Maxwell wrote to Peter Guthrie Tait on 11 December 1867. Maxwell and Tait were old friends, the two having been together at school, at the University of Edinburgh, and at Cambridge. Throughout their adult lives they carried on a lively and amusing correspondence—sometimes, because they were thrifty Scots, by means of postcards which required only a halfpenny stamp instead of the penny stamp then needed for a letter. The point of this particular letter written by Maxwell was to show how, in principle (but hardly ever in practice), the second law could be violated.

Maxwell considered a vessel divided into two compartments, A and B, separated by a partition which had a hole in it that could be opened or closed by 'a slide without mass.' (Fig. 2.21). The gas in A was at a higher temperature than the gas in B. By this time Maxwell was well aware of the fact that in a gas at a given temperature the molecules will be moving with a variety of speeds; some will be moving fast, others slowly. Earlier Maxwell had, in fact, worked out a mathematical treatment of the distribution of speeds, and he is famous for this; it is called the *Maxwell distribution*. In a gas at a higher temperature, the average speed of the molecules will be greater than if the gas is cooler, but there still is a distribution of speeds.

Maxwell imagined 'a finite being', later called a demon, who knew the speeds of all the molecules. This creature would open the hole for an approaching molecule in A when its speed was low, and would allow a molecule from B to pass through the hole into A only when it was moving fast. As a result of this process, said Maxwell, 'the hot system has got hotter and the cold colder and yet no work has been done'. Of course, Maxwell did not imagine that his 'finite being' could exist; he emphasized that his intention in

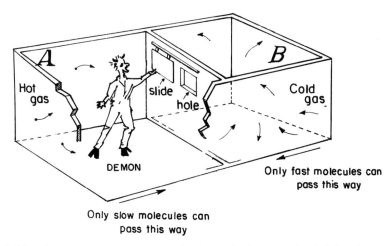

**Fig. 2.21** Schematic representation of Maxwell's demon. The left-hand compartment, A, contains gas that is hotter than the gas in B, which means that the molecules in A are on the average moving faster than the molecules in B. The demon allows only fast-moving molecules to move from right to left, and only slow molecules to move from left to right. The hot gas therefore gets hotter, and the cold gas cooler! Maxwell used this imaginary device to explain the second law of thermodynamics.

inventing it had simply been to provide us with an understanding of why the second law of thermodynamics applies; it is just a matter of probability. In a letter to J W Strutt (later Lord Rayleigh), written in December 1870, Maxwell commented that

> The 2nd law of thermodynamics has the same degree of truth as the statement that if you throw a tumblerful of water into the sea, you cannot get the same tumblerful out again (i.e, exactly the same molecules as before).

Maxwell's idea of a doorkeeper who could bring about a violation of the second law was soon taken up by others. It was Kelvin who first referred to the creature as a demon, and he later playfully endowed it 'with arms and legs—two hands and ten fingers suffice.'

It is interesting to note that earlier discussions of the second law made no reference to a process that today is often used as a good example of the application of the law, namely the mixing of gases at constant temperature. Maxwell's demon could reverse such a process; air, for example, could be separated into oxygen and nitrogen by the use of a demon who would allow only oxygen molecules to pass in one direction and only nitrogen molecules to pass in the other. The earlier workers thought of the second law as relating only to engines, where there is a passage of heat from a higher temperature to a lower one, and a mixing process is not obviously related to such transfers of heat.

Maxwell's ideas on the relationship of entropy and probability, and his treatment of the distribution of molecular speeds, were greatly extended by the Austrian physicist Ludwig Boltzmann (1844–1906; Fig. 2.22). In 1871 Boltzmann published a paper of great importance in which he created an entirely new branch of science, a branch that was later called statistical mechanics. In that paper Boltzmann showed how all the properties of a system can be calculated from a particular function, which can be obtained from fundamental data about molecules. In 1877 he published his famous relationship between entropy ($S$) and the natural logarithm (ln) of the probability, $W$, in terms of a constant $k_B$ which has come to be known as the *Boltzmann constant*:

$$S = k_B \ln W$$

This equation is engraved on Boltzmann's tombstone in Vienna.

Although Boltzmann made great contributions to science he became increasingly despondent in his later years. The reason for this was partly because some scientists at the end of the nineteenth century were expressing the opinion that atoms do not really exist, but are just a convenient fiction. One of these scientists was the great physical chemist Wilhelm Ostwald (1853–1932; Fig. 2.23), with whom Boltzmann was on friendly terms. These attacks struck at the roots of Boltzmann's entire life's work. He ably defended the atomic theory against its critics, and received strong support in many quarters. In spite of this his distress became more acute, and he made several unsuccessful suicide attempts. Finally, in 1906, on holiday with his wife and daughter and while they were out swimming, he hanged himself from the window of his hotel room. It is a sad irony that soon afterwards the French

**Fig. 2.22**   Ludwig Boltzmann (1844–1906), who extended thermodynamics into a new branch of science, *statistical mechanics*.

**Fig. 2.23** Friedrich Wilhelm Ostwald (1853–1932), who played a key role in the development of the science of physical chemistry. Of German ancestry, he was born in Riga, Latvia, then a province of the Russian empire, and much of his career was as professor of physical chemistry at the University of Leipzig. He lucidly explained the new principles of thermodynamics in his well-known textbooks, and made important contributions in a wide range of topics. He was awarded the 1909 Nobel Prize for chemistry.

physicist Jean Perrin (1870–1942) produced compelling evidence for the real existence of atoms—work that convinced even the sceptical Ostwald—and that Perrin's work was greatly influenced by Boltzmann's own contributions to the statistics of molecular motions.

## THE WIDE SCOPE OF THERMODYNAMICS

When Kelvin first invented the word thermodynamics in 1849 he chiefly had in mind its applications to steam engines and other machines in which heat is converted into mechanical work. Over the years, however, it became apparent that the subject has much wider applications. We have already seen an example of this, in that entropy provides an explanation for the mixing of gases, where a heat transfer is not necessarily involved.

It is now realized that thermodynamics is concerned in a fundamental way with whether any process, physical or chemical, occurs naturally or not. A simple example is provided by the chemical reaction between hydrogen gas and oxygen gas with the formation of water:

$$2H_2 + O_2 \rightleftharpoons 2H_2O$$

It is well known that at ordinary temperatures a mixture of hydrogen and oxygen gases is rapidly converted into water, for example when a flame is applied to the mixture. Water at ordinary temperatures, on the other hand, does not spontaneously decompose into the gases. The reason the reaction goes in one direction and not in the other is that the process from left to right involves a large increase in entropy. This is mainly because the formation of water from hydrogen and oxygen gives out a great deal of heat, which heats up the surroundings and greatly increases the entropy of the surroundings. Chemists have found it convenient to think in terms of *equilibrium constants* for reactions, the equilibrium constant being approximately the ratio of the product of the concentration of products to the product of the concentration of reacting substances. Thus for the hydrogen–oxygen reaction the equilibrium constant is

$$K = \frac{[H_2O]^2}{[H_2]^2 \, [O_2]}$$

At ordinary temperatures this quantity is exceedingly large, even though the reaction between hydrogen and oxygen is very slow when not induced by a flame or a spark.

It should be emphasized that thermodynamics by itself has nothing to say about the rates of chemical reactions (which we will consider briefly at the end of this chapter). The reaction between hydrogen and oxygen is actually extremely slow; if a mixture of the gases is left undisturbed, nothing observable will happen over billions of years. A spark or a flame will, however, cause the reaction to go explosively.

We have seen that the large increase in entropy when hydrogen and oxygen combine to form water is largely due to the heat evolved, which causes an increase in the entropy of the surroundings. It is inconvenient to have to consider entropy changes in the surroundings, and efforts were made to discover an alternative thermodynamic function, relating only to a chemical system, which would relate to the tendency of a process to be spontaneous. Success was first achieved by the American theoretical physicist Josiah Willard Gibbs (1839–1903; Fig. 2.24), who in the 1870s introduced the concept of *free energy*. The importance of this concept is that the free energy change in a reaction, which is a function of the entropy change and the heat evolved, must *decrease* if the reaction is to occur spontaneously. Between 1873 and 1878 Gibbs developed the whole subject of thermodynamics in a masterly and rigorous fashion. However, he was not able to make his ideas understandable. Also, Gibbs made no reference to the useful concept of equilibrium constant, perhaps because, having little contact with chemists, he had never heard of it. As a result of these communicational deficiencies only a few scientists (one of whom was Maxwell) had much idea of what Gibbs was saying, and only in the twentieth century did his work receive recognition.

**Fig. 2.24** Josiah Willard Gibbs (1839–1903), who developed many fundamental aspects of thermodynamics in the 1870s, but whose work was so abstract and obscurely presented that it had little impact in its time.

Because of this lack of appreciation of Gibbs's work, much of the same ground was covered independently by others, particularly by von Helmholtz, Max Planck, and van't Hoff.

Hermann von Helmholtz (1821–1894; Fig. 2.25), born in Berlin, took a medical degree in 1842 at the Friedrich-Wilhelm-Institüt in Berlin, and then served as an army surgeon. In 1848 he became professor of physiology at Königsberg, transferring to Heidelberg in 1858. In 1871 he moved to the University of Berlin as professor of physics. He had a remarkably original mind and covered a wide range of science, including physiological acoustics, physiological optics, thermodynamics, hydrodynamics, and electrodynamics. His early studies in physiology led him to consider how chemical energy, heat, and work are interconverted in living systems. It was mentioned earlier that in 1847, shortly after the work of Mayer and Joule, he discussed in detail the principle of the conservation of energy.

For some years after 1847 he devoted his attention mainly to physiological problems and to some aspects of electrochemistry. In 1851 he invented the ophthalmoscope and worked out a mathematical theory of it. In about 1880 he returned to the subject of thermodynamics and in 1882 he put forward an explanation of free energy that was slightly different from that of Gibbs.

Helmholtz achieved a position of high popular esteem, similar to that of his close friend Kelvin, and like him he received many honours. Some idea of his distinction is provided by the fact that he was offered professorships at both Oxford and Cambridge. He was offered the Oxford chair of experimental

**Fig. 2.25**   Hermann von Helmholtz (1821–1894), who played a role of great importance in the development and application of thermodynamics.

philosophy in 1865 when he was visiting Oxford, but he turned it down, preferring to remain at Heidelberg where the facilities were better. As it turned out, his refusal had a far-reaching effect on the development of physics at Oxford. Until that time Oxford had done rather better than Cambridge in all of the sciences, but in place of Helmholtz the appointment went to Robert Bellamy Clifton (1836–1921), who had been a Wrangler at Cambridge, and had served for a short period as the first professor of natural philosophy at the newly-formed Owens College in Manchester (which later evolved into the University of Manchester). Unfortunately, Clifton was of the opinion that 'the wish to do research betrays a certain restlessness of mind'. Since he himself did not suffer from that mental affliction, research in physics languished at Oxford during his 56 year tenure of the professorship. Clifton and his staff were conscientious in the teaching of physics, but a university science department cannot achieve more than mediocrity unless some first-class basic research is carried out. This is a remarkable example of how one person can sometimes have a strong effect on the development of a particular discipline at a university—sometimes, as in this case, a detrimental one, but sometimes a good one.

The offer of the Cambridge chair of experimental philosophy was made to Helmholtz in 1871, when the chair was created, and work on the construction of the Cavendish Laboratory had begun. The post had first been offered to Kelvin, but he was unwilling to leave Glasgow where he had established a distinguished school of research. It was at Kelvin's suggestion that the post was then offered to Helmholtz, but he had just accepted the chair at Berlin, and he also declined. The consequences to Cambridge were not as serious as they were to Oxford, as the Cambridge appointment went to Maxwell who, although he was not as good a teacher as Kelvin and Helmholtz, continued to carry out—in the remaining eight years of his too short life— research and scientific writing of distinction (see Chapter 5). Since Maxwell's succesors were also of great distinction, Cambridge became and remained a world centre for research in physics.

Max Planck (1858–1947), who is chiefly remembered today for his great work on the quantum theory (Chapter 7), also did important work in thermodynamics. Indeed, during his earlier years, at the Universities of Munich, Kiel, and Berlin, most of his efforts went into thermodynamics, including the theory of solutions. His work that led to the quantum theory in 1900 was thermodynamic in character. His achievements in thermodynamics, although always of high quality, were a considerable disappointment to him since time and time again he found that what he did had already been done by others, particularly by Gibbs, Helmholtz, and van't Hoff. He devoted some attention to chemical problems, but having been trained as a physicist he lacked the chemical insights that van't Hoff in particular applied to such great advantage. His book *Vorlesungen über Thermodynamik*, first published in 1897, did exert a considerable influence, giving an exceptionally clear and systematic treatment of the subject. It appeared in many editions (the 11th edition in 1966) and was translated into many languages, including English, Russian, and Japanese.

Jacobus Henricus van't Hoff (1852–1911; Fig. 2.26) was born in Rotterdam and studied in The Netherlands, Germany, and France, obtaining his PhD degree from the University of Amsterdam in 1874. In the meantime he had published a pamphlet, later recognized to be of great importance, on the shapes of organic molecules. After holding a rather unsatisfactory position at the State Veterinary School in Utrecht, he was in 1878 appointed professor of chemistry, mineralogy, and geology at the University of Amsterdam. There he embarked on his pioneering research in physical chemistry. In 1896 he moved to the University of Berlin, and in 1901 he was awarded the first Nobel Prize in chemistry, 'in recognition of the extraordinary value of his discovery of the laws of chemical dynamics and of the osmotic pressure in solutions'.

The thermodynamic work of van't Hoff, carried out for the most part at the University of Amsterdam, was in marked contrast to that of Gibbs and Helmholtz. During the period he was in Amsterdam van't Hoff carried out a comprehensive programme of research in thermodynamics and kinetics, which is concerned with the rates and mechanisms of chemical reactions. All of this work is conveniently summarized in his *Études de dynamique chimique*, which appeared in 1884, and in the second edition, in German, which was published in 1896. His thermodynamic work was less rigorous than that of Gibbs and Helmholtz, his concern being more with arriving at relationships that would be useful in the interpretation of experimental results. He

**Fig. 2.26** Jacobus Henricus van't Hoff (1852–1911), who derived many thermo-dynamic relationships, and applied them to practical chemical problems.

posed various chemical problems, such as the way a chemical equilibrium is shifted by various factors, and he derived simple equations that explained the results.

Van't Hoff's approach to thermodynamics was that of a practical chemist interested in understanding chemical reactions, and he was less concerned with scientific rigour than in arriving at simple relationships applicable to lab-oratory work. He was a modest and unassuming man who never made prior-ity claims; indeed in his later books he sometimes gave credit to others for things that he had first discovered himself. He tended to proceed independ-ently; although later recognizing the importance of the work of Clausius, Helmholtz, and Gibbs, he preferred to use the equations of thermodynamics that he had derived himself. He never, for example, made use of the concept of entropy, although he referred to it in his books. Instead of free energy he preferred to think of the work done by a chemical reaction.

Perhaps the most important contribution to thermodynamics in the book was the treatment of equilibrium constants. Van't Hoff was one of the first to recognize the dynamical nature of chemical reactions. Previously it had been generally assumed that when a chemical system is at equilibrium, all motion has stopped, but van't Hoff recognized that chemical change is still taking place, at equal rates in the two opposite directions. The idea was not original with him: it had been first suggested by the British chemist Alexander William Williamson (1824–1904) and had been accepted by Clausius.

Another who made important contributions to thermodynamics was Walther Nernst (1864–1941; Fig. 2.27). He was born in Brieson, West

**Fig. 2.27** Walther Nernst (1864–1941), who applied thermodynamics to electrochemical problems, and derived a heat theorem which has been called the third law of thermodynamics.

Prussia (now Wabrzezno, Poland), and obtained his doctorate from the University of Würzburg in 1887. He lectured first at Leipzig, and in 1891 became professor of physical chemistry at Göttingen, where he established a distinguished institute of physical chemistry. Early in his career Nernst did work of particular importance on the application of thermodynamics to electrochemical cells. Indeed, during the 1880s and 1890s he almost single-handedly transformed electrochemistry from a subject in which there was little understanding of anything beyond the basic principles into one in which all the thermodynamic aspects were well understood.

In 1905 he became professor of physical chemistry at the University of Berlin, and after 1922 he was at the same time president of the Physikalisch-technische Reichsanstalt of Berlin-Charlottenburg. By the time he took up his appointment in Berlin his interests had shifted in the direction of the behaviour of solids at temperatures close to the absolute zero, and for many years his laboratory was devoted to investigations of that kind. In 1906 he formulated an important theorem, '*Nernst's heat theorem*', about the behaviour of crystals at the absolute zero. This theorem is often referred to as the third law of thermodynamics, although Nernst himself did not favour this description.

Nernst ranks with van't Hoff, Wilhelm Ostwald, and Svante Arrhenius as one of the great pioneers of physical chemistry, and his work covered an even wider range than that of the other three. He won the Nobel Prize for chemistry in 1920 'in recognition of his thermochemical work', and his work on the heat theorem has often been regarded as his most important work.

His work in electrochemistry was also of fundamental importance and of lasting value.

Nernst exerted a wide influence in many ways. His book *Theoretische Chemie*, which first appeared in 1893 and went through many editions and translations, was the leading book in physical chemistry until the 1920s; in its preface Nernst said that his object was to present 'all that the physicist must know of chemistry, and all that the chemist must know of physics.' Many who later did important work in physics and chemistry gained much from their early association with Nernst; among them were the American chemists G N Lewis and Irving Langmuir.

Nernst had an unusual personality, and Einstein, himself hardly conventional, said that 'he was so original that I have never met anyone who resembled him in any essential way.' Nernst's character showed many contradictions: he could be extremely generous and helpful, and sometimes devastatingly sarcastic. He had many personal eccentricities, and many amusing stories are told about him. His own office and laboratory were always in a state of extreme chaos, which his colleagues appropriately referred to as a state of maximum entropy. To approach him at his desk one had to negotiate winding corridors through waist-high piles of books. Amazingly, if he wished to find a particular book himself he was able to push his hand into the confusion and extract it at once. In his laboratory he would sometimes carry out a quick experiment, perhaps with a galvanometer perched precariously on a pile of books, and miraculously obtain a reliable answer.

Nernst was devoted to his wife and family, with whom he had a close relationship. He left all the family arrangements to his wife Emma, who assumed many additional responsibilities. He dictated all his papers to her, and she also inspected the starched cuffs of his shirts before they were laundered; he was in the habit of making notes on the cuffs, and these she transcribed. He had no sympathy with research students who spent time taking exercise. The British physicist F A Lindemann (later professor at Oxford and the future Lord Cherwell) enjoyed playing tennis with his brother when both were working with Nernst. Nernst indignantly asked them why they had to hit that silly little ball from one to the other; their father was comfortably off, surely he could buy a ball for each of them.

Nernst was by no means noted for his modesty; his students reported that in his lectures he presented physical chemistry in a way that left the impression that he had done all the work himself. In 1937 he was awarded an honorary DSc degree by Oxford, and I attended the lecture he gave on that occasion; although his English was reasonably good, he gave it in German. He ended the lecture by saying that it had taken three people to formulate the first law of thermodynamics (perhaps he meant Rumford, Mayer, and Joule?), two people to work out the second law (presumably Kelvin and Clausius), but he had been obliged to formulate the third law all by himself. He added that it followed by extrapolation that there could never be a fourth law.

One story about Nernst was told to me shortly after the event by a senior German physical chemist, who knew the people involved. Nernst died in 1941 at his manorial estate near Bad Muskau in East Prussia, and was buried there. When the Russians annexed East Prussia his remains were moved to Berlin, where there was a second interment. Later his family decided that Nernst should be buried in Göttingen where he had been professor for most of his career, and there was a third interment. Present at all three interment ceremonies were several distinguished scientists, including the physical chemist Karl Bonhoefer (1899–1957). When one of them commented to Bonhoefer that he was getting a little tired of these interment ceremonies, Bonhoefer replied 'As far as I am concerned, we can't bury Nernst too often'.

Important contributions to thermodynamics were also made by the American chemist Gilbert Newton Lewis (1875–1946; Fig. 2.28), who was born in West Newton, Massachusetts. After obtaining his PhD degree at Harvard in 1897 he worked with Wilhelm Ostwald at Leipzig and with Nernst at Göttingen. For some years he taught at the Massachusetts Institute of Technology, and made important contributions to thermodynamics, being particularly concerned with what are called 'non-ideal systems'. Certain simple equations apply approximately to gases, one of them being Boyle's law, according to which the product of the pressure and volume of a gas is a constant at a fixed temperature. Gases that do not obey the laws, and liquids

**Fig. 2.28** Gilbert Newton Lewis (1875–1946), who made pioneering contributions to thermodynamics in 1900–1907, and in his book with Merle Randall, *Thermodynamics and the free energy of chemical substances* (1923). He also did important work on the theory of the chemical bond. (Photograph courtesy of Professor Michael Kasha, who took the photograph on Lewis's seventieth birthday).

that show deviations from other simple relationships, are said to be *non-ideal*. Lewis interpreted these deviations from ideality in terms of important new concepts called *fugacity* and *activity*. It is of interest that in his earlier thermodynamic work Lewis followed the form of thermodynamics that had been formulated by van't Hoff. Later, particularly in his important book on thermodynamics, he based his treatment more on the contributions of Willard Gibbs.

In 1912 Lewis moved to the University of California where he was killed accidentally in 1946, as a result of the escape of hydrogen cyanide vapour from apparatus in his laboratory. His work on the chemical bond is considered in Chapter 8.

## The fullness of time

In our discussion of thermodynamics I have so far avoided saying anything about the speed with which processes occur. This was a deliberate omission, for a very good reason. Thermodynamics by its very nature is not concerned with the speeds of processes, but only with the direction they take. We have seen that the second law is an arrow of time, pointing in the direction that is to be followed. But it does not command that its instructions be instantly obeyed; on the contrary, it is satisfied if they are obeyed in the fullness of time—and sometimes the time is extremely full. The second law is satisfied even if the changes occur over thousands of millions of years—only as long as they occur in the right direction.

We considered earlier the reaction between hydrogen and oxygen gases, and it again provides us with a good example. We can keep the mixture in a bottle for any length of time at ordinary temperatures, and will never be able to detect any change. The reaction under those conditions is indeed so slow that no one could possibly measure its rate, simply because we do not live long enough. From what we know of the way the reaction proceeds we can estimate that only after several thousand millions of years would anything detectable have happened. And yet from the second law, together with data about the reaction, we can calculate that after an enormous period of time at ordinary temperatures the reaction really does go in the direction of water formation.

At first this may seem puzzling, and perhaps unsatisfactory. What is the use of a law of thermodynamics that orders something to happen, and then doesn't care if there is such a dragging of feet that nothing seems to happen in a billion years? A sergeant major who ordered a private to shine his boots would hardly be pleased, and might easily be mildly irritated, if the response was 'Rightyho, they'll be done in a year or so'. Physical and chemical systems may be even more insolently insubordinate than our private, since they may follow the second law's command at such a snail's pace that nothing detectable has happened in a hundred years—perhaps not in a million years. We might think that if something is done as slowly as that, it might as well not be done at all. Does the second law really help by insisting on a direction, without imposing any speed requirement?

Yes it does, and we have already seen the answer earlier in this chapter. In discussing the hydrogen–oxygen reaction we mentioned that if we put a lighted match to the mixture, it explodes. It would not do that if the second-law direction were not the direction of water formation. In other words, the second law has told us something important about the reaction. It has told us that there may be ways of speeding the reaction up. Putting a light to it is one way. Another way is to heat the whole mixture—it will explode at about 500°C—but anyone trying this should be very careful, as the explosion can be violent and destructive. Another way is to introduce certain powdered solids, such as platinum or palladium, into the gas mixture.

The branch of science that deals with the rates of chemical reactions is known as *chemical kinetics* or *chemical dynamics*. It would require another long chapter to go into this topic, and here we must be content to cover just a few important points. It is convenient to think of any chemical reaction, like the one between hydrogen and oxygen, in terms of motion along a reaction channel. We can visualize this as a trough which connects the reacting substances (the hydrogen and oxygen gases) and the products (water). It is the shape of the trough that is important from the point of view of the speed of reaction.

All that the second law of thermodynamics has told us about this reaction is where the trough begins, and where it ends. We know that the energy of the product state (water) is much lower than the energy of the initial state (hydrogen and oxygen), as heat is given off to the environment when the reaction occurs. Figure 2.29 is a diagram which shows the energy levels of the initial, intermediate, and final states. (Strictly speaking we should be talking about free energies here, but the point can be made by referring simply to energy.)

The lines connecting the initial and final states of the system show us the way in which the troughs connect the two states. Line 1 goes through a maximum, which means that we have a trough that at first goes uphill as it leaves the initial state; it then curves downwards into the final state. We say that there is an energy barrier to reaction. Line 2 has no maximum; as we move along the trough we are going downhill all the way.

Line 1 represents—in a very simplified way—the kind of situation we have if the gases are just left together at room temperature. Because the trough goes over a rather high hill, it is very rare for any of the hydrogen and oxygen molecules to react with one another. Of course, if we have a dish of water, and insert a curved tube in it, the water will *never* run out. With a system like the mixture of hydrogen and oxygen the system is, however, a little different. There is a distribution of energy, in that a few molecules are moving with much more than the average speed, and a few are moving with lower speeds.

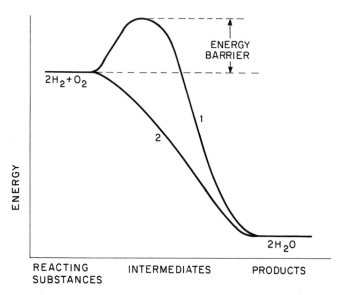

**Fig. 2.29**   An energy diagram for the reaction $2H_2 + O_2 \rightarrow 2H_2O$, showing the energy levels of the initial, intermediate, and final states. It is convenient to visualize the reaction as a motion along a reaction channel, or trough, connecting the initial and final states. Line 1 shows the path of a channel that initially rises, then passes through a maximum, and finally descends into the final state. Line 2 shows an alternative reaction channel in which there is no maximum; the channel descends all the way. Channel 1 corresponds to the situation at ordinary temperatures, with the gas mixture undisturbed. Because the barrier to reaction is quite high, the reaction is exceedingly slow. Channel 2 represents a new reaction path, made possible if a flame is applied to the mixture, or if certain substances are added to it. The two alternative channels are useful also for explaining an uncatalysed reaction (channel 1) and a catalysed reaction (channel 2). For a catalysed reaction there is often a low barrier rather than no barrier at all.

Occasionally, molecules will come together with enough energy to get over the barrier. In this reaction, at ordinary temperatures this is exceedingly rare, which is why the reaction is so incredibly slow. If we raise the temperature, the chance of two molecules colliding with enough energy to react is increased. As we have seen, if we raise the temperature to about 500°C, the mixture will explode.

Line 2 represents the situation if we have done something special, like putting a flame to the mixture, or introducing powdered palladium metal. Another reaction path is now possible for the system, and it is one in which there is no barrier to reaction—the trough goes downhill all the way. What the flame does is to introduce a few atoms into the system. Some of these are hydrogen atoms, which react extremely rapidly with oxygen molecules:

$$H + O_2 \rightarrow OH + O$$

Oxygen atoms react very rapidly with hydrogen molecules:

$$O + H_2 \rightarrow OH + H$$

Each of these reactions has produced OH, which is called the hydroxyl radical, and is very reactive. It reacts with $H_2$, for example:

$$OH + H_2 \rightarrow H_2O + H$$

(The heavy arrows in these equations indicate that these reactions are believed to occur in a single stage; we say that they are elementary). Since each of the two OH radicals does this we will have $2H_2O$ (+ 2H). Notice that we started with one H atom, and that after these reactions have occurred we have three H atoms in addition to the two water molecules! Each one of the three H atoms can perform the same trick over again, and we will then have nine H atoms and six $H_2O$ molecules—in addition to the two we got earlier.

It is easy to see that the reaction gets faster and faster, and since each individual reaction is fast it turns out that within about a millionth of a second there has been almost complete conversion into water. In other words, the mixture has exploded; since heat is given out the bottle gets warm, and there is a flash of light. This kind of reaction, in which a cycle of reactions occurs many times, is called a *chain reaction*. In Chapter 9 we shall refer to chain reactions again, from the point of view of how they came to be discovered. A chain reaction, by the way, does not always lead to an explosion, or even to a very fast reaction. It has done so in the case of the hydrogen–oxygen reaction because the chain process had a special feature. In each step of the cycle more hydrogen atoms are produced than we started with, so that the reaction went faster and faster.

It was mentioned that the reaction can also be made to go very fast by the addition of something like powdered palladium metal, and the explanation is similar. On the surface of some solids, hydrogen molecules easily dissociate into hydrogen atoms, and these initiate the same kind of chain reaction that we discussed. This effect of metals has been known for a long time. In 1817 Humphry Davy was working on the miners' safety lamp that first made him famous. He noticed that if he took a mixture of coal gas and oxygen and introduced a hot (but not red-hot) piece of platinum or palladium into it, it then did become red-hot. There was no explosion of the gases, but rapid reaction was obviously occurring. Later the German chemist Johann Wolfgang Döbereiner (1790–1849), on the basis of Davy's discovery, designed a hydrogen-air lamp that could be ignited by contact with a special form of platinum. Such lamps had some popularity until friction matches of the modern type became available later in the century. Döbereiner made no money out of his invention, as he had taken out no patent.

A schematic diagram like that in Fig. 2.29 can be drawn for any reaction, and is useful for illustrating various ideas in chemical kinetics. One of these is the idea of catalysis, which is the speeding up of a reaction by an added substance which is itself unchanged in the reaction. We have just seen one type of *catalysis*, involving a metal introduced into the hydrogen–oxygen mixture; this is a rather special type of catalysis, in which the metal initiates the reaction, which continues in the gas mixture. In the usual type of catalysis by a solid, the reaction occurs entirely on the surface of the solid. One of the best-known examples is the Haber–Bosch catalyst used for the production of ammonia from nitrogen and hydrogen:

$$N_2 + 3H_2 \rightarrow 2NH_3$$

During World War I this reaction was particularly important to the Germans, who needed ammonia for the production of explosives. About four-fifths of our atmosphere is nitrogen, $N_2$, but molecular nitrogen is quite inert and cannot be converted into explosives unless it is first converted into a more re-active form, such as ammonia. The reaction with hydrogen to form ammonia is referred to as the fixation of nitrogen. The difficulty about the reaction is that at ordinary temperatures the reaction is exceedingly slow. One could speed it up by increasing the pressure, but it was still too slow. The usual method of raising the temperature did more harm than good, as it shifted the position of equilibrium in such a way that little ammonia was formed. (This is a direct consequence of the second law: the total entropy must increase for a reaction to proceed, and for this reaction a high temperature causes ammonia to decompose into nitrogen and hydrogen).

The only practicable way to obtain ammonia, therefore, was to use high pressures and find some way of speeding up the reaction without using too high a temperature. At the Kaiser Wilhelm Institüt in Berlin, Fritz Haber (1868–1934), with his colleague Carl Bosch (1874–1940) who was his brother-in-law, devoted their attention to this problem. It was known that iron was a catalyst, and they found that treating the iron with small amounts of certain other chemicals brought about a great improvement in the efficiency of the catalyst. This was of great practical importance to the Germans in their production of explosives. The award to Haber of the 1918 Nobel Prize for this work led to some criticism, particularly as Haber had also worked on poison gases, which were used by the Germans during the war. In 1933 Haber, who was Jewish, resigned his position and left Nazi Germany, dying in Switzerland in the following year. Bosch shared the 1931 Nobel Prize, particularly for his work on reactions at high pressures.

There are other types of catalysts. Acids and alkalis are often catalysts, and are widely used in industrial work. Digestive and metabolic processes in animals and plants are brought about by special catalysts called enzymes. All of them are large molecules, and practically all are proteins (Chapter 8).

Catalytic behaviour can be understood in terms of a diagram like Fig. 2.29. Line 1 now represents the shape of the reaction channel for the uncatalysed reaction. It goes uphill at first, and the reaction is slow. Addition of the cata-lyst introduces another mechanism for the reaction, with a lower barrier, and the reaction proceeds more rapidly. Note that if an added substance allowed the reaction to proceed along a path having a higher barrier, the rate would be unaffected, as the reaction would continue to occur along the original path. Chemists used to talk about 'negative catalysts', but there is really no such thing. There are substances that reduce the rates of reactions, and these are best called inhibitors. They can act in two ways. Some of them interfere with the action of a catalyst already present in the system. Other inhibitors act by removing active intermediates from the reaction system; the active intermediates might, for example, be the hydrogen atoms which we saw to be important in the explosion between hydrogen and oxygen.

Reaction barriers, which chemical kineticists call *activation energies*, are of very great importance. We can measure the height of a reaction barrier by studying the way in which temperature affects the rate of the reaction. We noted that as the temperature is raised it is easier for the reacting molecules to surmount the barrier. Some reactions have zero activation energy (i.e., there is no barrier), and temperature then has hardly any effect on the rate. If, on the other hand, the activation energy is large, the rate is greatly affected by the temperature. For many of the ordinary chemical processes that we are familiar with in daily life, such as those involved in cooking, the rate of reaction roughly doubles if we raise the temperature by ten degrees Celsius. It is for this reason that oven temperatures have to be controlled fairly carefully. The digestive and metabolic processes that occur in the body are also quite sensitive to temperature, which is why hyperthermia (too high a body temperature) and hypothermia (too low a temperature) can have such serious effects.

The mathematical relationship between the rate of a reaction and the temperature was first suggested in 1884 by van't Hoff, and his discussion was extended in 1889 by Svante Arrhenius. The equation they arrived at is usually known as the *Arrhenius equation*. It applies to a remarkably wide range of chemical and physical processes. Chemists have applied it successfully to an enormous number of chemical reactions. A less conventional example of an Arrhenius plot is shown in Fig. 2.30. The test of the equation is whether the points lie on a straight line, which they do in this case. In some country areas the local people do make rough estimates of the air temperature by counting the rate of chirping of crickets. Naturally they do not make Arrhenius plots (perhaps the odd physical chemist has done so) but use some kind of rule-of-thumb formula. The Arrhenius equation has been shown also to apply to the rate of flashing of fireflies, and to the rate of creeping of ants.

From the steepness of an Arrhenius plot we can calculate the height of the energy barrier to reaction. There is one rather simple and straightforward conclusion that can be drawn from barrier heights. Physical processes, such as the flow of a liquid and the deformation of a solid, always have rather low energy barriers to reaction. Chemical processes, on the other hand, often have much higher barriers. By a chemical process we mean a process in which chemical substances are changed into other substances, so that there is a

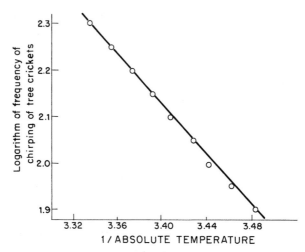

**Fig. 2.30**  An Arrhenius plot for the rate of chirping of tree crickets. The logarithm of the rate is plotted against the reciprocal of the absolute temperature. The height of the energy barrier can be calculated from the steepness of the line. A steep line means that the barrier is high; since the system has difficulty surmounting the barrier, the rate is greatly increased by raising the temperature. If there is no barrier (line 2 in Fig. 2.29) temperature has hardly any effect. A straight line from this type of plot is found for many processes. Arrhenius plots are important in the study of the mechanisms of processes of all kinds.

breaking of chemical bonds and a rearrangement of atoms. Energy is required to break chemical bonds, which is the reason that there is a fairly high energy barrier.

If, therefore, the steepness of an Arrhenius plot tells us that there is a fairly high energy barrier, we can be fairly sure that some chemical reaction is controlling the process that is being studied. This is the case, for example, for the chirping of crickets (Fig. 2.30), the flashing of fireflies, and the creeping of ants. We conclude, therefore, that the rates of these processes are controlled by some chemical reaction occurring in the creatures.

The Arrhenius equation even applies to psychological processes, such as subjective time—our impression of how rapidly time passes. Shakespeare was aware of this subjective time. In *As You Like It* he wrote

> Time travels in divers paces with divers persons. I'll tell you who Time ambles withal, who Time trots withal, who time gallops withal, and who he stands still withal.

A person with a high temperature tends to think that others are reacting slowly—they are too slow to bring them cups of coffee, for example. An abnormally low temperature has the opposite effect. With this idea in mind the American biochemist Hudson Hoagland made studies of hospital patients whose body temperatures varied over a few degrees. He had them count at a rate of what they believed to be one a second, and found that the rate of counting was greater in the case of persons with a high temperature. He then made an Arrhenius plot of his results, and obtained a satisfactory straight line. The height of the barrier was typical of the barrier heights obtained for chemical reactions in the body. It thus seems that our sense of subjective time is controlled by some chemical reaction in the body—perhaps a reaction occurring in the brain.

Rates of forgetting have been measured over a range of temperature by Heinz von Foerstler. Persons were asked to remember a string of nonsensical syllables, and measurements were made of the times for which they were able to keep this material in their minds. Again, the Arrhenius equation was obeyed, and the height of the barrier was much the same as obtained in the counting experiments. Again, the evidence is that processes occurring in the brain are controlled by chemical reactions.

It is hardly necessary to add that the existence of energy barriers is essential for the universe to be unfolding the way it does. If there were no barriers, and the second law were obeyed with military promptness, there would at once be chaos. Suppose that the world were as it now is, and the barriers were all removed. Every tree would at once react with the oxygen of the atmosphere, and the same would happen to our bodies. The universe as we know it is therefore as much controlled by the laws of chemical dynamics as by the laws of thermodynamics.

THREE

# Daguerre, Talbot, and the legacy of photography

It is hard to think of any invention that has changed our lives more substantially, or a technique that has contributed more to the progress of science, than photography. Modern newspapers contain many photographs, some transmitted by radio over large distances. Before photography became possible, the only pictures possible, including portraits, had to be made by hand, with much room for error. Many of the more recent techniques, such as television and photocopying, are extensions of photography, and could not have existed if photography had not been invented.

We can conveniently restrict the word photography to procedures in which cameras are used to produce fixed images. Related techniques, such as television, we can call *photoimaging*, by which we mean creating an image by photochemical means. The term photoimaging can also be applied to some of the pioneering work done before 1839, the year in which photography was announced almost simultaneously in France and England.

To a great extent the invention of photograpy was an empirical one, owing little to pure science. Before photography was introduced in 1839 there had been a few relevant scientific observations. In the eighteenth century it was discovered that silver salts are blackened under the influence of light. The first important work in photochemical imaging was done in the last two decades of that century by Elizabeth Fulhame, about whom almost nothing is known. She was probably the wife of an English or perhaps Irish doctor, Thomas Fulhame, and lived from about the middle of the eighteenth century until some time in the nineteenth century. In 1794 she published privately a remarkable book *An essay on combustion* (Fig. 3.1) in which she made two pioneering contributions; one was concerned with photochemical imaging, the other with what came to be called catalysis. She was the first to be successful in creating patterns on dyed materials. She did this by impregnating cloths with silver and gold salts, and exposing them to light. In her book Elizabeth Fulhame said that she had begun her scientific work in about 1780, so that perhaps she was born about the middle of the century. In 1810 she was reported to have been an honorary member of the Philadelphia Chemical Society, but that is the last we hear of her.

In 1800 the distinguished chemist Joseph Priestley (1733–1804), then residing in Pennsylvania, published a book in which he mentioned that he had met Elizabeth Fulhame in London in 1783, the year before he emigrated to

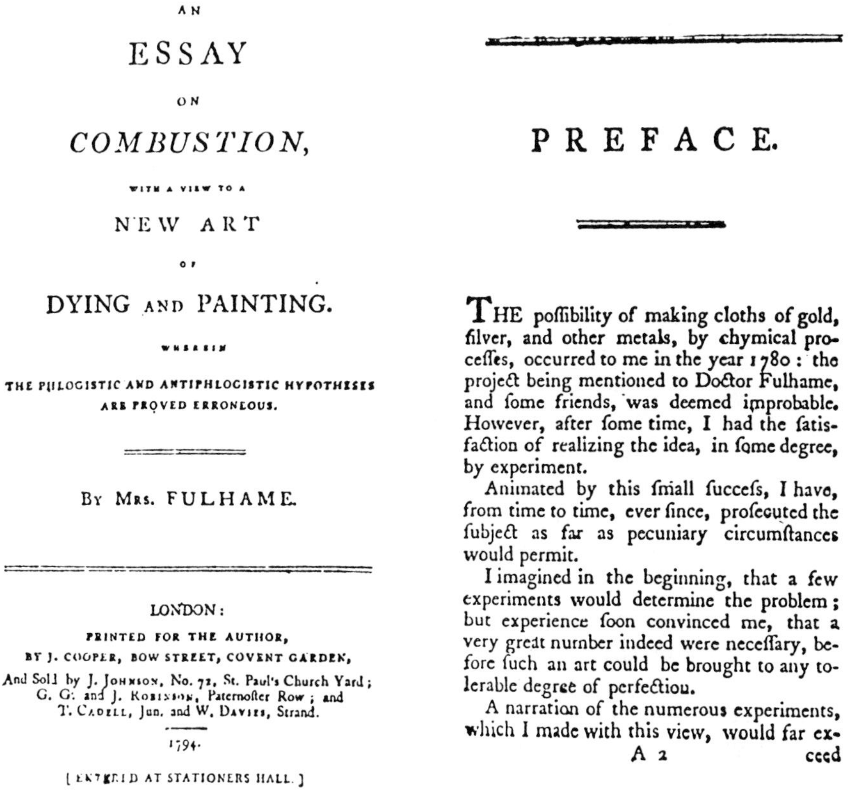

AN

# ESSAY

ON

## COMBUSTION,

WITH A VIEW TO A

### NEW ART

OF

### DYING AND PAINTING.

WHEREIN

THE PHLOGISTIC AND ANTIPHLOGISTIC HYPOTHESES
ARE PROVED ERRONEOUS.

By Mrs. FULHAME.

LONDON:

PRINTED FOR THE AUTHOR,
BY J. COOPER, BOW STREET, COVENT GARDEN,

And Sold by J. JOHNSON, No. 72, St. Paul's Church Yard;
G. G. and J. ROBINSON, Paternoster Row; and
T. CADELL, Jun. and W. DAVIES, Strand.

1794.

[ ENTERED AT STATIONERS HALL. ]

## PREFACE.

THE possibility of making cloths of gold, silver, and other metals, by chymical processes, occurred to me in the year 1780: the project being mentioned to Doctor Fulhame, and some friends, was deemed improbable. However, after some time, I had the satisfaction of realizing the idea, in some degree, by experiment.

Animated by this small success, I have, from time to time, ever since, prosecuted the subject as far as pecuniary circumstances would permit.

I imagined in the beginning, that a few experiments would determine the problem; but experience soon convinced me, that a very great number indeed were necessary, before such an art could be brought to any tolerable degree of perfection.

A narration of the numerous experiments, which I made with this view, would far exceed

A 2                              ceed

(a)                              (b)

**Fig. 3.1**  (a) The title page of Elizabeth Fulhame's book. (b) The first page of the Preface of the book, in which reference is made to the possibility of making 'cloths of gold, silver, and other metals'. This she did by impregnating the cloths with salts of the metals, and exposing them to light. This is believed to be the first time photo-imaging had been achieved. Note the reference in the Preface to Dr Fulhame. This may be Dr Thomas Fulhame who obtained an MD degree from the University of Edinburgh in 1784.

America. It seems likely, in view of a comment in the Preface of Mrs Fulhame's book, that it was this meeting that encouraged her to publish it. After settling in Pennsylvania, Priestley became a member of the Philadelphia Chemical Society, and perhaps it was he who proposed her for membership.

During the next few decades more work was done on photochemical imaging, and it all contributed to the final success of photographic techniques. In 1798 Count Rumford (Benjamin Thompson), whom we met in the last chapter (Fig. 2.13), published an account of experiments along the same lines as those of 'the ingenious and lively Mrs Fulhame', as he called her, saying that his results agreed entirely with hers, and that her work had inspired his own. At about the same time Thomas Wedgwood (1771–1805), a son of the famous potter Josiah Wedgwood (1730–1795), carried out similar investigations, the results of which were included in a paper that Humphry Davy published in 1802 on his own studies of photochemical imaging. This paper made no mention of Mrs Fulhame, but presumably Wedgwood and Davy were aware of her work.

## NIÉPCE AND DAGUERRE

The investigations mentioned so far created images by placing flat objects in contact with sensitized materials. Thus, although contributing to the development of photography, they cannot be classified as photography since they made no use of cameras. The first to have used a camera to secure an image was Joseph Nicéphore Niépce (1765–1833; Fig. 3.2), who achieved success in about 1816, although the oldest specimen of his work that still exists was made a decade later. One of the light-sensitive materials he used was bitumen, and his interest was in photoengraving as much as in photography. In 1829 Niépce entered into partnership with Louis Jacques Mandé Daguerre (1787–1851), who had already been making experiments along similar lines. Four years later Niépce died; Daguerre continued the work, and by about 1837 made his great achievement, the *daguerreotype* as it came to be called.

**Fig. 3.2**  Joseph Nicéphore Niépce (1765–1833), who is believed to have taken the first photograph, in about 1816. From 1829 until his death in 1833 he was in partnership with Daguerre. Niépce's scientific knowledge was much greater than that of Daguerre, and he contributed a great deal to Daguerre's ultimate success in 1839.

Daguerre (Fig. 3.3), like Niépce, had no training in science. He was a painter of theatrical scenery and a showman. One of his achievements was to invent the diorama—small scale scenery in front of which three-dimensional figures are displayed. His greatest achievement was to invent a practical way— although as it turned out not the most convenient way—of taking a photograph. A daguerreotype is created from a silver-plated sheet of copper, sensitized by exposure to iodine so that light-sensitive silver iodide is deposited. Exposure to light causes the silver iodide to be reduced to silver to an extent depending on the intensity of the light. The exposed plate, which as yet bears no visible image, is next brought into contact with mercury vapour; this combines with the silver, forming what is called an *amalgam*, and the image is then visible. The plate is then placed in a strong solution of sodium chloride (common salt), which renders the unchanged silver iodide fairly insensitive to light. In the resulting daguerreotype the regions that were exposed to light have been converted into silver amalgam, which is pale in tone, while the unexposed surfaces consists of the bare mirror surface, which is dark. Some of the early daguerreotypes can still be seen today in various museums, and are remarkably clear. The main disadvantage of the method is that since daguerreotypes are created on metal it is not possible to make copies of them directly. Another disadvantage is that, unless special mirrors

**Fig. 3.3** Louis Jacques Mandé Daguerre (1787–1851), from a daguerreotype made in 1848 by J E Mayall. Daguerre was trained as a painter, but was primarily a showman, and his elaborate stage designs in the 1820s won him a wide reputation. In 1922 he opened his Diorama for the display of large panoramic views, which were enhanced by dramatic changes in lighting. He perfected his daguerreotype technique in about 1837, and announced it in 1839.

are used, they are laterally reversed; if one parts one's hair on the left, the daguerreotype shows it parted on the right.

Daguerre brought his invention to the attention of several well-known French scientists, particularly François Arago (1786–1863) who was very influential; he had been appointed Director of the Paris Observatory, was perpetual secretary of the Académie des Sciences, and was an elected member of the Chambre des Députés. Arago carried out experiments of his own, and confirmed Daguerre's conclusions. On 7 January 1839 he presented the technique at a meeting of the Académie des Sciences in Paris, but provided little technical information, so that it would have been impossible for anyone to reproduce the procedures.

Arago also made an appeal in the Chambre des Députés for state support for Daguerre's further investigations, and was successful; Daguerre was awarded a state pension for the rest of his life. In August 1839 Daguerre also tried to come to some arrangement with the British government, to avoid taking out a British patent to protect his invention. In this, however, he was unsuccessful, and he was forced to take out a British patent. Details of his process were then made public for the first time.

## WILLIAM HENRY FOX TALBOT

In Britain in the meantime investigations with the same objective, but with the use of quite different techniques, had been pursued by William Henry Fox Talbot (1800–1877; Fig. 3.4). Talbot's background was very different from that of Daguerre. He was a man of independent means, and had been educated at Cambridge where he studied mathematics and a certain amount of science. He became an enthusiastic amateur scientist of broad interests, including mathematics, chemistry, botany, optics, art, archaeology, and linguistics. He had carried out some research in spectroscopy, and had published a few papers, as a result of which he was elected a Fellow of the Royal Society in 1832.

In about 1833 Talbot deposited silver chloride on paper, and obtained images by placing an object such as a leaf in contact with it and exposing it to sunlight. He then made the paper fairly insensitive to further exposure to light by washing with a strong salt solution. He was also able to prepare positive prints from negatives; the prints, of course, did not suffer from lateral reversal. Later he was successful in making negatives and then positives by the use of small box cameras. Because he used a paper base his photogenic drawings, as he called them, were much less sharply defined than daguerreotypes, but ultimately his method was more useful as it led, after a number of refinements, to the techniques in use today. Since a negative was involved, any number of copies could be made.

Talbot had so many interests that he tended to jump from one subject to another. He often dropped important work that he had started, and this he

**Fig. 3.4** William Henry Fox Talbot (1800–1877), from a daguerreotye taken by Antoine Claudet in 1844. His interests were broad; he was a mathematician, archaeologist, linguist, and chemist. Having invented a photographic technique in the 1830s he lost interest and immersed himself in antiquarian research. On hearing of Daguerre's announcement in 1839, he quickly resumed his photographic work.

did with his pioneering photographic experiments. After carrying them out in the middle 1830s, he became more interested in other matters, such as antiquarian research. However, on hearing of Daguerre's announcement in January 1839 he at once realized that his own contribution was in danger of being scooped, and he decided to take immediate action.

Talbot first announced his method on 25 January 1839, at one of the regular Friday evening meetings of the Royal Institution, with Michael Faraday present and with his enthusiastic cooperation. He exhibited samples of his work, including a photograph of his own house taken in 1835. In a paper presented on 21 February 1839 sufficient details were given to enable others to repeat the experiments, and a full account of this paper later appeared in the *Proceedings of the Royal Society*. Although Daguerre has priority in announcing his technique, Talbot has priority in giving a full scientific account of his method of taking a photograph by means of a camera.

Another action taken by Talbot was to write letters to Arago and other French scientists, claiming priority over Daguerre. However, after discussions between them, the French scientists continued to uphold Daguerre's priority. This was reasonable, as Talbot had failed to make any announcement of his work until Daguerre had done so. In the end, however, it was Talbot's method, after many refinements, that prevailed. Daguerre's method gave better images than Talbot's, but had the important disadvantage of not providing copies. This difficulty could not be overcome, and the daguerreotype is now of only historical interest. It must be recognized, however, that Daguerre's work had the effect of stimulating further work in the field of photography. If he had not made his announcement in 1839, would we ever have learnt of Talbot's work? Photography might not have evolved for several more decades.

## SIR JOHN HERSCHEL

At about the same time that Daguerre and Talbot announced their techniques, Sir John Herschel (1792–1871; Fig. 3.5) was carrying out similar studies. He was the only son of the famous astronomer Sir William Herschel (1738–1822; Fig. 3.6), who discovered the planet Uranus and two of its satellites, and the rotation of Saturn's rings. He was also the nephew of the famous astronomer Caroline Herschel (1750–1848) who discovered eight comets and many star clusters. John Herschel was educated at Cambridge, where he was Senior Wrangler in 1813. He became a Fellow of St John's College, Cambridge, and missed by one vote being elected professor of

**Fig. 3.5**   Sir John Frederick William Herschel, Bart (1792–1871), from an oil painting of him by Christian Albrecht Jensen, painted in 1834. At this time Herschel was actively experimenting with a variety of photographic techniques, and had already made some coloured photographs of spectra. He was highly skilled in chemistry, but his main work was in astronomy, in which he followed in the footsteps of his father Sir William Herschel and of his aunt Caroline Herschel.

**Fig. 3.6** Sir William Herschel (1738–1822), from a portrait painted in 1785. He was a refugee from Germany who first earned his living in England as an organist, music teacher, and composer; he became interested in astronomy, began to construct telescopes, and made important contributions to spectroscopy.

chemistry at Cambridge. He was a scientist of great versatility and originality, and did outstanding work in a number of fields, including astronomy. He was elected a Fellow of the Royal Society in 1813, was knighted in 1831, and was created a baronet in 1838. He followed in Newton's footsteps by becoming Master of the Mint in 1850. Today he is not as well remembered as his father, but in his time he was held in high popular esteem. When he died in 1871 he was mourned by the whole nation, and was buried beside Newton in Westminster Abbey.

John Herschel became personally acquainted with Talbot in 1824, when they were both visiting Munich, and there learned of his photographic achievements. On hearing, in January 1839, of the announcement made by Daguerre he at once began a detailed investigation of a number of related problems. He was particularly interested in trying to improve Talbot's techniques. On 14 March 1839 he communicated to the Royal Society a paper in which he introduced the procedure, still used today, of 'fixing' negatives with a solution of sodium thiosulphate, which he called 'hyposulphite', a name preserved in the word 'hypo', still employed by photographers. At this meeting Herschel exhibited a number of samples of the photographs he had taken, one being of the famous telescope that his father had built at Slough (it is fortunate that he took the photograph as the telescope was demolished a few years later).

On 5 March 1840 Herschel presented to the Royal Society a pioneering paper entitled 'On the chemical action of the rays of the solar spectrum on preparations of silver and other substances, both metallic and non-metallic; and on some photographic processes.' This paper described a detailed investigation of the photographic effects of different spectral regions, including the ultraviolet and infrared. Herschel reported further work along the same lines in subsequent publications.

It seems to have been Herschel who first used, in March 1839, the word 'photograph', instead of 'photogenic drawing' which Talbot had used. He was also the first to refer to 'negatives' and 'positives'. Herschel also gave us the word 'snapshot', which is commonly used today. Previously the word 'snap-shot' had been used by hunters to refer to a hurried shot that they would take at a bird or rapidly moving animal, without taking deliberate aim. In an article in the *Photographic News* in 1860 Herschel referred to 'the possibility of taking a photograph—as it were by a snap-shot—of taking a picture in a tenth of a second of time...'. At the time, because of the insensitivity of photographic plates, long exposures, of up to a minute even for portraits, had to be used. Herschel was looking forward to the possiblilty of more sensitive materials, so that much shorter exposures were possible. His word certainly caught on, as it is now so widely used. Today it refers not so much to a photograph with a short exposure—since practically all exposures are less than a tenth of a second—as to one taken at short notice, with no posing of the subjects, and often without their knowledge.

In 1842 Herschel described a procedure for producing a 'cyanotype'—or as we should say blueprint. Blueprints were still made, particularly by architects, until the 1950s, when they were replaced by xerography (from the Greek *xeros*, dry, xerography being a dry process). All of Herschel's work was done in the spirit of pure scientific enquiry. He never competed with his friend Talbot, as he might easily have done; instead he always tried to help him. He never took out a patent or profited in any way from his work, and he rarely took a photograph for pictorial purposes; most of his experimental images were contact copies of engravings, made with the object of studying the chemical aspects of the process.

It was the method initiated by Talbot that ultimately became the photography that is practised today. In 1840 Talbot introduced the *calotype*, a refinement of his photogenic drawings, and patented the process on 8 February 1841, describing it to the Royal Society on 10 June. Much of the success of the Talbot method is due to the improvements made by Herschel. As early as March 1840 Herschel had described to the Royal Society a method of precipitating on glass a coating of photosensitive material, in this way making it possible to improve the quality of the positive prints that could be made.

In 1851 Frederick Scott Archer (1813–1857) introduced a wet-plate process in which a film of collodion was attached to a plate of clear glass.

Since there was now no paper base the images were much clearer, and this method soon became widely used both by scientists and by artistic photographers. Dry plates were introduced into photography in the 1870s, and were much more convenient than the wet plates. One manufacturer of dry plates was the American inventor George Eastman (1854–1932), who in about 1878 succeeded in attaching the light-sensitive emulsions to strips of paper, and later to celluloid, which could be rolled on spools. This new technique was of great practical value to all photographers. Later Eastman founded the Kodak Company, which through its research laboratories and enlightened business practices has exerted such a wide influence on the improvement of photographic techniques. The word Kodak has no etymology; it was invented by Eastman out of thin air, its merits being that it is short, can hardly be mispronounced or misspelt, and does not resemble any other trade name.

By 1850 or so a good deal of photography was being carried out by professionals and by amateurs. At first patents presented something of a problem, but there were so many technical improvements during the 1840s that the patents began to give little protection. Daguerreotypy was protected by patents in the United Kingdom but not in North America, where the technique was much used for a few years. Methods based on Talbot's invention soon gained ascendancy everywhere, because of its much greater convenience.

## EDWIN LAND

Important contributions to photography were made from the 1930s and for many years by the American scientist and inventor Edwin Herbert Land (1909–1991; Fig. 3.7). Land is remarkable for the fact that he was a highly original pure scientist, an extremely skilful technologist, an effective scientific administrator, and at the same time an excellent man of business. His first important contribution was in 1932 when he invented the plastic material called *polaroid*, which is so widely used today. To understand its function we must know a little about the vibrations that occur in light (Fig. 3.8). In ordinary light these vibrations are in all directions at right angles to the direction in which the light is travelling (Fig. 3.8(a)). For a long time it has been known that certain crystals, such as tourmaline and calcite (also known as Iceland spar) were able to cause light to be polarized, which means that the vibrations are confined to one direction (Fig. 3.8(b)). In particular, William Nicol (1768–1851), at the University of Edinburgh, cemented two crystals of Iceland spar together with Canada balsam. For many years the resulting Nicol prisms were widely used for the investigation of polarized light and its interaction with matter, but were somewhat inconvenient to make and to use.

Land's idea was to produce a material that could be made easily in large quantities. He had this idea while still a freshman at Harvard, and he decided

**Fig. 3.7**    Edwin Herbert Land (1909–1991), who is best known for his Polaroid Land Camera, in which the picture is developed inside the camera in a very short time. Land also invented polaroid, and made fundamental contributions to the basic theory of colour vision.

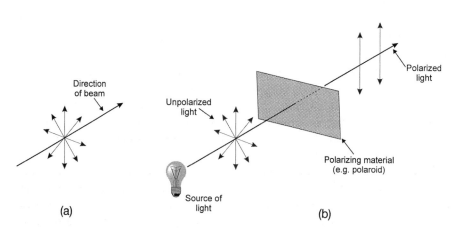

**Fig. 3.8**    (a) The vibrations that occur in ordinary light. These vibrations are at right angles to the direction of propagation of the light. (b) The production of plane-polarized light by passing ordinary light through a polarizing material. The vibration is now in one direction only (up and down in this diagram).

to leave the University while still in his first year and continue his education at the New York Public Library—as Thomas Edison had done several decades earlier. When he and his wife Helen were reading in the library they came across an interesting observation made in 1852 by an English physician, Dr William Bird Herapath. He had noticed that dogs treated with quinine had microscopic crystals in their urine, and that these crystals would polarize light. Seen by themselves the crystals were completely transparent, but if two of them were crossed, the region of their intersection was black. The substance became known as *herapathite*, and was later found to be iodoquinine sulphate.

Land decided to line up suitable tiny crystals of herapathite and immobilize them in a transparent plastic, so as to produce a plastic polarizing material. He later said that his original purpose had been to reduce car accidents at night, caused by the glare of headlights, which at the time could not be dipped. His idea was that headlights could be fitted with his polaroid, and that windshields (windscreens) would be fitted with the same material oriented at an angle so as to reduce the glare. It was later found that there are many other uses for the material.

He set up his own private laboratory to achieve his purpose, and took out his first patent on polaroid in 1930 when he was twenty. During the course of his early investigations he effectively overcame in an original way what is a common dilemma for many research workers today. To get support for one's research one has to convince others that one will ultimately be successful, but one can only do so by actually carrying out the research. Land's problem was that to orientate his herapathite crystals he needed to use an extremely powerful electromagnet, and there was a suitable one in the physics department of Columbia University in New York. However, on applying to the chairman of the department for permission to use the magnet he was informed that it would be granted only if he could provide evidence that the experiments would produce successful results. He could do that only by doing the experiments. Later, having carried out the experiments with the successful production of promising material, Land reapplied for permission, and this was then freely granted; he was given a key to the room and could carry out his experiments with greater convenience.

What the department chairman fortunately did not know at the time was how Land had done his experiments. He had in fact been using the Columbia electromagnet, which was in a locked room on the sixth floor of the physics building. On a number of occasions Land had gone up to the sixth floor in the evening when few people were about, had climbed out of a window and had scrambled along a ledge on the outside of the building, finally climbing through another window to gain access to the electromagnet. Receiving the key to the room merely gave him a more convenient and safer way of doing the experiments.

Land later returned to Harvard, still technically a freshman but giving some physics lectures. He never took his Harvard degree, but at the end of his career he had about twenty honorary doctorates, including one from Harvard awarded in 1957.

He established the Polaroid Corporation in Boston in 1932. Polaroid's use in sunglasses is based on the fact that reflected light undergoes some polarization. If one looks at a horizontal surface, such as a body of water or a paved road, the reflected light has more vibrations in the horizontal direction than in the vertical direction. Lenses fitted with polaroid so oriented that they cut out the horizontal vibrations will therefore cut out much of the polarized light, and will reduce the intensity of the unpolarized light by removing its horizontal component. The result is a great reduction in the glare, which means that one can see objects more clearly because of the greater contrast with their surroundings.

Polaroid's application to sunglasses came about in an interesting way. One of Land's colleagues took some polaroid film on a trout fishing expedition, and found that on putting it to his eyes in a suitable way he was less bothered by the glare of the sun's reflection and could see the fish more easily. At first polaroid sunglasses were available only in sports stores, for use by fisherman, but they were soon found to be useful for a wide variety of purposes, particularly for skiing or driving a car.

Polaroid also found an important use as a filter in photography, particularly to enhance the contrast between the clouds and the sky. The reason the sky is blue is that there is scattering of the light from the sun, the blue end of the spectrum undergoing more scattering than the red end. The scattering is due partly to dust particles in the atmosphere, but also to random fluctuations in the density of the atmosphere. The scattered light is partially polarized to various degrees depending on the position of the sun in relation to the region of sky that is being observed. At all points the polarization is quite substantial, so that a polaroid filter effectively reduces the amount of light coming from the sky, and greatly improves the quality of a photograph of clouds.

In 1947 Land produced his famous Polaroid Land Camera, which produces prints within a minute or so of taking the photograph. Land later recounted how this invention came about. In 1944 he was in Santa Fe, New Mexico, with his three-year-old daughter, and when he took a picture of her she asked why she could not see it at once. This set him thinking, and while walking around the city he considered whether it would be possible to develop and print film more rapidly. By this time he was already thoroughly familiar with the technical details of photography, and within an hour had been able to outline in his mind the way to proceed.

The technique is extremely ingenious, and in bare outline is as follows. The camera contains film consisting of negative film and positive paper, with sealed containers of chemicals in between. After the photograph is taken the pack is passed through a roller and the chemicals are released; the final positive print automatically appears in a very short time. Extension of the method to colour photography, in 1963, is mentioned later.

The original polaroid cameras went on sale for $89.75; today they can be obtained at an even lower price. One of Land's maxims was that to be commercially successful a device must retail at less than $100; if it costs more it will be bought only by the wealthy, and there are too few of them. Land himself, of course, became extremely wealthy as a result of his many inventions.

During World War II Land invented many important devices, including infrared polarizers and dark-adaptation goggles. Later he designed high-altitude optical surveillance systems for use in planes and satellites. On one occasion he gained the support of President Eisenhower, an enthusiastic golfer, by explaining a surveillance system not in technical terms but by demonstrating how well it could detect a golf ball at 2000 yards.

## COLOUR PHOTOGRAPHY

Today we think of colour photography as something that became possible in the present century, and indeed commercial coloured prints did appear only in the 1950s. More than a century earlier, however, some colour photography had been carried out. The first to be successful was John Herschel, who added various dyes to his photographic paper films. Inevitably, by modern standards, the quality of his coloured prints left a good deal to be desired.

Another scientist to have been successful in colour photography was Alexandre Edmond Becquerel (1820–1891). He had been interested in Daguerre's invention from the time of its first announcement in 1839, when he was only 19 years of age. By 1845 Becquerel had recorded coloured solar spectra on daguerreotype plates; to produce them he built up layers of various chlorides and other chemicals on the plates. Becquerel later became professor of physics at the Conservatoire and then director of the Musée d'Histoire Naturelle, succeeding his father Antoine César Becquerel (1788–1878) who had done important early work in electrochemistry; since Antoine remained active to a ripe old age and died only 13 years before his son he was able to enjoy Edmond's successes. Edmond's own son, Antoine Henri Becquerel (1852–1908), shared with the Curies the 1903 Nobel Prize for physics, for his discovery of radioactivity.

The investigations just mentioned, by John Herschel and Becquerel, were empirical, and were not closely related to the theory of colour and of colour perception. The work that eventually led to the widespread use of colour photography was firmly based on colour theory. We should therefore now consider some of the principles involved.

The first important work on the nature of colour was carried out in 1676 by Isaac Newton (1642–1727), in rooms at Trinity College, Cambridge. Ingenious detective work by Lord Adrian (1889–1977), who was Master of Trinity from 1951 to 1956, has revealed that Newton probably did the optical experiments in two different rooms, both in Great Court: the first experiments in a room with a window facing west between the Master's Lodge and the Chapel, and later experiments in a room between the Great Gate and the Chapel.

Before Newton's time it had been noticed that colours can be formed from white light, but there had been no careful investigation of the matter. Newton allowed a ray of sunlight to pass through a hole in the shutter of a dark room, and then through a prism where it was split into a band or spectrum of coloured light (Fig. 3.9); a rainbow is a familiar example of such a spectrum, and it is often said that there are seven colours in the spectrum, sometimes identified as red, orange, yellow, green, blue, indigo, and violet.

Newton then did something even more important: he showed that the band of light could be reconstituted into white light by passing it through another prism (Fig. 3.9). He explained his results in terms of the refraction of light, which was well known in his time, and is familiar to everyone. If a stick is placed partially in water it appears bent, which means that the light is bent when it passes from water to air or vice versa. The action of a lens, such as a spectacle lens, depends on the refraction of light by glass.

Newton's conclusion that white light is composed of seven basic colours is entirely reasonable, but artists have long been aware of the fact that any desired hue can be obtained by combining pigments in just three primary colours; for example, red, green, and blue pigments can be used as the primary colours. Important scientific work based on the idea of three primary colours was carried out in the early nineteenth century by Thomas Young (1773–1829). Young was one of the most talented persons who ever lived; he is comparable in intellect with Newton, Maxwell, and Einstein, although he did not achieve as much as any of them. The reason was that his interests were too broad, and that he never quite finished anything he did. Towards the end of his life he said that he had never spent an idle day in his life, but

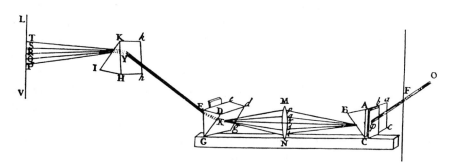

**Fig. 3.9**  The illustration of Newton's spectrum experiment that appeared in his *Optics* (1704). Sunlight enters the room through a hole F in the window shutter. After passing through the first prism it produces the spectrum pqrst. The lens MN focuses it at X where it enters a second prism. There the rays are reconstituted into white light EY. A third prism produces a spectrum again. Newton emphasized that the reconstituted beam EY had, as far as he could determine, the same properties as the original beam.

admitted that his main contributions had been to make 'acute suggestions', and to leave it to others to work out the details.

Young was an infant prodigy who could read at the age of two, studied Latin at six, and while still in his teens had a working knowledge of more than a dozen languages, including Hebrew, Chaldean, Arabic, and Turkish. He studied medicine at the Universities of London, Edinburgh and Göttingen, and then practised medicine, and wrote medical treatises. In 1801 he became the first professor of natural philosophy at the Royal Institution, but was not a successful lecturer. For this reason, and because his relations with the Institution's managers became strained, he resigned in 1803. His main scientific contributions were to clarify the understanding of vision, to discover the interference of light, and to outline a wave theory of light (these topics are considered further in Chapter 5). Before continuing to develop his wave theory, however, he returned to his long-standing interest in languages and hieroglyphics. (It was he who provided the key to deciphering the Egyptian hieroglyphics on the famous Rosetta Stone, but that work was completed by the French scholar Jean François Champollion (1790–1832), to whom the credit is usually given.)

Young's first contribution to physiological optics was to show that the lens of the eye changes its focal length, and so makes a contribution to the eye's ability to accommodate, i.e., to focus on objects at various distances. Previously it had been thought that accommodation was due solely to changes in the curvature of the cornea (which in fact do not occur) and to changes in the depth of the eyeball. Young then went on to study colour vision. He concluded that the eye contains three colour receptors, corresponding to red, yellow, and blue, and that the eye recognizes colours by the superposition of images from these receptors. When lights of these three colours are mixed together in suitable proportions the result is white light.

James Clerk Maxwell, whose career is considered in some detail in Chapter 5, took up the subject where Young had dropped it. In 1855 he presented to the Royal Society of Edinburgh a paper entitled 'Experiments on colour, as perceived by the eye, with remarks on colour-blindness'. He demonstrated what had become his favourite instrument in this field, a specially designed colour top, having a flat surface to which he could attach coloured sectors of various sizes (Fig. 3.10); this was an improvement over the device that Young had used for mixing colours. Maxwell's article, largely experimental, is a model of thoroughness, and marks the beginning of the science of quantitative colorimetry.

Maxwell's investigations on colour showed that although red, yellow and blue make an acceptable set of primary colours, a better set is provided by red, green, and blue. Maxwell distinguished clearly, for the first time, between *hue* (or colour), *tint* (or saturation), and *shade* (lightness or intensity). The hue, which is defined by the wavelength of the light, simply means what we call the colour; the hue may be red, yellow-green or blue, for

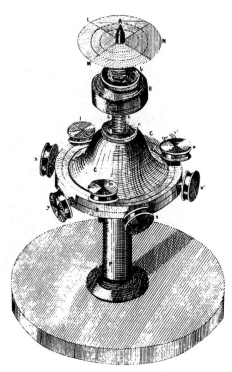

**Fig. 3.10** A colour top as used by James Clerk Maxwell. Tinted papers could be inserted into adjustable sectors, so that when the top was spinning colour mixtures could be compared.

example. The term tint or saturation has to do with the extent to which the hue has been influenced by the addition of other hues. If, for example, blue is the only hue present, the saturation is said to be 100%; the tint is a pure tint. The shade or intensity is related to the amount of light that reaches the eye. If an object is illuminated by white light and it reflects all the incident light it has maximum brightness and appears white. If it reflects no light, it appears black. In the case of a self-luminous object, the shade can be called the brightness.

Maxwell used a number of observers for his investigations on colour, one of whom was his wife, who assisted him in many other ways. Some of his observers suffered from colour-blindness. All of his results convinced him of the essential correctness of Young's three-receptor theory.

His procedure was to obtain matches between various mixtures of colours, and to relate the compound colours to the primary ones by means of equations. He constructed colour diagrams consisting of equilateral triangles, with the primary colours at the angular points. Any colour produced from a mixture of only two primaries was represented by a point on the side of the triangle. If three primary colours were involved the point was within the diagram.

In 1858 Maxwell abandoned the colour top and arranged for the construction of a colour box (Fig. 3.11), with which he could combine light of different colours; he later constructed other colour boxes based on the same principle. When Maxwell was professor at King's College, London, he worked on one of these colour boxes in the attic of his house in Kensington. When his neighbours observed him constantly staring into this box, which was painted black and looked like a coffin, he gained quite a reputation as an eccentric. His wife and several others assisted him in making observations with the box. In 1860 he presented to the Royal Society a major paper, 'On the theory of compound colours, and the relations of the colours of the spectrum', which was later published in the *Philosophical Transactions*. In it he established which colours had to be added or subtracted to produce any compound colour.

In 1855 Maxwell discussed the implications of his conclusions with respect to the possibility of colour photography. In his own words:

> Let a plate of red glass be placed before the camera, and an impression [i.e., a photograph] taken. The positive of this will be transparent wherever the red light has been abundant in the landscape, and opaque where it has been wanting.... Let this operation be repeated with a green and a violet glass, and by means of three magic lanterns (which we now call projectors) let the three images be thrown on the screen. The colour at any point on the screen will then depend on that of the corresponding point on the landscape and...a complete copy of the landscape will be thrown on the screen.

On 17 May 1861 Maxwell made a demonstration of the effect, at the Royal Institution, in the presence of Michael Faraday among many others. The object he photographed was 'a bow made of ribbon, striped with various colours'; the colours were red, green, and blue, and the ribbon was tied into a

**Fig. 3.11** Maxwell's colour box, built in 1860. It allowed coloured lights to be mixed. A similar box, painted black and looking like a coffin, attracted the puzzled attention of neighbours when the Maxwells resided in Kensington.

rosette. Three black and white photographs were taken, through red, green, and blue filters. The negatives were made on wet collodion containing silver iodide, and from them glass positives were prepared. Three projectors, lighted by red, green, and blue lamps, were then used to produce superimposed coloured images from the corresponding three black and white positives. The result was a reasonably good replica of the ribbon. As will be seen in Chapter 5, there are some rather interesting aspects of the demonstration (it only worked by good luck). Maxwell is sometimes said to have produced the first coloured photograph, but it seems better to call it the first photoimage, as no permanent record had been obtained.

What is important about Maxwell's demonstration is that he had established the basic principle behind a particular type of colour photography, the *additive* principle. The colours seen on the screen had been obtained by addition of the primary colours in the right proportions. Actually Maxwell had rather simplified matters by using a ribbon with only the three primary colours red, green, and blue. If he had included yellow stripes on the ribbon, they would have been reproduced by a mixture of red and green light. Maxwell was well aware of that, and he had done many experiments to demonstrate how colours combine by addition. He was being rather cautious in avoiding all but the primary colours.

The additive principle is appropriate for the projection of an image on a screen, where one is producing the colour by combining lights coming from the projector. It is the additive principle that is used in television, the colour being produced by tiny phosphors which glow with red, green, and blue light when activated by electron guns which rapidly scan the screen. However, when one looks at a coloured picture on an opaque background, such as a painting or a coloured photographic print (as opposed to a transparency), the situation is different. The light that one sees on the picture is supplied by the light in the room, which is more or less white light, a mixture of all the colours of the spectrum. When the white light strikes the picture, the dyes subtract light from it, and the reflected light has a colour that is complementary to the colour of the dye. For example, if white light strikes a dye that subtracts the orange-red light from it, the reflected light has a blue-green colour, known as cyan. If for the three primary colours we select orange-red, green, and violet, their complementary colours are as follows:

| Primary | Complementary |
| --- | --- |
| Orange-red | Blue-green (cyan) |
| Green | Blue-red (magenta) |
| Violet | Red-green (yellow) |

It follows that the colour of a body depends on the colour of the light that falls on it. An object that appears green in white light also appears green in green light. In light with which it does not interact, however, it appears black. A piece of glass that appears blue in white light may do so because it absorbs red light but reflects and transmits green light as well as blue light (Fig. 3.12). A piece of glass may appear yellow because it transmits red and green, but not blue light. Suppose that, as shown in Fig. 3.12, beams of red, green, and blue light are mixed together to give white light, and are first passed through blue glass. The red light is absorbed, and the light that continues is blue-green. If this is passed through yellow glass the blue is absorbed, and the light that passes through is green.

Much the same applies if light strikes the surface of a solid and is reflected. Some of the light is absorbed at the surface, and the reflected light has the colour of the original light minus the colour absorbed. It follows that if we blend together particles of blue and yellow glass, the mixture appears green. The same argument illustrated in Fig. 3.12 applies, and the resulting colour has been obtained by a *subtractive* process. The mixing of pigments by a painter is another example of a subtractive process. A green paint, for example, may be obtained by mixing yellow and blue paints. The same principle applies to obtaining a photographic print, as opposed to a transparency.

The principle of subtractive photography was first put forward in the 1860, shortly after Maxwell performed his demonstration, by the Frenchman Louis Ducos du Hauron (1837–1920). In 1869 he made a specific proposal for producing a coloured print by a subtractive process. He suggested that three negatives could be made of the same subject, by orange-red light, by green light, and by violet light. Prints would then be made from the three negatives, on transparent paper, using the dyes of the complementary colours. For example, the negative made with orange-red light would be printed with a cyan (blue-green) dye. The negative made in green light would be printed with a magenta (blue-red) dye, and the negative in violet light with a yellow (red-green) dye. When the three negatives were carefully superimposed, the result would be a print in the correct colours.

The subtractive process has an important advantage over the additive one. The additive process involves the use of filters, which often transmit as little as

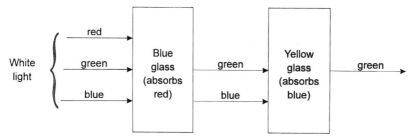

**Fig. 3.12**  White light passing first through blue glass, which absorbs the red so that what passes through is blue-green. When this passes through yellow glass the blue is absorbed, and what remains is green. This subtractive principle applies to the mixing of pigments, and to the production of a photographic print (as opposed to a transparency, or a television image, where the additive principle applies.)

10% of the light of the appropriate colour falling on them. A number of special cameras were devised, by du Hauron and others, for producing three separate negatives from a single exposure. Much effort also went into combining the three positive prints to obtain a single positive print.

Many people tried to devise an alternative procedure for colour photography, avoiding the need to combine separate positive prints. Again, Ducos du Hauron did pioneering work. In 1869 he proposed a screen made up of fine lines in the three complementary colours. By the selective blocking of the lines of the screen any colour could be produced. It was many years before this idea could be used in a practicable method of colour photography.

The idea was first used effectively by two brothers, Auguste Marie Louis Nicholas Lumière (1862–1954) and Louis Jean Lumière (1864–1948), who made many important contributions to photography. In 1907 they introduced their Autochrome plates involving the colour additive method. Their single-plate process was the first practical colour system to reach the public. Their method involved the use of a mosaic, of red, green, and blue filters made up of dyed starch particles. This was supported on a glass plate or film. An emulsion was attached to the mosaic, and when the plate was exposed the light passed through the plate to the emulsion. The plate was then reversed to a positive which could be projected on to a screen or viewed by transmitted light. The colours were very satisfactory, and the method could be used by professional photographers, but was rather too complicated for use by amateurs. A disadvantage of the process was that the exposure times required were much longer than for a black and white photograph. This additive system has now been much improved.

A considerable number of subtractive colour processes, producing prints on paper, have been put forward, the first in about 1902. Perhaps the most successful of these was the Kodacolor system, introduced in 1942. It involved the superposition of three films in each of which was a sensitive emulsion; one of the films contained red particles, another had green particles, and the third blue particles. Over the years there have been steady improvements in the colours obtained and the speed of the films.

Even before Maxwell did his work it had been recognized that it is not necessary for all three primary colours to be present to produce white light. Only two colours need be present, but these must be properly chosen. Particularly important contributions to colour theory were made by Edwin Land. The classic theories of Thomas Young and James Clerk Maxwell were concerned with isolating small regions of a picture and dealing with the colour characteristics of those regions. It is well known, however, that one's perception of a colour is strongly affected by neighbouring colours. A coloured photograph or painting may therefore be completely acceptable even though the colours of individual small regions of it may not exactly correspond to the original. Land paid particular attention to this matter, and developed a theory, his *retinex* theory, that took these perceptions into account.

Land's practical conclusion from his basic studies was that one could simplify colour photography by going to a two-colour, instead of a three-colour, system; this was technically much simpler, and the results were just as satisfactory. He was then able to produce in 1963 a colour film which could be developed inside the Polaroid Land Camera in a very short time.

## ASTRONOMICAL SPECTROSCOPY

One branch of science that was completely transformed by the invention of photography was astronomy. No doubt it was partly for this reason that John Herschel, the son and nephew of great astronomers and himself a distinguished astronomer, devoted much attention to the improvement of photographic techniques. Before the invention of photography, astronomers had to use drawings in their scientific papers, but photographic records are obviously more accurate and reliable.

The first person to take an astronomical photograph was John William Draper (1811–1882; Fig. 3.13), who was born in St Helens, Lancashire, the son of a Methodist preacher. In 1829 he began premedical studies at the University of London (later called University College) and was granted a 'certificate of honours in chemistry'. In 1832 Draper left England for Virginia and in 1838 was appointed professor of chemistry and physiology at New York University, where he remained until his death. For a period he was President of the New York University School of Medicine.

**Fig. 3.13** John William Draper (1811–1882). Born and educated in England, he did most of his scientific work at New York University.

Draper was another photographic pioneer. Even before Daguerre's announcement in 1839 he had taken temporary photographs by coating surfaces with potassium bromide and other materials. In December 1839 he was successful, using Daguerre's technique, in obtaining an excellent portrait of his sister Dorothy Catharine, with an exposure of 65 seconds. He announced this achievement in the *Philosophical Magazine* of March 1840; the portrait still survives, and is probably the oldest photographic portrait.

In 1841 Draper also took a daguerreotype of the moon, the first astronomical photograph to be taken. The first photograph—again a daguerreotype—of the sun, was taken in 1841 by two Frenchmen, Hippolyte Fizeau and Léon Foucault. We will meet these two remarkable investigators again, particularly in Chapter 5 in connection with their measurements of the speed of light.

Some of Foucault's work on the photography of stars is of special interest. Such photography is difficult, because of the low light intensities. Long exposures are required, and to compensate for the apparent movement of a star on account of the earth's rotation it is necessary to use a clockwork-driven siderostat to keep the telescope accurately trained on the star. Foucault used a clock which was driven by a pendulum, and he noticed in 1851 that over a period of time there was a continuous change in the plane of oscillation of the pendulum. In fact, the plane rotated completely, through 360°, in one day, and Foucault realized that this was a convincing demonstration of the rotation of the earth.

Later in 1851 Foucault was able to persuade the French authorities to allow him to install a large pendulum in the Panthéon in Paris, so that the public could see the demonstration. Foucault pendulums are today installed in many of the large museums around the world. Foucault, besides being a scientist of great originality, was also a remarkable mechanical genius. In addition to his pendulum he invented the modern form of the gyroscope, and showed how it too can demonstrate the rotation of the earth.

Much astronomical photography has been concerned with spectra, which will now be discussed.

## PHOTOGRAPHY OF SPECTRA

The introduction of the technique of photography into spectroscopy gave rise to an amusing misunderstanding. The January 1908 issue of The *British Journal of Photography* included the surprising comment that C E K Mees of the Eastman Kodak Company in Rochester, New York, had designed a photographic plate 'eminently suited to photograph spooks, and ghosts, and astral beings, as they are now styled in the most refined spiritualistic circles'. Mees was in fact concerned with the more mundane task of photographing the spectra of atoms and molecules; he had no interest in spectres or even American specters. Charles Edward Kenneth Mees (1882–1960), born and educated in England, was the founding director of the Kodak Research Laboratories in Rochester, New York, and carried out much pioneering work on photography. His name was so closely associated with the company that it was widely believed that his middle initials stood for Eastman Kodak.

The invention of photography completely transformed the subject of spectroscopy, which is related to the splitting of light into constituent colours. First we should say something about spectroscopy before the invention of photography. As was mentioned earlier (Fig. 3.9), Isaac Newton in 1672 showed that a prism splits white light into its constituent colours, and that a second prism reconstitutes it into white light. Visible light, however, comprises only a part of the entire spectrum; there is a region of invisible 'light', which we now refer to as radiation, on both sides of the visible spectrum. The region beyond the red is known as the infrared, and was discovered by Sir William Herschel (Fig. 3.6) in 1800. He used thermometers with blackened bulbs to measure the heating powers of various parts of the spectrum (Fig. 3.14). He found a steady increase in heating power from the violet to the red, but then noticed that there was heating even when the bulb was placed in the dark region beyond the red. In this way he discovered the infrared spectrum.

The ultraviolet region was discovered in 1801 by the German scientist Johann Wilhelm Ritter (1776–1810), who observed the blackening of silver

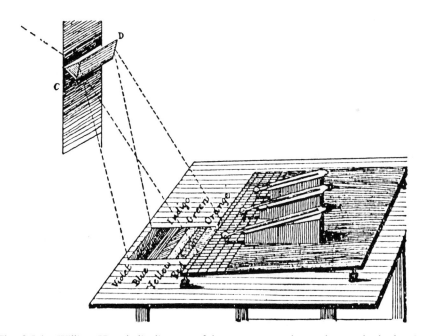

**Fig. 3.14** William Herschel's diagram of the arrangement he used to study the heating power of various regions of the spectrum (see text for details). He discovered that there was heating beyond the red end of the spectrum, and so discovered the infrared.

chloride brought about by radiation beyond the violet end of the spectrum. A year later the English chemist William Hyde Wollaston (1766–1828) independently obtained evidence of the ultraviolet spectrum.

Wollaston also observed something of great importance that Newton had unaccountably missed—or at any rate had not reported—namely dark lines in the spectrum of sunlight passed through a prism. Wollaston mistakenly regarded these lines as boundaries between the coloured bands, and he believed that there were four colours in the spectrum. In 1817 these dark lines were rediscovered and properly interpreted by Joseph von Fraunhofer (1787–1826), and the lines are now known as *Fraunhofer lines*. Fraunhofer was a Bavarian lens manufacturer who studied the optical properties of glasses of various colours. In doing so he saw a pair of bright yellow lines in the spectrum of the flame he was using. He then passed sunlight through a prism and observed a number of black lines. In later experiments he saw lines in the spectra of the moon, the planets and the stars. He observed that the lines found with light from various sources were not always in the same parts of the spectrum. This and other evidence convinced him that the lines were not due to his instrumentation but were characteristic of the light that he was studying and of the substances through which it passed.

To designate the more prominent spectral lines Fraunhofer suggested the use of capital letters from A to H, A being in the red and H in the violet, and this notation is often used today; the lines in the yellow region of the spectrum, now known to be due to sodium, are called D lines. Fraunhofer would probably have made many more fruitful investigations had it not been for his early death.

Even before these fundamental studies had been made, a few chemists had used coloured lines in spectra for the identification of materials placed in flames. In 1752 the Scotsman Thomas Melvill (1726–1753), who had studied divinity at the University of Glasgow, examined the flame spectra of various salts, and must be regarded as the founder of the technique of flame spectroscopy; his promising investigations were cut short by his early death in Geneva. Similar identifications were later made by two of the great pioneers of photography, Sir John Herschel and William Henry Fox Talbot.

An important contribution was made in 1834 by the Scottish physicist David Brewster (1781–1868) who did much work in optics and other branches of physics; in 1816 he achieved popular fame with his invention of the kaleidoscope; this was a tube containing mirrors and coloured glass which produced colourful and symmetrical designs. Brewster observed the Fraunhofer lines when white light is passed through various gases, and suggested for the first time a method of chemical analysis in which 'substances might be characterized by their action on different parts of the spectrum.' His studies were extended by John Frederic Daniell (1790–1845), who in 1831 became the first professor of chemistry at King's College, London. Some of his spectroscopic work was done in collaboration with William Allen Miller (1817–1872) who succeeded him as professor at King's College.

In 1852 fluorescence was discovered by George Gabriel Stokes (1819–1903), who from 1849 until his death 54 years later was Lucasian professor at Cambridge—a position that Newton had held. He observed that a solution of quinine sulphate, when viewed by transmitted light, has a blue colour, and he showed that this is due to the absorption of ultraviolet radiation, the light emitted being visible because it is of a lower frequency. The word fluorescence, which Stokes invented, has no etymological connection with light. It derives from the name of the fluorescent mineral fluorspar, the name of which comes from the Latin *fluor*, flowing, and the German *Spat*, which means spar, which is a particular kind of crystal which is easily cleaved. Stokes developed a technique for studying ultraviolet light by causing it to produce fluorescence in a solution of a blue dye. This procedure was used for a time until it was replaced by photographic methods.

The study of spectroscopy was made more convenient by the invention, by Bunsen and Kirchhoff, of the spectroscope. Robert Bunsen (1811–1899) and Gustav Robert Kirchhoff (1824–1887) had worked together at the University of Breslau. In 1852 Bunsen became professor of chemistry at Heidelberg, where he began research on combustion and flames, in the course of it using the famous laboratory burner which he perhaps did not invent but which now bears his name. In 1854 Kirchhoff joined Bunsen at Heidelberg as professor of physics, and suggested to him that his studies on the effects of salts on the colour of flames would be much improved by the use of a prism to produce the spectra. The two then collaborated on the construction of a spectroscope (Fig. 3.15) which although simple was invaluable for its purpose.

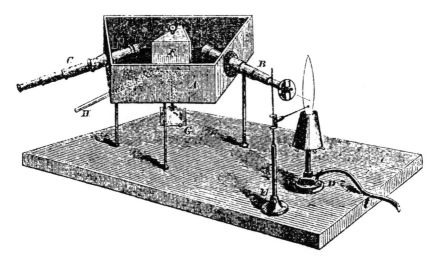

**Fig. 3.15** Bunsen and Kirchhoff's diagram of the spectroscope they devised in the 1850s. They published this diagram in 1860. They viewed their spectra through the telescope C. Soon they arranged for a camera to take photographs of the spectra.

Bunsen and Kirchhoff both did work of great importance, and both were concerned with the discovery of new chemical elements by means of spectroscopy. The story is told that on one occasion Kirchhoff met his banker who said that he had heard some silly story about a scientist who claimed to have found gold in the sun. When Kirchhoff admitted that he was the scientist, the banker was still sceptical, saying that in any case no one could bring gold down from the sun. Later Kirchhoff was awarded the Rumford medal by the Royal Society, which carried with it a cash award in gold sovereigns. Kirchhoff then had the great satisfaction of giving these to his banker to be added to his account, saying that he had brought them down from the sun.

Besides being a great scientist, Robert Bunsen exerted a considerable influence on the progress of science by his lectures and through the many people who came to work with him in his laboratories and who later became eminent scientists themselves. Bunsen never married, but he apparently proposed to one young lady and was accepted. He then became so absorbed in his research that he forgot about his successful proposal, and proposed to her again. This time she indignantly showed him the door. It was probably just as well that he did not marry, as he had many eccentricities, including a tendency not to wash very often.

After photography was introduced in 1839, spectroscopists found that their work was greatly assisted by its use. Before it became possible to take photographs, spectroscopy had to be done entirely by visual observation, and this had serious disadvantages: there was no permanent record which could be checked by others, and only the visible region of the spectrum could be studied. In the scientific journals the early spectra had to be presented as wood-cuts or etchings prepared from hand-drawn sketches.

John William Draper was successful in photographing the solar spectrum in about 1842. The technique he used for photographing spectra was daguerreotypy, and a drawing of a daguerreotype of the solar spectrum appears in a paper he published in the *Philosophical Magazine* in 1843. Other important contributions of Draper, made in the 1840s, include photography in the infrared and ultraviolet. He was one of the first to show that Fraunhofer lines exist outside the visible region of the spectrum. In 1843 he also took a daguerreotype of a spectrum taken using a diffraction grating instead of a prism. A diffraction grating, which gives sharper spectra than a prism, consists of very closely spaced indented parallel lines on glass. As early as 1823 Fraunhofer had ruled some gratings, but they were not very effective. The grating used by Draper had been made for him by the United States Mint, and his photograph of a spectrum was probably the first ever to be taken using a diffraction grating. The main pioneer of gratings was the American physicist Henry Augustus Rowland (1844–1901), who towards the end of the 1870s ruled gratings of high quality with lines much closer together than previously possible; he achieved nearly 15 000 lines per inch. His gratings brought about a considerable improvement in the quality of spectra.

Another scientist to have been highly successful in photographing spectra was Alexandre Edmond Becquerel (1820–1891), whom we met previously in connection with colour photography. By 1843 he had made an excellent daguerreotype of the solar spectrum, and engravings made from it were later published in his book *La lumière, ses causes et ses effets*, which appeared in 1867. Some of his engravings are reproduced as Fig. 3.16; the visible region includes prominent Fraunhofer lines designated A to H, while lines L to R and many others are in the ultraviolet.

As photographic techniques improved, efforts began to be made to photograph the spectra of stars. Some of the earliest attempts to do this were carried out by William Huggins (1824–1910), an amateur astronomer who established an observatory at his house in Tulse Hill, now in Greater London but then in the country. It happened that W Allen Miller of King's College was his neighbour and Huggins enlisted his help, especially with the photographic aspects. At first the siderostat they used to keep the telescope trained on the stars was inadequate, and the spectra were blurred. After Miller's death in 1870, Huggins had obtained a much improved siderostat, and was able to obtain excellent spectra of stars. At first he used wet collodion plates, and then dry plates which were much more sensitive and allowed shorter exposures. In many of his investigations Huggins collaborated with his wife, the former Margaret Lindsay Murray (1848–1915). He was knighted in 1902 and awarded the Order of Merit in 1902, the year the order was instituted. The Order of Merit is one of the most exclusive of British honours.

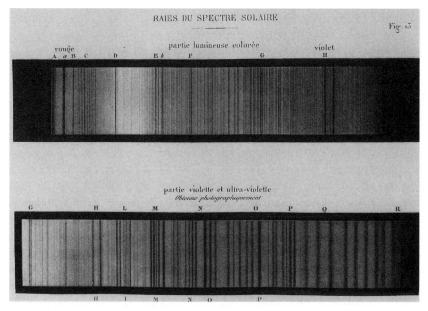

**Fig. 3.16** Engravings based on daguerreotypes of solar spectra made by Edmond Becquerel in 1842. Lines L to R are Fraunhofer lines in the ultraviolet.

Henry Draper (1837–1882) also recorded the spectra of many stars. He was the son of John Draper and followed closely in his father's footsteps. He studied medicine at New York University, obtaining his MD degree in 1858. From 1860 until his death he held various professorships at New York University. Henry Draper became one of the leading astronomers of his time. His wealthy and enthusiastic wife, Anna Mary Palmer, assisted him in much of his research, and they established an observatory in their house on Madison Avenue. They were particularly successful in taking astronomical photographs. Draper obtained satisfactory photographs of stellar spectra in 1877, shortly after this was first achieved by William Huggins. Henry Draper's career was cut short at the age of 45 when he succumbed to double pleurisy, contracted on a hunting trip in the Rocky Mountains when he was exposed overnight to low temperatures with no shelter.

The photography of spectra also played an important part in the discovery of the rare earths and the noble gases. The story of the discovery of helium is an interesting but involved one, and in brief is as follows. On 18 August 1868 a total eclipse of the sun was visible in India, and a number of scientists went there to make observations of the solar prominences (the solar corona). One of those who examined photographs of the spectra was Joseph Norman Lockyer (1836–1920), who although a civil servant at the War Office had already in his spare time done valuable work in astronomical spectroscopy. Having no previous experience of laboratory spectroscopy, and realizing that this was important for the interpretation of astronomical spectra, he turned to Edward Frankland (1825–1899) who had recently become professor at the Royal College of Chemistry which was then in London's Oxford Street. Frankland was already a distinguished organic chemist, and he allowed Lockyer the use of a room in his laboratories and part-time use of some research assistants.

Lockyer was particularly interested in one of the Fraunhofer lines in the yellow region of solar spectra that had been obtained during the eclipse in India; it had been called the $D_3$ line. It was already known that there was a prominent D line in the spectrum of sodium; when a sodium salt is put into a flame the flame becomes yellow, and this is because of emission of the sodium D line. It was also known that the D line on closer inspection was really two lines close together, called the $D_1$ and $D_2$ lines. The $D_3$ line found by Lockyer in the spectrum of the sun, however, could not be obtained from any substance available in the laboratory, and Lockyer boldly suggested that it was caused by a new element, found in the sun but apparently not on earth. He gave this new element the name helium, from the Greek *helios*, the sun.

Frankland, who had shown great interest in Lockyer's work, did not accept this suggestion, and was caused some embarrassment in 1871 when the president of the British Association (Sir William Thomson, later to be Lord Kelvin) announced that Frankland and Lockyer had discovered a new element. Frankland did not think it appropriate to make a public disclaimer, but he wrote to Lockyer—the letter still exists—making it clear that he did not believe in the new element. Partly because he respected Frankland's opinion Lockyer himself said little in public about his proposed new element, and even his book *Chemistry of the sun* (1887) did not mention it.

Years later, in 1895, William Ramsay (1852–1916), who with Lord Rayleigh had already discovered the element argon, began to investigate the gas produced by the mineral cleveite (this mineral is radioactive, and emits $\alpha$ particles, found to be helium gas). On examining its spectrum he found a line which he remembered to have been mentioned in a lecture by Lockyer that he had attended many years previously. He recalled that Lockyer had called it the $D_3$ line, and had suggested that it might relate to a new element. Ramsay sent a sample of cleveite to Lockyer, who was then director of the Solar Physics Laboratory in South Kensington, and Lockyer at once turned over his whole laboratory to the study of the material. The existence of helium was confirmed and soon afterwards, in 1897, Lockyer was knighted; Ramsay was knighted in 1902 and in 1904 received the Nobel Prize for chemistry 'in recognition of his discovery of the indifferent gaseous basic materials in the atmosphere....'

Lockyer continued to make important contributions to astronomy and to spectroscopy, making much use of photographic methods. He was particularly interested in the solar corona. Perhaps his most important achievement was his founding in 1869 of the journal *Nature*, which he edited for the first fifty years of its distinguished existence; *Nature* is still one of the most important scientific journals. Lockyer's personality was a complex one, in that he was highly opinionated and engaged in acrimonious quarrels with many of his colleagues. James Clerk Maxwell, an enthusiastic versifier, composed the following ditty about him:

> And Lockyer, and Lockyer,
> Gets cockier and cockier,
> For he thinks he's the owner
> Of the solar corona.

## PHOTOMECHANICAL PRINTING

Until photography was invented, books could be illustrated only by means of wood-cuts or engravings. These have the disadvantage that subjectivity is involved, a particularly serious matter in a scientific publication.

Soon after photography was introduced, many efforts were made to include photographs in books, periodicals, and newspapers. The first book to have been produced photographically was published privately by Anna Adkins (1797–1871), an amateur biologist. She was the daughter of John George Children (1777–1852), an active early photographer whom we shall meet again in Chapter 4 in connection with work on batteries. It happened that as a Vice-President of the Royal Society, Children had presided at the meeting of 21 February 1839 at which Talbot had first described his technique of making his 'photogenic drawings'. Anna Adkins learned of the cyanotype or blueprint method directly from its inventor, Sir John Herschel, with whom, and with Talbot, she was on friendly terms. Her book, *Photographs of British algae: cyanotype impressions*, was issued in parts over a period of ten years from 1843. Each page was a unique contact negative obtained by the cyanotype (blueprint) technique.

In the 1840s some books were illustrated by simply pasting photographs into them. The first book of this type was Talbot's famous *The pencil of nature*, which described his pioneering work on photography; the first part of it appeared in October 1843. Such a procedure was inefficient, and many efforts were made to convert a photograph into an *intaglio* plate, which is a plate used for printing in which the ink is held in indentations. Many procedures have been used.

One of them was *photogravure*, which was based on a technique invented by Talbot in 1852. He treated a specially prepared negative chemically in such a way that the parts exposed to light became hardened. Etching with acid then produced a surface in which the regions corresponding to shadow were indented, and when the plate was inked the ink went into the indentations. In order for the indented parts to hold ink, Talbot had to give them a texture, which he achieved using layers of muslin or lace. These layers were the proto-type of the modern half-tone screens.


## CINEMATOGRAPHY

Cinematography depends for its success on the persistence of vision. In the second century the Alexandrian astronomer and mathematician Ptolemy (*c.* 90–168 AD) described how a disc, one sector of which is coloured, appears to be uniformly coloured if it is spun sufficiently rapidly. Little notice seems to have been taken of this observation for many centuries. In 1824 an observation of a different kind was made by Peter Marc Roget (1779–1869; Fig. 3.17). Roget is best remembered today for his *Thesaurus of English words and phrases*, written in his retirement and first published in 1852; it has been revised many times, first by his son and then by his grandson, and is still widely used not only where the English language is used but also in versions in many other languages.

Roget's scientific career was of some distinction. He was born in the Soho district of London where his father, a Huguenot refugee, was pastor of a French Protestant church. He obtained a medical degree from the University of Edinburgh, and practised medicine first in Manchester and then for many years in London. He carried out a great deal of scientific work, and wrote many articles and books on physiology, electricity, and magnetism. From 1827 to 1848 he was Secretary of the Royal Society.

**Fig. 3.17** Peter Marc Roget (1779–1869), who is today chiefly remembered for his *Thesaurus of English words and phrases*. He practised medicine in Manchester and London, and was appointed Fullerian professsor of physiology at the Royal Institution in 1833, the same year that Faraday became Fullerian professor of chemistry. In 1824 he rediscovered the persistence of vision, and demonstrated some of its applications. For this contribution he has been called the grandfather of cinematography.

Roget rediscovered the persistence of vision, and brought it to popular attention, so that it is not unreasonable that he has been called the grandfather of cinematography. He described observations he had made in the basement kitchen of his London house, while looking through a vertical Venetian blind at a moving carriage wheel. He noticed that the slats of the blinds broke up the motion of the spokes into what appeared to be still pictures. Also, the spokes of the wheels appeared curved instead of straight. He pursued his observations (leaving his wife waiting to have breakfast with him), and gave a street vendor a shilling to move his cart back and forth outside the window.

Roget reported these observations to the Royal Society on 9 December 1824 in a paper entitled 'Explanation of an optical deception in the appearance of the spokes of a wheel seen through vertical apertures'. The published version of this paper contained meticulously drawn plates which greatly clarified his argument. Figure 3.18 may help to explain his interpretation of his observations. In Roget's paper he referred to

the illusion that occurs when a bright object is wheeled rapidly round in a circle, giving rise to a line of light throughout the whole circumference; namely, that an impression made by a pencil of rays on the retina, if sufficiently vivid, will remain for a certain time after the cause has ceased.

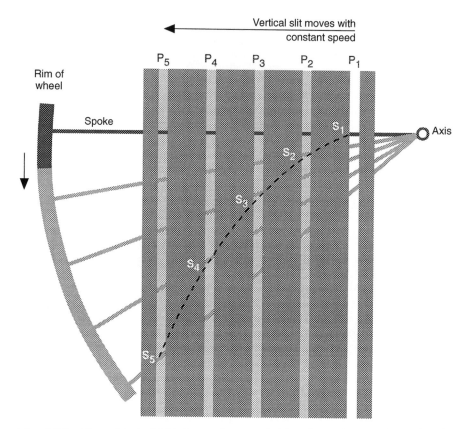

**Fig. 3.18** To explain in 1824 the persistence of vision, Roget imagined a spoke of a wheel turning at constant speed, but at first with its axis stationary. An opaque partition is in front of the spoke, and it has one vertical slit moving from right to left at constant speed. Suppose that at a certain time the spoke is vertical and the slit is at position $P_1$; a portion of the spoke is seen at position $S_1$. After a period of time the slit has moved to position $P_2$, and the spoke is seen at $S_2$. After equal intervals of time the slit is at positions $P_3$, $P_4$, and $P_5$, and the spoke is seen at positions $S_3$, $S_4$, and $S_5$. An observer will thus see the spoke at positions $S_1$, $S_2$, $S_3$, $S_4$, and $S_5$, and it will appear to be curved, as represented by the dashed line. If instead the spoke is observed through vertical Venetian blinds, the situation is only slightly different. If there were no persistence of vision, the observer would see the entire spoke at the same instant of time, and it would appear straight. In fact, however, an observer looking at a turning spoke through a Venetian blind has to scan an area, and sees different parts of it at slightly different times. The spoke will therefore appear to be curved. The effect is related to the fact that wheels shown cinematographically often appear to rotate in the wrong direction.

Roget's paper attracted much attention. Michael Faraday made reference to it several times, and constructed a disc to demonstrate the effect. Another suggestion that practical use might be made of the persistence of vision was made in 1826 by J A Paris (1785–1856). He constructed a 'thaumatrope' (from the Greek *thauma*, a wonder or marvel, and *trope*, turning). This was a cardboard disc on each side of which two distinct objects were painted—for example a tree in winter with no foliage on one side, and on the other side the foliage. When the disk was spun about an axis the tree was seen in full leaf. In the same year Sir John Herschel described another thaumotrope which worked on the same principle as that devised by Paris. The Austrian scientist Simon R von Stampfer (1792–1864) developed a mathematical treatment of Roget's observation, and took out a patent on 'stroboscopic discs' in 1833. A stroboscope (from the Greek *strobos*, a whirling around) works on the same principle as a thaumotrope, producing the illusion of motion when one looks through openings in a rotating disc.

Another man who became interested in the persistence of vision at about the same time was Joseph Antoine Ferdinand Plateau ((1801–1883), who was one of the best known Belgian scientists of the nineteenth century. He was professor of physics at the University of Ghent from 1835 until his retirement, and was particularly interested in optics and in liquid surfaces; he originated the technique of producing soap films, and discovered the tiny second drop, named after him, which always follows the main drop that falls from a surface. In 1829 he carried out an experiment in physiological optics in which he stared into the sun for 25 seconds. He was blinded for several days, but his sight recovered partially. Later his sight deteriorated, and by 1841 he was totally blind. In spite of this he continued to do research for many years, aided by his colleagues.

He also worked on the persistence of vision, and devised a stroboscope, which he called a *phénakistiscope*, consisting of pictures of a dancer placed round a wheel. When the wheel was turned, the dancer was seen to execute a turn. Plateau sent one of his phénakistiscopes as a present to Michael Faraday.

A hundred years or so after Roget presented his 1824 paper, the centenary of his contribution to cinematography was celebrated by a number of newspapers and periodicals, including *The Times* of London (19 March 1929), the *Illustrated London News* (19 August 1922) and the *Photographic Journal* (December 1924).

Beginning in 1872, Eadweard Muybridge (1830–1904; Fig. 3.19) carried out work that may reasonably be said to have led to cinematography. He made remarkable photographs, using short exposures, of trotting and galloping horses. Muybridge was born Edward James Muggeridge in Kingston-on-Thames, near London. He disliked his name so much that he changed it to something that was closer (or so he thought) to the original Anglo-Saxon form. In 1852 he emigrated to California where he worked at first as a bookseller. After receiving a severe blow to his head, which appears to have affected his personality, he became a professional photographer, and soon achieved great success. His extensive photographic work, involving a good deal of travel under difficult circumstances, is all the more remarkable in view

**Fig. 3.19**  Eadweard Muybridge (1830–1904). Few clear pictures of him seem to exist, since he was usually photographed in the course of his work with rapidly moving people or animals, at exposures of 1/1000th of a second or less. He was a frenetic individual, and probably never kept still long enough to have a portrait photograph taken. The photograph shown here shows that, at least during his stay in America, he had adopted a Wild West hat in addition to his propensity to shoot dead his wife's alleged lovers.

of the great clumsiness of photographic techniques at the time. The wet collodion plates had to be prepared shortly before taking a photograph—an awkward task in a tent under the heat of the California sun—and the plates had to be developed, fixed, washed, and dried right away.

In 1872 the millionaire ex-Governor of California, Amasa Leland Stanford (1824–1893), got into an argument with a friend about the gait of a rapidly moving horse. (Leland Stanford and his wife are remembered today as the founders and endowers, in 1891, of Stanford University in Palo Alto, California.) The French investigator Étienne Jules Marey (whose dates are the same as Muybridge's, 1830–1904), another pioneer photographer, had made some interesting experiments in which a rubber bulb was attached to each hoof of a horse. Tubes attached to them led to four pens resting on a revolving paper drum on the horse's saddle, so that when a hoof touched the ground an ink record was made. Marey's work seemed to show that at times all four hooves were off the ground. Stanford accepted this conclusion but got into an argument (according to some accounts it was a $25 000 bet) with a millionaire friend who thought otherwise. Stanford then commissioned Muybridge to settle the matter by photographic means.

The project proved difficult, and success was not achieved until about 1877. The sensitivity of photographic film is given by its ASA value, and even the slowest modern film has an ASA value of 25 or so. With such film and a shutter speed of 1/1000th of a second, the question could easily be settled today. The wet collodion plates then in use had ASA values of about 0.1, and shutter speeds were much slower. Muybridge's initial attempts failed, and he achieved success later only through the design of greatly increasing shutter speeds, and by the use of more sensitive films.

In the earlier attempts, at Stanford's vast estate in Palo Alto, California, Muybridge devised an ingenious arrangement involving a camera set up beside the horse track. This was lined with bedsheets, contributed by neighbours, so that the photographs would be more clearly defined, and there was a white backing. Later the arrangements were much improved.

The experiments were interrupted by two incidents. In 1873 Muybridge was commissioned to take photographs in Northern California of the Modoc Indian War, one of the many attempts by the settlers to exterminate the Indians. The second interruption is probably unique in the history of science and technology. In 1874 Muybridge's wife bore a child which, since he was photographing the war at the relevant time, he believed to have been fathered by another expatriate Englishman, Major Harry Larkyns. In October 1874 Muybridge went to Larkyns's house, shot him dead, and was arrested and charged with murder. He was held in gaol until his trial in February 1875, during which time his hair and beard became snow white. At the trial he made no attempt to deny his action, and no defence of insanity was offered. The California jury, all-male (of course) and apparently less prejudiced against an English photographer than an English major, found him not guilty, with the comment that they would have acted in the same way in the circumstances.

This incident seems to have interfered in no way with the success of Muybridge's career. He continued his work at Stanford's estate, steadily improving his techniques. He began to use gelatine dry plates, which were twenty times more sensitive than wet collodion. He devised cameras with exposure times of 1/1000th to 1/750th of a second, and they were activated either by clockwork or by means of trip wires stretched across the track; these were broken by the moving horse. His photographs, although somewhat blurred, were clear enough to show conclusively that at times the horse did have all feet off the ground (Fig. 3.20). His work showed incidentally that artists' depictions of moving horses, in a 'rocking horse' attitude, were unrealistic.

Later Muybridge discovered that he had another talent that was to take him all over America and Europe; he was an excellent lecturer—interesting, lucid, and humorous. Perhaps his most memorable lecture was at a Monday Evening Discourse at the Royal Institution in London, on 13 March 1882. The Prince of Wales (later to be King Edward VII) took the chair at the meeting. Also present were the Prince's wife Princess Alexandra and three of the Queen's daughters, Prime Minister Gladstone, the poet Alfred (later Lord) Tennyson, Sir Frederick Leighton (the President of the Royal Academy,

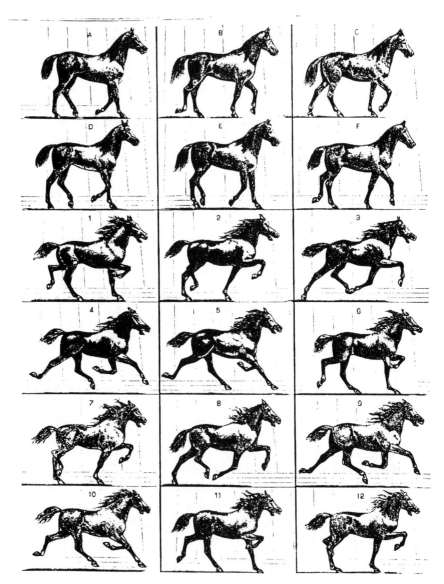

**Fig. 3.20** A set of photographs of moving horses, taken by Muybridge on Leland Stanford's estate in California. This set was reproduced, as an engraving, on the cover of the 19 October 1878 issue of *Scientific American*.

perhaps anxious to increase his skill at painting galloping horses), and the biologist Thomas Henry Huxley. The title of the lecture was 'The attitudes of animals in motion, illustrated with the zoopraxiscope'. One wonders how many of the august persons present knew that Muybridge had committed a crime for which he would certainly have been hanged in England.

Muybridge continued to take photographs of various kinds, including many of humans performing acrobatics. One consequence of his high-speed photography is that it became possible to improve stroboscopes, which previously had used drawings; now sufficiently well-defined photographs could be used. Muybridge made a device which he first called a 'zoogyroscope' (from the Greek *gyro*, circle). Regrettably, he later referred to it by the awkward name 'zoopraxiscope', derived from the Greek *zoo*, animal, and *praxis*, action. This device, illustrated in Fig. 3.21, gave the illusion of animation by means of a disc on which there were a number of photographs, for example of a galloping horse. In about 1890 he got the idea of combining his zoopraxiscope with the phonograph recently designed by Thomas Edison. In 1893, at the World Columbian Exhibition in Chicago, he installed as a sideshow what he called the Zoopraxographical Hall. In it he projected images of galloping animals to the accompaniment of sound from a phonograph, in this way achieving a rudimentary kind of 'talkie'. For some reason the demonstration

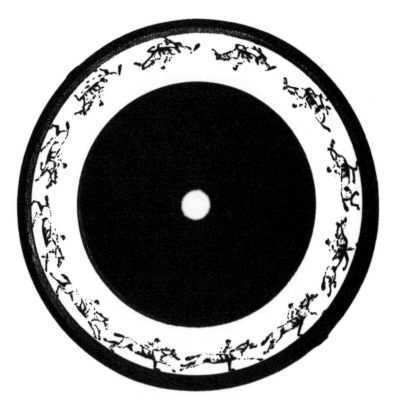

**Fig. 3.21** A 'zoopraxiscope' disc, of the type exhibited by Muybridge at the World Columbian Exhibition in Chicago in 1893. Some of his discs were more complicated, having several rings of horses, so that the projector had to move in an appropriate way.

was not popular (perhaps the name put people off), and he made only $200. The Zoopraxographical Hall has been called—with hardly adequate justification—the world's first motion picture theatre. It is more appropriate to call Muybridge the stepfather of cinematography, just as we have called Roget its grandfather.

Later Muybridge returned to his native Kingston-on-Thames, where he died. He left his zoopraxiscope, his slides, and many notebooks to the Kingston Public Library and the Kingston Museum, but some of them have been transferred to the Science Museum in South Kensington, London.

In the 1880s Thomas Alva Edison (1847–1931) began to take an interest in doing for visual images what in 1877 he had done for sound by his invention of the phonograph. He came of a remarkable family. His father, Samuel Ogden Edison, was born in Nova Scotia, received little education, and could scarcely write his name. He was of a restless disposition, and when in 1837 the fiery Scotsman William Lyon Mackenzie, mayor of Toronto, launched his plot to overthrow the Canadian government, Sam Edison enthusiastically joined in, leading a band of agitators towards Toronto. The rebellion was quickly crushed, and a number of its leaders hanged. Fortunately for technology, Sam narrowly escaped capture by fleeing to Port Huron, Michigan, and it was there that Thomas Edison was born ten years later.

He and his assistants experimented with various devices that were forerunners of cinematography. Much of the enthusiasm for this work came from his assistant William Kennedy Laurie Dickson (1860–1935), a man of French–Scottish descent who had travelled from England in his teens to work with Edison. In 1887 they opened in West Orange, New Jersey, a workshop officially called the Kinetographic Theater, but which on account of its appearance became known as the Black Maria. Early in 1889 the Theater was visited by Eadweard Muybridge, and he and Edison discussed in detail the possibility of making motion pictures. Edison and his assistants, particularly Dickson, took pictures of various events on long strips of celluloid film, and also constructed equipment for exhibiting the films. They also had the idea of mounting spirally 42 000 tiny photographic images on a phonograph cylinder, with the idea that the soundtrack could be played and the images displayed at the same time. It is not certain whether this clumsy arrangement was ever constructed, but it is not surprising that it came to nothing.

In 1891 they produced a 'kinetoscope', a nickel-in-the-slot device in which a person could peer through a peephole and see about 15 seconds of action on a succession of film frames exposed at a speed of 15 a second. On 9 May 1893 Edison gave a public demonstration, at the Brooklyn Institute of Arts and Science, in which he projected a film, 30 seconds in length, on a screen. The quality of the film, however, was poor.

In 1894 a 'Kinetoscope Parlor' was opened at 1158 Broadway, New York. In it a customer could look through a slit and see a variety of films, lit from behind by a lamp and moved by an electric motor; the speed had been increased to 46 frames a second, and each show lasted about one minute. Edison personally had little to do with this enterprise; he delegated its operation to two Canadian brothers, Andrew and Edwin Holland of Ottawa, who had earlier worked with Edison on his phonograph. The Kinetoscope Parlor had ten machines, arranged in two rows of five; an 'attractive young woman' was in attendance, and customers paid her 25 cents to see one row, and 50 cents to see both. One of the machines showed the famous strongman Eugene Sandow, scantily clad and swelling his rippling muscles in all directions. Another machine showed Fred Ott, one of Edison's assistants who enjoyed being a ham actor, performing an elaborate minute-long sneeze; a copyright was quickly taken out on this star performance. There were also Scottish dances, can-can dances, and fencing matches. The Parlor drew large crowds, and Holland Brothers made a considerable profit. Edison's company manufactured a number of these kinetoscopes, and they were established in Atlantic City, Chicago, and elsewhere.

Because Edison's technique involved driving the film continuously, projection on to a screen was unsatisfactory, the result being very jerky. Dickson tried to get Edison interested in improving the technique, and in particular in developing the idea of projecting moving pictures to large audiences. Edison, however, was rather surprisingly content to keep to the peephole technique. Because of this difference of opinion, Dickson left Edison's employ in 1895 and cofounded a syndicate which became the American Mutascope Company, which for a time was the Edison Company's chief competitor. Dickson returned to England in 1897.

The final and most important step in cinematography was taken in France by two brothers, Auguste Marie Louis Nicholas Lumière (1862–1954) and Louis Jean Lumière (1864–1948). They were both chemists and excellent technicians, and were manufacturers of photographic materials. In 1895 they invented the first successful instrument that performed the function of a ciné camera and of a ciné projector, driving the film by means of perforations at its edges. Most important of all, they applied the principle of intermittent movement to the projection of the film, so that the picture observed would be smooth running. Edison had rejected this idea, and his results were jerky.

To the Lumières, rather than to Edison, must go the credit for producing the first satisfactory moving picture, and also for the word 'cinématographie' which they introduced. It is interesting that having achieved such success they regarded their invention as 'an invention without a future'; they thought it to be of scientific interest only, never to be popular with the public. Edison tried to block the progress of cinematography by litigation, at first with some success, but within a few years the moving-picture industry had become well established in a number of countries, and highly successful with the public.

## HIGHER SPEED PHOTOGRAPHY

By modern standards the plates and film used in early photography were of low sensitivity, and long exposure times had to be used. Persons having their portraits taken often had to sit completely still for a minute or more. Much effort went into the improvement of film 'speeds', It is estimated that since the early efforts of Daguerre and Talbot, speeds of plates and films have been increased by a factor of over 20 million.

With fast film it is possible to take photographs of rapidly moving objects. The problem then is to use sufficiently short exposure times. This can be approached in two ways: by devising cameras having shutters which remain open for very short periods of time, and by using intense flashes of light of short duration.

Various techniques were devised for reducing shutter speeds. One of the most famous of these was achieved by the German photographer Ottomar Anschütz (1846–1907), who like Eadweard Muybridge was interested in photographing rapidly moving animals. In 1888 he patented his focal plane shutter which could remain open for speeds as short as $1/1000$th of a second. We have already noted the high speed shutters used by Muybridge at about the same time for his photographs of moving animals and people. The ones he used were less efficient than those of Anschütz.

There is, however, a limit to the speed with which shutters can be made to move. Much shorter exposure times than can be achieved by shutters can be achieved by short flashes of light. Again, there has been enormous progress over the years. As early as 1852 Talbot used electric flashes of perhaps $1/100\,000$th of a second duration, by discharging a capacitor through a spark gap. However, the technique was complicated and was mainly used by scientists in their laboratories; it could not be used conveniently by commercial photographers.

The physicist Ernst Mach (1838–1916), at Charles University in Prague, used the electric flash technique in the 1880s to photograph moving projectiles, meteorites, explosions, and sound waves. He also did important theoretical work on projectiles moving faster than the speed of sound—that is, on the subject of supersonics. Mach was born in Brno, which is now in the Czech Republic; it was then part of the Austro–Hungarian empire. As a result of his work on supersonics his name is still in common use; the Mach wave is the wave behind a travelling projectile, and the Mach number is the ratio of the speed of a projectile to the speed of sound in the surrounding medium. The Mach angle is the angle the shock wave makes with the direction of motion of a projectile. It is of interest that Mach himself did not think that his work on supersonics was as important as his ideas about the philosophy of science. Today we recognize that he went badly wrong in a number of his ideas, such as his scepticism about the existence of atoms and his rejection of the use of probability theory in science. His supersonic theory, however, is still valid today.

It was only later that electric flashes came into general use by photographers. Since the 1880s photographers for the most part made use of magnesium flash powder, which emitted a cloud of acrid white smoke as well as light. An additional disadvantage was that the light was not of good quality, so that the photographs were unsatisfactory. Since these flashes were not of short duration the exposure times still had to be controlled by the camera shutter. There was much improvement following the invention, in Germany by Paul Vierkötter, of the flash bulb, in which the magnesium was encased in glass so that fumes were not emitted. After about 1940 developments in electronics made possible flash units that were much more convenient, and were synchronized with the movement of the shutter in the camera. Kodak, for example, introduced their Kodatron electronic flash unit in 1940.

Such electronic flash units could easily be used by amateur photographers. Previously a number of techniques for producing flashes of short duration had been used in scientific work. I have already mentioned the flash systems used by Ernst Mach in his work on supersonics. In the 1930s Harold Edgerton (1903–1990) of the Massachusetts Institute of Technology began to perfect flash systems, and over a period of many years took numerous remarkable photographs of projectiles and of people and animals in rapid motion. Many of his photographs were of high artistic quality as well as being of scientific importance. He often produced a number of flashes in rapid succession, recording the images on a single film so that the character of the motion could be clearly seen. He used the term stroboscopic photography (from the Greek *strobos*, a whirling) for such multiple exposures. His exposure times were often less than a microsecond (a millionth of a second), and he could take as many as a hundred photographs in a second.

Flashes of short duration are also produced for a different purpose. Certain chemical reactions, called photochemical reactions, are brought about when light interacts with chemical systems; for example, if a mixture of hydrogen gas and chlorine gas is exposed to a bright light, reaction occurs at once, sometimes with a loud noise, to produce hydrogen chloride (hydrochloric acid). The word photolysis is used to refer to such reactions. In the late 1940s much effort went into the production of flashes of short duration to investigate fast reactions of this kind. This technique of flash photolysis was first developed at Cambridge University by George (now Lord) Porter (b. 1920) and Ronald George Wreyford Norrish (1897–1978), their first paper on the subject appearing in 1949. For this work they were awarded Nobel Prizes in 1967, together with Manfred Eigen (b. 1927) who had developed other techniques that are important for the investigation of rapid reactions. Shortly after the work of Porter and Norrish was done, flash photolysis equipment was built and used by Gerhard Herzberg (b. 1904) and Donald Allan Ramsay (b. 1922) at the National Research Council in Ottawa, Canada, and by Norman Ralph Davidson (b. 1916) at the California Institute of Technology in Pasadena.

In Porter's earliest experiments at Cambridge, in 1948, the duration of his flashes was about a millisecond (a thousandth of a second, $10^{-3}$ s) and it is a remarkable fact that during the next four decades the duration of the flash was reduced by nearly 12 powers of 10, to a few femtoseconds (a femtosecond is a thousand million millionth of a second, $10^{-15}$ s). As a result, techniques are now available for photographing the fastest chemical and physical processes.

## LASERS

By 1950 flashes of a microsecond ($10^{-6}$ s) duration had been produced, and that might have been the best that could be done if it had not been for the invention of the laser. We must now briefly digress to say something about lasers. The word laser is an acronym for light amplification by stimulated emission of radiation. Laser technology may be said to have originated in 1954 with the introduction by the American physicist Charles Hard Townes (b. 1915), at Columbia University, of the maser, which operates in a region of the spectrum which is far from the visible (see Chapter 5, particularly Fig. 5.16). In 1960 Theodore Harold Maiman (b. 1927), of the Hughes Research Laboratories in Miami, Florida, constructed a similar device which operates in the visible and ultraviolet regions of the spectrum. In this way he invented the laser. Maiman's device produced a laser pulse; the first continuously operating (c.w., or continuous wave) laser was constructed in 1961 by A Javan and his co-workers at the Bell Telephone Laboratories.

An important feature of a laser is that the beam is highly monochromatic, which means that it is confined to an extremely narrow region of the spectrum; we can say that it corresponds to a very pure colour, uncontaminated by other colours. A laser beam can also be of very high intensity. A laser beam is also what physicists call coherent. This means that all of the waves are in phase; in other words, they match each other and there is no interference between them. If all of the waves in a beam of light started exactly in phase they would remain in phase throughout the path of the light, and we would say that the coherence length was infinite. Since the waves never start out exactly in phase they get more and more out of phase as they proceed on their way. The coherence length is the distance they travel and remain more or less in phase. The coherence length of a laser beam is usually a matter of centimetres. By contrast, an ordinary incandescent light bulb has a coherence length that is shorter by a factor of many millions. For a light bulb that is no disadvantage, but there are many applications (such as holographs, to be considered later) where a long coherence length is essential.

By 1966 Porter, who in that year became Director of the Royal Institution in London, had developed a highly efficient laser flash system capable of dealing with processes occurring in a matter of nanoseconds (a nanosecond is

$10^{-9}$ s, or a thousand millionth of a second). This allowed him to photograph the course of many chemical processes that could not previously be studied.

Flashes of a picosecond duration (a million millionth of a second, $10^{-12}$ s) were achieved during the 1970s, notably by Charles Vernon Shank (b. 1943) and Peter M Rentzepis (b. 1934) at the AT & T Bell Laboratories in New Jersey, where great advances in laser technology have been made. During the 1980s, again largely due to the efforts of Shank and his colleagues, it became possible to achieve flashes of even shorter duration, of only a few femtoseconds ($10^{-15}$ s, or a thousand million millionth of a second). With flashes of such short duration photographs can be taken of the fastest motion of any kind.

## HOLOGRAPHY

Most of the many improvements in photographic techniques, including even television, have involved the same basic optical principles applied by Daguerre and Talbot in 1839. In 1947, however, a completely new principle was suggested by the Hungarian-born Dennis Gabor (1900–1979) of the Imperial College of Science and Technology in London. His interest was in improving the quality of certain scientific photographs, and his idea was to make a more detailed record of the light waves that were reflected from objects. He had no thought at the time of making three-dimensional images, but that is what his idea led to.

His idea could not be put into practice at first, as it was necessary for the light to be coherent—in other words the light waves must match each other. Moreover, the coherence length of the light must be comparable to the size of the object being photographed. This requirement is not satisfied by ordinary light, but the technique became possible with the introduction of lasers in 1960. Two years later lasers were applied to Gabor's principle by Emmett Norman Leith (b. 1927) and Juris Upatnieks (b. 1936) of the University of Michigan. They were successful in producing what has come to be known as a holograph or a hologram (from the Greek *holos*, whole). (The word holograph had been used earlier to mean a document which is entirely in the handwriting of the person whose signature it bears.)

The essential procedure in recording a holograph is shown in Fig. 3.22(a). No camera is used; instead the light from the laser beam is reflected by the object on to a photographic plate. Laser light is also reflected on to the plate by means of a suitably placed mirror. This light is referred to as the reference beam, and it interferes with the light that is reflected from the object, so that what is recorded is an interference pattern.

When this procedure is used without any refinements the resulting hologram bears no resemblance to the original object; it is just an irregular arrangement of lights and shades with no recognizable pattern. In order for the eye to recognize the object, laser light has to be used again, in the arrangement shown in Fig. 3.22(b). The hologram, in the form of a transparency, is irradiated with the laser light, and an image is formed, and is seen behind the transparency. The image has the quality of depth, in that there is an apparent change when it is viewed from different positions. Three-dimensional vision arises partly from the fact that each of our two eyes sees objects slightly differently. Since the hologram, unlike an ordinary photograph, has this quality of

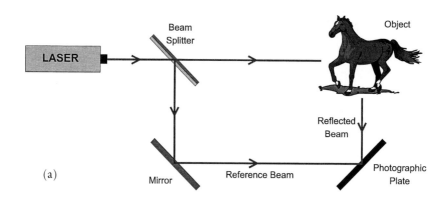

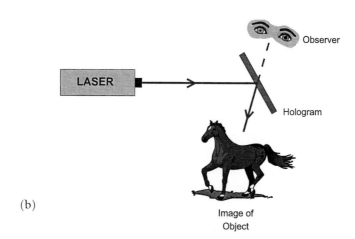

**Fig. 3.22** The principles involved in creating and viewing a hologram. (a) To form a hologram, a laser beam is split into two beams, one of which falls on the object to be photographed, and is reflected on to a photographic plate. The other beam, the reference beam, is reflected by a mirror and also falls on the photographic plate. The plate records an interference pattern which, unless special techniques are used, cannot be identified until laser light of the same kind falls on it. (b) The object becomes visible on the plate when laser light falls on it. A virtual image is formed, which can be seen by the observer, and it appears to be three-dimensional. (A virtual image is the type of image we see on looking into a mirror; it cannot be displayed on a screen. A real image can be displayed on a screen; we create one whenever we use a projector.)

depth it appears to be three-dimensional. A further sense of depth is given by the fact that when one's eyes have one part of a hologram in focus, the foreground and background are not in focus, so that the eyes have to adjust to them.

The first holograms had to be viewed using laser light, but later advances in technology made it possible to view specially prepared holograms in ordinary light. Holograms are widely used today; they have appeared on the covers of periodicals, such as *National Geographic*, and rather crude holograms appear on many credit cards and articles of jewellery.

## OTHER SPECIALIZED PHOTOGRAPHIC TECHNIQUES

Many other modifications of photography are in use today. To give some idea of the vast range of applications, a few of them will be mentioned and briefly defined:

*Digital photography* is a modern technique in which a special camera produces a focused image on a photosensitive device which is a mosaic of tiny units. A computer chip digitizes the data which can then be stored on a disc, displayed on a video screen, and printed.

*Electron microscopy* involves taking pictures by an apparatus in which electrons interact with matter, in such a way that individual atoms can be seen.

*False-colour photography* is a technique in which, for convenience, the colours recorded are deliberately changed. In infrared photography, for example, the film used is often made in such a way that regions in the infrared, which would be invisible to the eye, have a variety of colours, so that different regions can be distinguished by visual inspection. The technique of false-colour photography is also used in a CT scan (see *Tomography* below), to provide a clear distinction between different spectral regions.

*Infrared photography* is photography with the use of film sensitive mainly to infrared radiation. What is recorded on infrared film is very different from what is recorded on film sensitive to visible light. Infrared photography is particularly important in aerial photography. For example, the chlorophyll stored in plants strongly reflects infrared radiation, so that ecologists can use the technique to detect changes in the health of vegetation. Forensic experts use infrared photography to examine powder burns, and in other ways.

*Keratography* is photography that reveals the shape of the cornea of the eye; this is done by measuring the shape of the circular rings of an object reflected in it.

*Panoramic photography* involves the use of a rotating camera which will photograph a scene over a wide angle.

*Photographic dosimetry* involves the wearing of an emulsion plate by a person who might be exposed to radiation. The darkening of the plate indicates the dose received.

*Radiography* is photography involving X-rays (Chapter 7); the rays from a small source pass through the body on to a photographic plate, without the use of any lenses.

*Retinography* is photography of the retina of the eye.

*Schlieren photography* is related to variations in the optical properties of a gas or liquid, and is used to show the flow patterns in a moving fluid.

*Thermography* is the photography of the temperature patterns emitted by an object.

*Time-lapse photography* involves taking photographs at fixed intervals of time, and then projecting the film at a rate that speeds up the action.

*Tomography* is X-ray photography of the plane section of an object, the images of other planes being blurred out. A common example is *computed tomography*, or *CT scan*, used for obtaining cross-sectional views of an internal body structure.

*Ultraviolet photography* is photography using special film sensitive to ultraviolet light. Since ultraviolet light penetrates matter more than ordinary light, ultraviolet photography is valuable for examining underlayers of oil paintings, and for examining documents for forgery.

# FOUR

# Michael Faraday and electric power

The career of Michael Faraday provides us with an excellent example of how pure research, done as a result of scientific curiosity and with little regard for any possible applications, may nevertheless have far-reaching practical consequences. In some of his work Faraday did show interest in practical matters, but in much of it, particularly that on electricity and magnetism, he was mainly motivated by curiosity and a desire to formulate a theory of the behaviour of electric currents and of magnets.

Modern society is highly dependent on electric power, used for heating, lighting, and communication. The great transformation that has occurred since the early nineteenth century was greatly influenced by Michael Faraday's researches. In 1821 he was the first person to convert the energy of an electrical current into mechanical energy. Even more important, his discovery of electromagnetic induction in 1831 led to the efficient conversion of mechanical energy into electrical energy, and directly led to the vast modern electrical industry in which electricity is distributed on a large scale throughout the world. Also, and again of great importance, Faraday's unconventional ideas about electric and magnetic fields led James Clerk Maxwell to formulate his great theory of electromagnetism (Chapter 5), and to recognize that light is a form of electromagnetic radiation. The development of radio techniques, with all their consequences, was soon to follow the publication of Maxwell's ideas.

## EARLY WORK ON ELECTRICITY

Before we can understand Faraday's researches on electricity we must know a little about some of the earlier work on the subject. The existence of what came to be called electricity was known to the ancient Greeks. Thales of Miletus ($c$.636–$c$.546 BC), statesman, mathematician, philosopher, and engineer, made many important contributions to knowledge. He discovered that amber, when rubbed, acquires the property of attracting small pieces of pith or cork. Many centuries later William Gilbert (1544–1603) extended these observations, showing that other substances, such as glass and sulphur, show the same attracting properties when rubbed. To explain this behaviour he coined the word 'electric' from the Greek ἤλεκτρον (electron), meaning amber. Gilbert, who became physician-in-ordinary to Queen Elizabeth and to King James I, also made pioneering investigations on magnetism.

The early work on electricity was concerned with static electricity. It was found possible to store electricity in what were called Leyden jars, invented in 1746. Various machines were invented, one of them shown in Fig. 4.1(a), for generating electric charge. A Leyden jar (Fig. 4.1(b)) is the prototype of

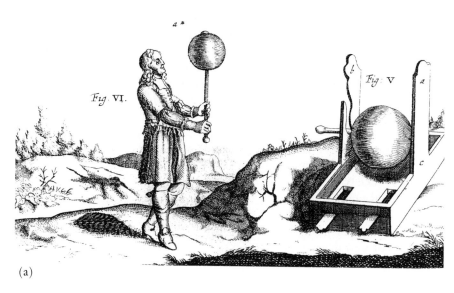

(a)

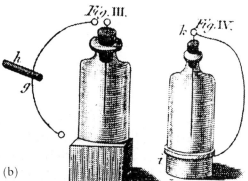

(b)

**Fig. 4.1** (a) A machine for generating electricity, as illustrated in Otto von Guericke's famous book *Experimenta nova*, 1672. The ball is made of sulphur, and it became charged if rotated while touched by the hand. The charge could be transferred to the insulated ball held by the man; it is shown attracting a feather, a. (b) Leyden jars, as illustrated in Benjamin Franklin's book *Observations on electricity, made at Philadelphia in America*, London, 1751. The jars are on metal stands, and a metal rod with a spherical top passes through the stopper. An electric charge transferred to the inner wire can be discharged by connecting the upper sphere to the metal base by means of a wire, g, held by an insulating handle, h (left-hand diagram, Fig. III).

what were called condensers and are now usually called capacitors, which are two pieces of metal close to one another but separated by an insulator, which does not allow electric charge to pass. If a capacitor is sufficiently highly charged with electricity, wires connected to the two plates will produce a spark if they are brought together; a person holding the wires may receive a dangerous electric shock. Capacitors are much used in research and technology today.

Electricity was at first regarded as little more than a curious and unimportant phenomenon. In 1729 Stephen Gray (1666–1736) distinguished between conductors and non-conductors (insulators) of electricity. Soon afterwards a two-fluid theory of electricity was formulated. Bodies that were not electrified were supposed to contain equal quantities of 'vitreous and resinous fluids'. An electrified body was supposed to have gained an additional quantity of one fluid and to have lost an equal amount of the other, the total amount of electrical fluid remaining the same. In those times the idea of an imponderable fluid was a popular one; heat, light, magnetism, and electricity were all commonly regarded as invisible and weightless fluids which could pass with ease from one body to another.

Important advances in the understanding of electricity were made by the great American scientist and statesman Benjamin Franklin (1706–1790; Fig. 4.2). He predicted, and confirmed experimentally in 1747, that if a

**Fig. 4.2** Benjamin Franklin (1706–1790), famous American scientist and statesman. He carried out work of great importance on electricity, and on the effect of oil in calming a rough sea. He was one of the three authors of the American Declaration of Independence, and in 1781 played an important role in negotiating the final peace with Britain.

sharp metal rod is connected to the earth ('earthed' or 'grounded'), and is pointed to an electrically charged body, the body becomes discharged. This observation led to Franklin's important invention of the lightning rod, one of which he installed at his own home in Philadelphia in 1749. Franklin proved, by his famous kite experiment in 1752, that thunder-clouds are due to the accumulation of electricity. In that same year he attached a bell to the lightning rod on his house to let him know of any electrification by passing clouds.

The lightning rod, which was never patented by Franklin, soon attracted wide attention. In May 1750 the *Gentleman's Magazine* (the first periodical ever to be issued, founded in London in 1731 and continued until 1914) published Franklin's suggestion of using a pointed metal rod for the protection of buildings. After a London church was struck by lightning, lightning rods were installed on St Paul's Cathedral in 1769. Lightning rods inspired some controversy. Some argued that their installation acted against the work of God; if an angry God wished to destroy a building, so be it. There was also controversy about whether lightning rods should be pointed or rounded. King George III, never the most stable of individuals, allowed himself to be persuaded by Tory politicians that the pointed rods were a plot by Franklin and other British colonists in America to destroy government buildings in England, and that rounded rods were preferable. Although a committee of the Royal Society considered the matter and recommended pointed rods, His Majesty ordered rounded rods on government buildings.

Franklin's investigations led him to reject the two-fluid theory in favour of a one-fluid theory of electricity, according to which there is a single electrical fluid which non-electrified bodies were considered to possess in a certain normal amount. A body having more than this amount was said by Franklin to be positively charged, while one deficient in the fluid was negatively charged. Many of Franklin's conclusions about electricity were set forth in his book, *Observations on electricity, made at Philadelphia in America*, which was published in London in 1751.

An important contribution was made by the French military engineer Charles Augustin de Coulomb (1736–1806), whose early work had been on the construction of military fortifications. In 1784 he constructed a torsion balance, consisting of a long bar hung at its centre from a wire. With it he showed that the force of electrical attraction is proportional to the product of the electrical charges, and inversely proportional to the square of the distance between them. He also showed that the same law applies to magnetic poles. This law is now known as *Coulomb's law*.

In 1776 Joseph Priestley (1733–1804) summarized all that was then known about electricity in a book entitled *The history and present state of electricity, with original experiments*. When this book appeared the really important work on electricity was still to come, the subject being completely transformed by the discovery of the electric current.

## THE ELECTRIC CURRENT

The first work that led to investigations of electric currents was done in the late eighteenth century by Luigi Aloisio Galvani (1737–1798), who described his findings in a book which appeared in 1791. Galvani found that when metals are inserted into frogs' legs, muscular contractions occurred, and he attributed the effects to 'animal electricity'. Others had made similar observations, and Galvani's interpretations of his results were incorrect, but his name has been preserved in a number of expressions. A flow of electricity is sometimes called 'galvanic electricity', and an electrochemical cell producing electricity is often called a 'galvanic cell'. Iron which has been treated electrically so that it is protected by a layer of zinc is called 'galvanized iron', and the expression is used even when electricity is not used for a similar type of protection. An instrument used for detecting and measuring current is called a galvanometer, and a person may be galvanized into action, not usually by application of electricity.

More important contributions were made by the Italian physicist Alessandro Volta (1745–1827) of the University of Padua. His great achievement was to isolate the physical phenomenon of electricity from its connection with living systems. He recognized that in Galvani's experiments the function of the frog's leg was merely to act as a detector of electricity, and he realized that the electricity was being produced from the metals and solutions present. In 1800 Volta invented the pile known by his name, consisting of a series of discs of two different metals, such as silver and zinc, separated by paper moistened with brine (Fig. 4.3). He found that a current of electricity was produced, and announced this discovery in a letter, dated 20 March 1800, to Sir Joseph Banks (1743–1820), the President of the Royal Society of which Volta was a Fellow. Banks arranged for the letter to be presented to the Society on 26 June 1800, and later in the year it appeared in the *Philosophical Transactions*. Volta explained the electricity produced by his piles simply in terms of contact between the metals and the solution, but this view was not generally accepted. It only later became clear that chemical reaction is occurring, and that the source of the electrical energy is the chemical energy released when the chemical processes occur. At the time, however, the idea of energy was not well understood, and the principle of conservation of energy (Chapter 2) was not generally accepted for another 40 years.

At first the relationship between static electricity and what was produced in a voltaic pile was not at all clear. It was only after more work had been done with electric batteries (to be considered later) that it was realized that what was being produced was a flow of the same kind of electricity that could be stored in Leyden jars. Only very much later was the mechanism of the flow understood—for example that the flow of electricity along a wire is a flow of electrons, the existence of which was not recognized until much later in the century.

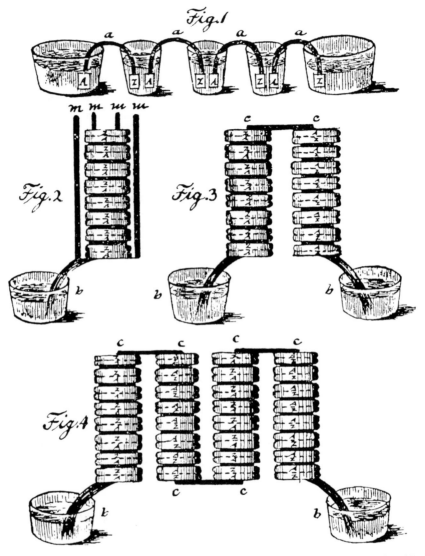

**Fig. 4.3**   Voltaic piles, as illustrated in Volta's first paper on the subject: *Philosophical Transactions*, **92**, 404–431 (1800). For his important contributions, the Emperor Napoleon created Volta a count in 1801.

## ELECTROLYSIS OF WATER

Even before Volta's letter appeared in print another paper was published which described experiments making use of the information presented by Volta, and using piles of the same kind. This paper showed that the electricity generated in the piles was capable of bringing about the production of hydrogen and oxygen gases from water. The author of this paper was William Nicholson (1753–1815), and the experiments had been done in collaboration with his friend Anthony Carlisle (1768–1840). Nicholson was a competent amateur scientist who had founded in 1797 the *Journal of Natural Philosophy, Chemistry and the Arts*; it was usually known as *Nicholson's Journal*. Carlisle was a fashionable London surgeon who was later knighted and became President of the College of Surgeons. On learning of Volta's work the two men constructed Voltaic piles of their own, often using half-crowns (a common silver coin in Britiain, still remembered by older people) as the silver discs; one of their piles, for example, consisted of '17 halfcrowns, with a like number of pieces of zinc, and of pasteboard, soaked in salt water....'. They then inserted wires from the two ends of their pile into a dish of water. When they used copper wires from the pile, hydrogen gas was evolved at one wire, and the other was oxidized. With either platinum or gold wires they found that hydrogen gas was evolved at one wire and oxygen gas at the other. Nicholson immediately announced the results in his journal, the paper appearing in July, 1800. This was before Volta's own paper was published.

Nicholson and Carlisle's discovery that electricity can cause water to produce hydrogen and oxygen created as great a stir as any scientific discovery ever made. The surprise was not so much at the fact that the gases were produced, but with where they were produced. It seemed to most people that if the gases were produced at all, they should be produced in one place. The puzzle was why the gases were produced separately at the two wires; in Nicholson's own words,

> It was with no little surprise that we found the hydrogen extricated at the contact with one wire, while the oxigen (*sic*) fixed itself in combination with the other wire at the distance of almost two inches.

We can look at the matter as follows: imagine the water at the wire at which hydrogen is evolved. Why is not the oxygen, also presumably formed from the decomposition of water, evolved at the same place? Why and how does it apparently burrow its way through the solution and appear only at the other wire?

Nicholson and Carlisle's experiments were at once repeated by many investigators. One of these was Johann Ritter (1776–1810), a brilliant but somewhat eccentric investigator at the University of Jena; his discovery in 1801 of ultraviolet radiation was mentioned in the last chapter. He concluded, also in 1801, that it was not possible for the gases hydrogen and oxygen to be produced from the decomposition of water, since there was no way they could travel through the solution to the wires that were attached to the voltaic pile. He then drew the provocative conclusion—and to emphasize it he set it in a single line of his article—that 'Water is an element.'

In his view, oxygen was water plus positive electricity, while hydrogen was water plus negative electricity.

These views at once attracted considerable attention, sometimes favourable and sometimes accompanied by indignation. The Institut de France set up a committee to look into the matter. Its chairman was Georges Cuvier (1769–1832), famous as a zoologist and paleontologist but also highly versed in the physical sciences. Cuvier was very fair-minded, and in his report which appeared later in 1801 he accepted Ritter's theory as a not unreasonable one. He also suggested other possibilities. One was that there actually was some unknown mechanism by which gases and other materials can travel invisibly through a solution.

The dilemma could not be solved satisfactorily for many years, until evidence was obtained that charged species, such as hydrogen ions ($H^+$) and hydroxyl ions ($OH^-$), exist in solution. The electrolysis of water could then be explained in terms of hydrogen ions moving towards one electrode, and hydroxyl ions towards the other.

Electrolysis will be mentioned again later in this chapter, in connection with the work of Michael Faraday on the subject.

## ELECTRIC BATTERIES

Volta's work, and that of Nicholson and Carlisle, stimulated many further investigations. An electric pile of particular interest was designed by George John Singer and described in his book *Elements of electricity and electrochemistry*, which was published in London in 1814; it is thought to be the first book dealing with the electric current. The diagram in the book that shows this pile is shown in Fig. 4.4. The pile consists of a large number of pairs of silver and zinc discs, separated from one another by manganese dioxide. What is remarkable about this pile is that one built in 1840, probably to the same specifications, is still in operation over a century and a half later. It was built in London by the instrument makers Watkin and Hill, and installed in Oxford by the Revd Robert Walker, who was a distinguished teacher of science and textbook writer. The pile, which is to be seen in the Clarendon Laboratory at Oxford, looks very much like the pile illustrated in Fig. 4.4, and its bell 'ceaselessly' rings. It has to be explained to those who go to see it that there is no perpetual motion; the bell is bound to stop eventually, and it seems possible that the clapper will wear out before the electrochemical energy runs out. It is estimated that the voltage is about two thousandths of a volt and the current about one thousand millionth ($10^{-9}$) of an ampere (1 nanoampere). Although the pile is referred to as a dry pile, it cannot be really dry as then it would not work. One of the reasons for its longevity is that the instrument makers sealed it well, so that the tiny amount of moisture present cannot escape.

It was soon realized that voltaic piles were awkward to use, and that it was more convenient to immerse rods or plates of two different metals in a bath

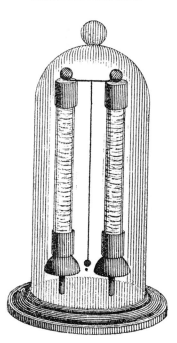

**Fig. 4.4** The voltaic pile designed by George John Singer and illustrated in his 1814 book. It consists of a large number of pairs of zinc and silver discs separated by manganese dioxide, which must be slightly damp in order for the cell to work. A cell constructed in 1840, perhaps to the same specifications, is still in operation, ringing the bell continuously, at the Clarendon Laboratory in Oxford.

of acid or brine (salt solution). Such a device was usually called a battery, although strictly speaking it should be called a cell; the word battery should be reserved for several cells connected together. Many different combinations of metals were used, and there was much rivalry among investigators.

Early in the nineteenth century a general feeling of euphoria resulted from the development of batteries, as it was thought that they would solve the problem of energy supply more effectively than steam engines. However, as we have seen in Chapter 2, Joule's work showed that batteries provided a source of energy that was not economical. The great development of the electrical industry had to wait for many decades, and batteries played a rather subsidiary role.

One of the early enthusiasts in the construction of batteries was John George Children (1777–1852), whom we met in the last chapter in connection with early work on photography. (His daughter Anna Adkins was the first to produce a book photographically.) John Children enjoyed making batteries of unusually large size. He gave an account of a large battery to the Royal Society in 1809, and of even larger ones in 1815. In one of his arrangements each cell had a copper plate 6 feet long and 2 feet 8 inches wide, and a zinc plate of the same size. By means of ropes, pulleys, and counterpoises these plates were lowered into acid, the nature of which was not specified; it was probably dilute sulphuric acid. Twenty-one of these cells were connected together electrically, the total volume of acid being 945 gallons. Children also constructed a single cell by connecting a number of plates together by means of lead straps; each of the resulting zinc and copper plates had a surface area of 1344 square feet. These enormous batteries were used by Children to produce large sparks and to fuse metals; it cannot be said that they led to scientific results of much importance.

Batteries of a more modest size were used at the Royal Institution by Humphry Davy (1778–1829) to isolate a number of chemical elements. On 6 October 1807 he electrolysed fused potash and obtained the element potassium, and a week later he prepared pure sodium by the electrolysis of fused soda. In the following year he isolated three more elements—calcium, strontium, and barium; he also obtained magnesium, but only in an impure form. On the basis of work of this kind Davy advanced the hypothesis that electrical attractions are responsible for the formation of chemical compounds—he turned out to be quite right, although the situation is much more complicated than he thought.

Also in 1808, Davy brought together two carbon rods which were connected to the poles of a battery, and produced a brilliant arc. This can be said to be the first lamp powered by electricity, but it was not for another half century that arc lamps could be used for public lighting. In the same year Davy launched an appeal for funds that would enable the Royal Institution to construct a battery of a large number of cells, capable of producing 5000 volts. (This unit, named in honour of Volta, was not introduced until later, so that this value of 5000 volts is an estimate.)

Later, several useful electric cells were devised by John Frederick Daniell (1790–1845), who was professor of chemistry at King's College, London. The college was on the Strand, next to Somerset House; it had a suitable lecture room and some laboratories for students, but no facilities for research, and Daniels had to do his research in a storeroom under the lecture room. One of Daniell's cells, constructed in 1836, consisted of an outer vessel containing copper sulphate solution, inside which was an ox-gullet bag containing a rod of amalgamated zinc in dilute sulphuric acid. It produced about 1.1 volts. In a later modification the ox-gullet was replaced by a porous pot which contained the acid and the zinc electrode. Daniell was a close friend of Faraday, and he first described this cell in letters to him. In 1839 he put together a battery of 70 of his cells and produced a brilliant electric arc which caused some skin blistering and eye injury to himself and others.

It was not at first realized that the source of the emf in these cells was the energy released in the chemical reactions that were taking place. As discussed in Chapter 2, understanding came in the 1840s, particularly with the work of Joule. In the later Daniell cell, zinc dissolves in the zinc sulphate solution, and copper is deposited in the copper sulphate; the chemical energy released provided the electrical energy.

In Chapter 2 we met William Robert Grove, barrister, scientist, and later judge, who made one of the first clear statements of the first law of thermodynamics. He also devised some useful electric cells. In 1838, when a professor at the London Institution, he devised a cell consisting of zinc in sulphuric acid and platinum in nitric acid, the two liquids being separated by porous material (see Fig. 4.5). This cell produced 1.8 to 2.0 volts—more than given by Daniell's cells—and cells of this type began to be used by Faraday in his demonstrations at the Royal Institution. In 1842 Daniell suggested that his own cells had inspired the Grove cell, and it is hard to see how Grove could not have profited from Daniell's work. However, Grove hotly denied the suggestion, and in 1842 and 1843 there was a sharp exchange of published letters between the two men.

Grove also devised what he called a gas voltaic battery, consisting of platinum electrodes immersed in acid solution, with hydrogen bubbled over one electrode and oxygen over the other. This cell was first described in any detail in a letter from Grove to Faraday in October, 1842. The source of the voltage

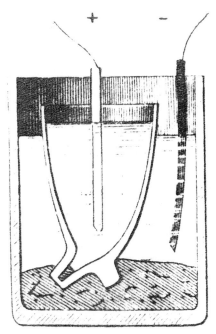

**Fig. 4.5** Grove's diagram of his primary cell. The positive pole was of zinc in dilute sulphuric acid, and the negative pole was of platinum in concentrated nitric acid. The two solutions were separated by use of the broken-off bowls of clay tobacco pipes. This cell gave a higher voltage (nearly 2 V) than any previous cell, and Faraday and others referred to it as Grove's intensity battery.

is the energy released by the combination of the two gases. Later Grove used other combinations of gases, such as hydrogen and chlorine, and oxygen and carbon monoxide. Cells of this type are today called *fuel cells*, and they are still the subject of much research to make them more effective for commercial use. Grove's first reference to his gas battery was in a paper in 1839.

Grove was a rather colourful individual who was fond of performing spectacular and sometimes dangerous experiments. At a lecture he gave to the Royal Society in May 1843 he connected together 50 of his fuel cells and caused the current to pass through five people holding hands, giving them all a shock. When he shorted the battery of cells, using charcoal poles, a brilliant spark was produced. He gave a total of 14 Discourses at the Royal Institution, and at one of them, on 13 March 1840, he managed to destroy Michael Faraday's pocket knife. Using a battery of 50 cells each having large metal plates Grove produced an arc about 3 cm long. Faraday lent him his knife to aid in the demonstration, and we are told that the 'large blade was instantly deflagrated'. We are not told what Faraday said about the incident; in view of his deeply religious nature we can assume that his comments had no need of expletive-deletion.

Later in the century came the introduction of dry cells, which are easier to transport than ones containing solutions or strong acids. As early as 1803 a dry cell had been constructed by Johann William Ritter, but the development of dry cells sprang from the work of the French engineer Georges Leclanché (1839–1882) who, however, did not construct a dry cell himself. In 1868 he devised a cell in which rods of zinc and carbon were dipped into a solution of ammonium chloride. Wet cells of this type were used well into the twentieth century for door-bells and other purposes where electricity is required only occasionally and for short periods. The familiar dry cells now used very widely are a development of the Leclanché cell.

The cells mentioned so far are called primary cells. They cannot be regenerated by electrical means but only by replacing the solutions or electrodes (as the 'poles' were later called by Faraday). A storage cell or accumulator, which could be recharged by the passage of a current, was first devised by Ritter in 1803, but the invention was far ahead of its time, and he lacked the facilities to develop it. The much later work of the Frenchman Raimond Louis Gaston Planté (1834–1889) led to the great development of storage batteries, which are used so widely today. Planté began his work in the 1860s and first demonstrated his batteries in 1879; they came into commercial use about two years later. The Planté cell consists of lead plates immersed in sulphuric acid. When the cell is charged the positive plate becomes plated with lead peroxide, and as the cell is discharged this process occurs in reverse. The Planté cells were greatly improved in 1881 by Camille Fauré, who coated the surfaces of the plates with lead dioxide, and there were various later technical improvements.

The invention of the storage battery inspired a second 'electrical euphoria' in the 1880s. It was thought that storage batteries would play an important part in the storage and distribution of energy, since rapidly moving streams could operate dynamos which would recharge the accumulators. By the late 1880s storage batteries were being used for such purposes as lighting railway carriages. A serious limitation to them, however, was their weight. As we shall see, the development of electric power followed an entirely different course, and storage batteries were destined to play only a subsidiary role—for example in automobiles. When electric automobiles become more efficient, storage batteries will be used to a much greater extent.

## OERSTED AND AMPERE: ELECTROMAGNETISM

In 1820 Hans Christian Oersted (1777–1851; Fig. 4.6(a)), professor of physics at the University of Copenhagen, brought a compass needle near to a wire connected to the poles of a battery (we know now that an electric current was passing along the wire, but at the time this was not clearly understood). He found that the needle tended to turn, and to turn in an unexpected direction; the poles of the magnet were not attracted or repelled by the current, but moved in a direction at right angles to the expected direction (Fig. 4.6(b)). When the direction of the current was reversed, the needle turned in the opposite direction.

A remarkable feature of this famous discovery is that it was made in front of a class of students. Oersted and his assistant had set up the experiment but

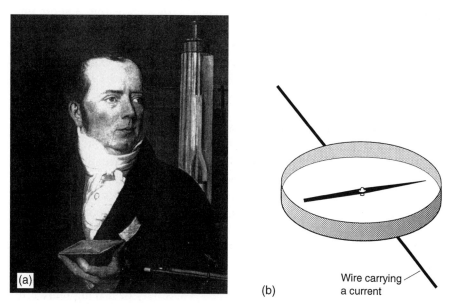

(a)

(b)

Wire carrying
a current

**Fig. 4.6** (a) Hans Christian Oersted (1777–1851). (b) Representation of Oersted experiment. The magnet tends to turn in the direction at right angles to the wire.

had not had time to try it out before the students arrived. Oersted first decided to defer the experiment until later, but during the lecture began to feel confident that it would work, and performed it successfully. Another remarkable feature is that, since the deflection of the needle had been small, Oersted was so little impressed by the result that he performed no further experiments on the subject for three months. Then, having confirmed the effect, he sent a four-page announcement of it, in Latin, to many leading scientific journals, and to a number of scientists. The announcement appeared, in various languages, in a considerable number of journals. Further publicity was given to the discovery by the fact that the distinguished scientific statesman François Arago (1786–1853) called attention to the discovery at a meeting of the Académie des Sciences in Paris on 4 September 1820. At first Arago was sceptical about the result, but a week later he repeated Oersted's experiment, and became convinced.

Oersted's experiment was the first to show a connection between electricity and magnetism, and can be called the birth of electromagnetism. At the time, in the scientific tradition established in particular by Newton, theories were formulated in terms of forces acting in straight lines between points; these were known as central forces, and they were believed to follow Coulomb's law, the force being inversely proportional to the square of the distance between the points.

The fact that the magnetized needle was subjected to a deflecting force (i.e., it moved towards a sideways direction) was therefore particularly surprising. It suggested a force acting not in a straight line but circularly, which most scientists thought to be unreasonable. However, within a short time many scientific investigators had confirmed the experimental result obtained by Oersted.

Within only a few days much progress, along both experimental and theoretical lines, was made by the French physicist André Marie Ampère (1775–1836; Fig. 4.7(a)). Ampère was a prodigious worker; his life was a tragic one and he deliberately drowned his sorrows in unremitting labour. At the age of eighteen he was forced to witness the guillotining of his father, to whom he was much attached. This experience left him highly traumatized, and for a long time he wandered listlessly. A happy marriage in 1799 raised his spirits, but the death of his wife four years later was a blow from which he never recovered. A second marriage to a woman who, with the connivance of her relatives, made his life unbearable, ended in divorce. Even his children by his first wife were a great trial to him, and he had constant financial problems.

Ampère was held in high regard by his colleagues, and he was elected a member of the exclusive Institut de France. At one of its meetings he failed to recognize the Emperor Napoleon, who with his usual good humour did not take offence. Instead he said that he knew a sure way of making himself recognized, and asked Ampère to lunch on the following day; Ampère unfortunately forgot to appear. There are many stories of his absentmindedness. On one occasion he suddenly had an idea and started to write equations in chalk on the back of a cab; just as he finished it sped away. Ampère kept a cat and had a hole of suitable size cut in the door so that it could go in and out. When he (or perhaps the cat) later acquired a kitten, he had a smaller hole cut in the door, not realizing until it was pointed out to him that the second hole was unnecessary. (This story, it must be admitted, has also been told about Newton and Bunsen, and perhaps is apocryphal.)

Ampère was in the audience at the Acadèmie des Sciences on 4 September 1820 when Arago announced Oersted's discovery. Ampère had an exceedingly agile mind, and although he had not worked previously on electricity, within a week he had made careful quantitative experiments and formulated a mathematical treatment. As early as the 18 and 25 September 1820 he gave reports to the Acadèmie, confirming Oersted's results and presenting some significant additional experimental results he had obtained himself. He also explained a hypothesis he had formulated on the relationship between electricity and magnetism. His work during the next few years gave strong support to the conclusions he had reached so quickly. The *Mémoire* he published on the subject in 1827 is regarded as a masterpiece of style and clarity, and has been called the *Principia* of electrodynamics; Maxwell later described it as 'perfect in form and unassailable in accuracy'. It is appropriate that the unit of electric current is called the ampere; it is one of the base units in the Système Internationale d'Unités (SI).

(a)

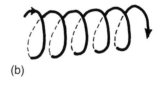

(b)

**Fig. 4.7** (a) André Marie Ampère (1775–1836). (b) Ampère's idea (1820) that a magnet is due to an electric current flowing through a coiled wire (which he called a solenoid).

Ampère made careful studies of the effects of electric currents on one another, and found that if currents travelled in the same direction along two parallel wires, there was repulsion between them; if the currents travelled in the opposite directions there was attraction. He also worked out a detailed mathematical treatment of the interactions, on the basis of the assumption that current-carrying elements of the wires interacted with one another according to the inverse square law. By adding up (using the mathematical procedure of integration) the effects of all the elements, he arrived at expressions that were consistent with the experimental results. This was a very impressive treatment, and because of it Maxwell in his *Treatise on electricity and magnetism* (1873) referred to Ampère as the 'Newton of electricity'. There is something of an irony about this, because Maxwell's own work caused the later development of electromagnetism to follow a completely different path. Ampère nevertheless deserves great credit for his experimental and theoretical work, as it played an important part in the later developments of the subject.

To explain the effect of an electric current on a magnet, Ampère supposed that magnetism arises from electricity moving in circular orbits round the axis of the magnet; that is, the planes of the orbits are at right angles to the magnetic axis. He coined the word solenoid to refer to a current moving in a small circuit and creating a magnet. He carried out experiments with solenoids in which wires were wound round glass tubes, and confirmed that when a current passed, a magnetic effect was obtained. He then developed an elegant mathematical treatment of the interactions between electric currents and the circular currents around the magnets, and was able to explain Oersted's results in terms of central forces.

Ampère's work was at once recognized by most investigators as an achievement of great importance, but a few objections were raised. It was pointed out that there was no experimental evidence for a flow of electricity round magnets, and no suggestion as to how it could arise. Volta had found that a current results when two dissimilar metals are present, but not if only one metal is present. Ampère later modified his theory to relate to the molecules in the magnets, suggesting that perpetual electric currents moved in orbits around them (we know today, of course, that electrons do move in orbits about the nuclei of atoms). At the time there was no understanding of the nature of electricity; the electron was not discovered until over half a century later. Ampère also formulated a theory of the magnetism of the earth in terms of electric currents moving in orbits.

## FARADAY'S EARLY RESEARCHES IN ELECTRICITY

Michael Faraday was particularly unhappy with Ampère's treatment. Because he knew little mathematics, and was ill-versed in physical theories, he simply could not understand the treatment. He was quite content to think of a circular force arising from a current flowing in a wire. Hardly anyone brought up on the physics of the time could accept such an idea, and yet it was the origin of Faraday's important concept of the electromagnetic field, which was to be the core of the later ideas of Faraday and Maxwell.

Michael Faraday (1791–1867; Fig. 4.8) was born to a Yorkshire family, his father James Faraday being a journeyman blacksmith. He was having difficulty making a living, and decided that the family should move south to London, which they did early in 1791. Since Michael was born on 22 September 1791 it is quite possible that Mrs Faraday was pregnant during the uncomfortable trip to London. Michael was born at Newington Butts, which was on the south of the River Thames, not far from London Bridge; the word Butts refers to the fact that the land was at one time reserved for the practice of archery. Newington Butts is now part of the London borough of Southwark.

James Faraday could do little more than provide the basic necessities of life for his wife and four children. For some years the family of six lived in a squalid single room. Michael Faraday's parents did, however, imbue in him strong religious principles which were to dominate his life. The family belonged to a strict and somewhat obscure Christian sect the members of which were known as the Glassites or more commonly as the Sandemanians. There seem to be no Sandemanians today, and even at their peak they numbered only about a thousand members in England and Scotland. In Faraday's own word, the sect was 'despised'. The actual founder of the sect was John Glas

**Fig. 4.8**  A photograph of Michael Faraday (1791–1867). Faraday was particularly interested in the discovery of the techniques of photography (Chapter 3), and gave much encouragement to Henry Fox Talbot, John Herschel, and others who were concerned in the work. As a result, Faraday himself was photographed on a great many occasions.

(1695–1773), who was born in Auchtermuchty, Scotland; Robert Sandeman (1718–1771) was his son-in-law. Glas was for a time a Presbyterian minister, but was deposed because of his unorthodox beliefs, and he formed a congregation based on strict adherence to the teachings of the Bible. The Sandemanians believed that the Bible, and the Bible alone, was a sufficient guide to personal behaviour. Although they were religious fundamentalists, the Sandemanians did not believe in a hell of fire and brimstone, but placed more emphasis on love and a sense of social duty. Faraday remained a loyal Sandemanian throughout his life. Some idea of the strictness of the sect is provided by the fact that he was on one occasion disciplined for being absent from a religious gathering. His excuse, over which he himself had agonized for some time, was that he had been summoned to see Queen Victoria, but it was deemed inadequate.

Michael Faraday's education, at an ordinary day school, was very rudimentary, consisting of little more than reading, writing, and simple arithmetic. In view of this it is all the more surprising that he was able to rise to a position of responsibility at an early age; we discuss this matter further, with more examples, in Chapter 9. At the age of 13 Faraday left school, and after delivering newspapers for a period entered into a seven-year apprenticeship with a good-natured bookbinder, George Riebaud, an émigré from France who had fled the excesses of the French Revolution. Ribeau encouraged Faraday to read the books he was binding, and an article on 'Electricity' in the *Encyclopædia Britannica* seems to have aroused his interest in science. Also, the entrancingly written book *Conversations on chemistry* made a particular impression on Faraday, and perhaps did more than anything else to make him enthusiastic about science; its author was Jane Marcet (1769–1858), herself a chemist of some distinction, and throughout his life Faraday spoke enthusiastically about her book. In 1812 Charles Dance (1755–1840), one of Riebaud's customers, gave Faraday a ticket to attend some of Sir Humphry Davy's lectures at the Royal Institution. In October 1812 when his apprenticeship expired, Faraday applied to Davy for a position, at the same time sending him notes he had prepared on his lectures. Davy was impressed, but at first could do nothing. In March 1813, however, when an assistant was dismissed for brawling, Faraday was appointed to his position at the Royal Institution, and was given two rooms on the top floor in which to live. The building was to be both his home and his laboratory for many years.

Later in that year Sir Humphry and Lady Davy paid an extended visit to France and Italy, and Faraday was invited to accompany them as scientific assistant. Although the experience was somewhat marred for Faraday by the fact that Lady Davy tended to treat him as a valet, he profited greatly from it, particularly from the opportunity of meeting many of the leading scientists of the time, including Volta, Arago, Cuvier, and Count Rumford. Although England and France were then at war, the travellers encountered no difficulties. They took with them scientific apparatus and carried out experimental research, performing demonstrations at the various places they visited. They experimented with the newly discovered element iodine, and from it prepared the explosive nitrogen triiodide, which today is still a favourite substance with lecturers on explosions.

At first Faraday's main work at the Royal Institution was in chemistry. By 1820 he had established a reputation as an analytical chemist, and from this work he was able to supplement his modest salary and also contribute to the support of the Royal Institution itself, which was not well endowed. Some of Faraday's chemical work was concerned with clays, and some was on metal alloys; he prepared several novel varieties of steel, from some of which he fabricated razors for himself and his friends. Occasionally Faraday appeared in court as an expert witness. One of these cases related to oil, and Faraday was led to investigate the properties of oils and gases which were beginning to be used for public heating and lighting. It was as a result of these studies that Faraday discovered benzene in 1825. Since this compound is the prototype of a vast number of organic compounds, this was a discovery of great importance. He also discovered a number of other important chemical compounds. One of these was the first recorded example of what is now called a clathrate compound—it was a compound in which a chlorine molecule is buried inside a group of water molecules. (*Clathri* is the Latin for a lattice or trellis, and the water molecules form a kind of trellis round the inner molecule). An example of a clathrate compound is found in permafrost, where water molecules trap methane ($CH_4$) molecules. Faraday was the first to liquefy a number of gases, such as ammonia and carbon dioxide.

In 1820, soon after Ampère had presented his interpretation of Oersted's result, Faraday's friend Richard Phillips, an editor of the *Philosophical Magazine*, persuaded him to look into the subject of electromagnetism. Like other editors of scientific journals, Phillips had been inundated with papers on the subject. Faraday accepted Phillips's suggestion rather reluctantly, as he considered that his skills were confined to chemistry, and his interests had been far from electromagnetism. Eventually he agreed to look into the work of Oersted and the later work of Ampère, and posterity must be grateful to Phillips for his gentle but persistent prodding.

Faraday at once repeated Oersted's experiments, and he noticed that when a small magnetic needle was moved round a wire carrying a current, one of the poles turned in a circle. He then speculated that a single magnetic pole, if it could exist, would move continuously around a wire as long as the current flowed. This led him to perform an experiment of great simplicity and also of great importance. In 1821 he attached a magnet upright to the bottom of a deep basin, and then filled the basin with mercury so that only the pole of the magnet was above the surface. A wire free to move was attached above the bowl and dipped into the mercury (see the right-hand bowl in Fig. 4.9). When Faraday passed a current through the wire and the magnet, the wire continuously rotated around the magnet. In an adaptation of the experiment

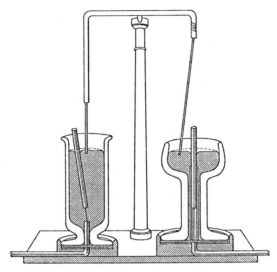

**Fig. 4.9** Faraday's published diagram illustrating his demonstration of electromagnetic rotation. This was the first time that electricity had been converted into useful energy.

(see the left bowl in Fig. 4.9), he caused the magnet to rotate around the wire. He also succeeded in rotating a wire by use of the magnetism of the earth. The great importance of these simple demonstrations is that electrical energy was being converted into mechanical energy for the first time. To Faraday the results implied that there were circular lines of force round the current-carrying wire, and he accepted this as a simple experimental fact. Almost everyone else concluded that the force could not be simple, but must be explained in some way in terms of central forces.

## ELECTROMAGNETS

For the next ten years Faraday worked only sporadically on electricity. During this period there were some important developments elsewhere in connection with electromagnets. In 1820, soon after Oersted had made his announcement, Arago showed that iron filings left in the neighbourhood of a wire carrying an electric current became magnetized, and that a steel needle became permanently magnetized if a current was passed through a coil of wire around it (Fig. 4.10(a)). An even more important observation was made in 1823 by William Sturgeon (1783–1850). Sturgeon was another man who had an unpromising background and education. His father was a not very hardworking Lancashire shoemaker who supplemented his income by poaching and other dubious activities. Sturgeon himself was first apprenticed to a shoemaker, but in 1804 enlisted in the Royal Artillery and attended some lectures, carrying out electrical experiments in his spare time. He developed his skills sufficiently well that after his discharge from the army he did some teaching, and in 1824 he was appointed lecturer in science at the East India Company Royal Military College at Addiscombe, a district of Croydon, near London. In 1825 he improved Arago's experiment involving the magnetization of a steel needle. Instead of using steel, Sturgeon placed a bar of soft iron in a solenoid (Fig. 4.10(b)), and found that the magnetism was greatly increased, but that the magnetization was not permanent, as it had been in Arago's experiment with a needle. Sturgeon used a coating of shellac to insulate the bare wires from the soft iron. It was Sturgeon who first suggested the word electromagnet for the device he had constructed.

Improvements to Sturgeon's electromagnet were made by Joseph Henry (1799–1878; Fig. 4.11), a great scientist who might be much better known today but for Faraday's work, because Henry did many of the things that Faraday did but always a little too late. Joseph Henry's grandfather had

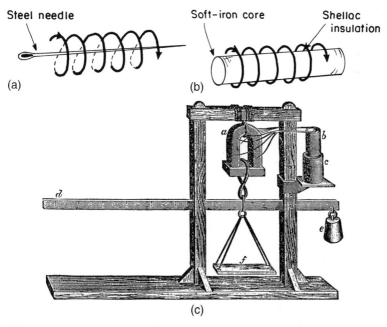

**Fig. 4.10**   (a) A representation of Arago's experiment of 1820 in which he produced permanent magnetism in a steel needle. (b) Sturgeon's electromagnet (1825); the word was suggested by him. He used a soft-iron core, in which case the magnetism was not permanent; the coil through which the current passed was insulated from the core by a coating of shellac. (c) Joseph Henry's diagram of the electromagnet he constructed in 1832; it could lift over 3000 lb. It still exists, and is in the Smithsonian Institution.

**Fig. 4.11** Joseph Henry (1799?–1878), who made important discoveries, particularly in electricity and magnetism. He perhaps deserves most credit for the electric telegraph. For a period he was professor of physics at Princeton, and later was the first director of the Smithsonian Institution in Washington. There is some doubt about his date of birth; his baptismal register records the date as 9 December 1797, but there is strong evidence from a family member that the year was 1799.

travelled from Scotland with his family in 1775 and had settled on a farm near Albany, New York. His name was William Hendrie, but feeling that this good Scottish name was not being pronounced to his satisfaction he changed it to Henry. This was later a matter of regret to Joseph Henry, who felt that Hendrie was more distinctive, and we must agree with him. He is often confused with other Henrys, particularly by chemists, who learn about Henry's law, which has to do with the solubilities of gases. But that was formulated by another and less distinguished Henry, William Henry (1774–1836), an English physician and chemist, who wrote a successful textbook of chemistry.

Joseph Henry was born in Albany, New York. His family was poor, but he was luckier in his education than either Faraday or Sturgeon. Like Faraday, he was introduced to books and the world of learning in an unusual way. One day his pet rabbit escaped, and he followed it on his hands and knees under the village library. Finding some loose boards he crawled into the library, and found a book that fascinated him. He frequently returned by the same route to the place that was giving him so much pleasure, but was eventually caught in the act. Happily, the authorities did not regard his action as a misdemeanour, but instead arranged for his admission to the library by a more convenient and conventional route. When he was about sixteen he became particularly interested in scientific books.

He was first apprenticed to a watchmaker and silversmith, and then at the age of twenty enrolled in the Albany Academy, at the same time supporting himself by doing a variety of jobs, such as surveying. He was soon recognized to be of unusual ability, and in about 1826 he was appointed professor of mathematics and natural philosophy at the Albany Academy. He remained there until 1832 when he was appointed professor at the College of New Jersey (now Princeton University). In 1846 he was appointed the first Secretary of the Smithsonian Institution, which had just been established in Washington as the result of a bequest by the eccentric Englishman James Louis Macie Smithson (1765–1829). Smithson was the illegitimate son of the 1st Duke of Northumberland, and in a fit of pique at the rejection of a paper that he submitted to the Royal Society he left all his money to be used for the foundation of an institution in a country he had never even visited.

Joseph Henry's interests in electricity and magnetism were similar to those of Faraday, but he was not as capable or persistent as Faraday, and time and again his discoveries came a little too late for him to be entitled to priority. Also, he was slow to publish his results, while Faraday was speedy. Faraday always expressed a great admiration for Henry, and the two had many pleasant meetings in 1837 when Henry visited the Royal Institution.

Henry's improvement of Sturgeon's technique for creating an electromagnet, carried out in 1826 when he was in Albany, was to insulate the wires before winding them on to the soft-iron core (Fig. 4.10(b)). This allowed many more turns, and therefore an increase in the magnetism produced. It is said that, unable to afford other material, Henry used his wife's petticoat, and perhaps her wedding dress, for the insulation of his wires. Henry also made the important discovery that the polarity of an electromagnet could be reversed by a reversal of the direction of the current. He published an account of some of his experiments on magnets in 1831.

Henry continued to improve his electromagnets, and investigated two different procedures for winding them. In some experiments he used one continuous wire, and he called the result an intensity magnet. Alternatively, he used several shorter wires connected electrically in parallel, and he called the result a quantity magnet. The intensity magnets were particularly effective for lifting heavy weights, and by 1832 he had constructed one that could lift a mass of over 3000 lb (Fig 4.10(c)). It is still in existence at the Smithsonian Institution in Washington.

Electromagnets were of great importance in themselves, but were especially significant in leading directly to the electric telegraph and the electric motor.

## THE ELECTRIC TELEGRAPH

Ask an American to name the person who first constructed a telegraph, and the most likely answer you will get is Samuel Morse. Ask anyone from the UK and you will almost certainly get a blank stare; if the person is unusually well-informed, you may hear the names Sir William Cooke and Sir Charles Wheatstone. Many people made contributions to the telegraph, but all the evidence I have found so far convinces me that the main credit should go instead to Joseph Henry, who set up simple telegraphs in Albany and later in Princeton, and who provided both Morse and Wheatstone with the basic ideas for making their long-distance telegraphs.

In 1831, just before moving to Princeton, Henry connected a mile of wire, strung round his room at the Albany Academy, to one of his intensity magnets. A small permanent magnet was pivoted near the magnet, and it would swing against a bell when the circuit was closed. In November 1832, after he had moved to Princeton, Henry stretched a wire from his laboratory in Nassau Hall (then and now the main academic building at Princeton) to his house, which was on the campus several hundred feet away; the wire was draped over the branches of the intervening elm trees. Henry completed the circuit through the earth by sinking the ends of the wire in two wells. In this way he was able to send signals from his laboratory to his wife, who heard the bell at their house. Henry never took out a patent on his telegraph, or indeed on anything else, as he felt strongly that in his position, with an established professorship, he should not profit personally from his scientific work.

In 1837 Henry obtained a year's leave of absence with full salary to visit laboratories in England and France. He was particularly happy about his reception in England, as his scientific distinction was well recognized there. He spent time with Faraday at the Royal Institution, and they carried out experiments together; some of these related to electrical induction and will be mentioned again later. He also met, sometimes at the Royal Institution, Charles Wheatstone, who was then at King's College, London. Together with William Cooke, Wheatstone had already been experimenting with a telegraph system, and Henry freely gave him all the advice he could. In that same year, 1837, a patent for a telegraph was taken out in London by Cooke and Wheatstone. This was the first telegraph patent taken out anywhere in the world. It is not clear to what extent this patent was influenced by the discussions with Henry, but there is no doubt that many of the later improvements to the British telegraph network owed much to Henry's ideas about electro-magnets and the use of relays.

Charles Wheatstone (1802–1875) was born in Gloucester and did not attend a university. He began his career as a musical instrument maker in London, and soon attracted attention with his experimental work on acoustics; he studied the vibrations in air brought about by a variety of musical instruments. In 1829 he invented the concertina, a type of accordion. In 1834 he was appointed professor of experimental physics at King's College, London, a position he held for the rest of his life. Later he invented the kaleidoscope and the stereoscope, an improved version of which is still used for viewing X-ray and aerial photographs in three dimensions. His name is remembered today in the Wheatstone bridge, still used by students of elementary physics; it was actually invented in 1833 by Samuel Hunter Christie (1794–1865), but Wheatstone modified it and made it popular.

Wheatstone was an excessively shy man who, although a professor, had an intense dislike of lecturing. The story is told that on 3 April 1846 he was due to give a Friday Evening Discourse at the Royal Institution, but got cold feet at the last moment and escaped from the building. In his place Faraday gave an improvised lecture in which he dealt with what Wheatstone had been expected to talk about, the title of Faraday's Discourse being 'Mr. Wheatstone's electromagnetic chronoscope'. After 40 minutes, Faraday ran dry as far as that subject was concerned, and for another 20 minutes he improvised, giving some of his ideas on the nature of light. Later, by invitation of the editor of the *Philosophical Magazine*, he published these ideas under the title 'Thoughts on ray-vibrations'. This paper was later treated with some ridicule, but it is interesting that when in 1864 Maxwell published a version of his famous theory of electromagnetic radiation (Chapter 5), he said that it was the same in substance as that proposed in 1846 by Faraday.

This incident led to a tradition at the Royal Institution that a person invited to give a Friday Evening Discourse is locked in a room for half an hour before the lecture is due to begin. A person who wishes to avoid giving a lecture must therefore either not turn up at all, or escape more than half an hour before the allotted time. The period of half an hour is no doubt considered to be enough time for someone else to prepare a lecture.

Sir John Meurig Thomas, who was Director of the Royal Institution from 1986 to 1991, tells this story in his delightful book *Michael Faraday and the Royal Institution*, published in 1991. He suggests, however, that the detail about Wheatstone's escape from the building before the lecture, although quite consistent with his character, is probably not true. There is no doubt that Faraday did give the lecture on that occasion, but that may have been by prearrangement. There is also no doubt that Faraday did on that evening explain his ideas about the nature of light, and that he published them later. Contemporary accounts, however, make no mention of Wheatstone's escape, which may have been invented later. For my part, I believe implicitly the story of the escape, whether it is true or not.

William Fothergill Cooke (1806–1879), who was Wheatstone's partner in the 1837 telegraph patent, was born in Ealing, near London, and studied medicine. The first telegraph that he and Wheatstone patented was an exceedingly clumsy affair, needing five lines to transmit all the letters of the alphabet. As early as 1838 Britain's Great Western Railway installed a telegraph system. Cooke and Wheatstone gradually improved their system, and in 1845 patented a system that transmitted along a single line. It is interesting that in 1858 Faraday gave a Friday Evening Discourse at the Royal Institution entitled 'On Wheatstone's electric telegraph in relation to science (Being an argument in favour of the full recognition of science as a branch of education)'.

The popularity of the telegraph in Britain was greatly enhanced in 1845 by its use in the arrest of a murderer. John Tawell, after a remarkable criminal career which led to his transportation to Australia, made himself rich in various dubious ways. On his return to England he dressed as a Quaker and posed as a respectable and benevolent man, but kept a mistress, Sarah Hadler, by whom he had two children. On 1 January 1845 she was living in Slough, west of London, and he murdered her with prussic acid, escaping by train to Paddington station in London. As a result of a telegram sent from Slough to Paddington he was arrested and charged with murder. His defence council, Fitzroy Kelly, who later became a judge, argued that Sarah Hadler had died as a result of eating a large number of apples, the pips of which contain the poison. Not surprisingly, this defence did not convince the jury, and Kelly was known thereafter as Apple-Pip Kelly.

Samuel Finley Breese Morse (1791–1872) was born in Massachusetts, graduated from Yale in 1810 and then studied painting in England. His early career was mainly as a painter and teacher, but he became interested in science with a view to inventing a telegraph. In 1835 he constructed one, but was disappointed to find that he could not transmit for more than about 40 feet. He was then told the secret of Henry's success, which involved using a much improved form of electromagnet. In 1839 Morse spent a few days at Princeton with Henry, who characteristically gave him every assistance, and even lent him several miles of wire. Henry also explained to Morse the usefulness of relays in the circuit when the signal is to be carried over long distances.

Morse tried to take out a patent on a telegraph in Europe, but was unsuccessful. He obtained a US patent in 1840, and in 1844 secured from the US Congress an appropriation of $30 000 to build a telegraph line between Washington and Baltimore. On 24 May 1844 he sent over it the message 'What hath God wrought?', which would seem a not very subtle way of calling attention to what he claimed to have accomplished himself. In 1855, quite improperly, he issued a public statement denying that he was indebted in any way to Joseph Henry for his invention. This was annoying and embarrassing to Henry, who had not wished to make any claim for himself, but felt that his integrity was under attack. He made no response himself, but appealed to the Regents of the Smithsonian Institution, who made a full investigation. Their report gave complete confirmation of the help that Henry had given to Morse, without which Morse could never have succeeded.

Morse is entitled to credit for the dot-and-dash code he invented in 1844, and which bears his name. He was sensible enough to design it in such a way that the most commonly used letters have the simplest representations. Thus e, the commonest letter, is given by a dot, and t, the next commonest, by a dash. This feature is not a virtue of braille, in which t is represented by a pattern of four dots, or of the semaphore in which t requires flags to be held in positions that can cause much discomfort to the signaller.

# ELECTRIC MOTORS

Electromagnets led, by a short step, not only to the electric telegraph but also to the electric motor in the form in which we know it today. However, in the previous century, in 1748, that quite remarkable genius Benjamin Franklin had constructed an electric motor using static electricity. Franklin attached a number of thimbles to the circumference of a wheel consisting of a circular wooden board, which lay horizontally and was pivoted at its centre. Two Leyden jars, one charged positively and the other negatively, were then placed on either side of the board, with wires attached to them, arranged so that they would touch the thimbles when the board rotated. If the board was set in motion with a small push, the thimbles would touch the wires and collect a charge; being repelled by the wires they would continue on their way. The wheel kept turning until the charges in the Leyden jars were depleted. Speeds of 12 to 15 revolutions per minute were achieved, and Franklin considered the possibility of employing such a device as a spit for the roasting of meat and game in front of a fire. A few motors of the same kind were later constructed by others. One was used to drive an orrery (a machine that simulates the rotation of the planets round the sun, first constructed for Charles Boyle, Third Earl of Orrery, a kinsman of Robert Boyle). Since these motors did not perform for long, because of the discharge of the Leyden jars, they were never more than conversation pieces.

When electromagnets were constructed in the second decade of the nineteenth century, it did not really require much further ingenuity to construct an electromagnetic motor. The first of these seems to have been made by Salvatore del Negro in 1830. In 1831 Joseph Henry reported on a motor in a paper entitled 'On a reciprocating motion produced by magnetic attraction and repulsion'. His device is illustrated in Fig. 4.12. A reciprocating motion of this kind is of less value than a circular motion, and this was soon achieved in a variety of ways. A schematic diagram of a simple motor is shown in Fig. 4.13. A permanent magnet is pivoted on an axle between the poles of an electromagnet. An arrangement known as a commutator is needed to reverse the direction of the current, and therefore the polarity of the electromagnet, when the magnet has made a half turn. Otherwise the magnet would stop after making a half turn; since the polarity of the electromagnet is reversed the rotating magnet is able to make another half turn. The arrangement shown can be improved in many ways; a rotating electromagnet is usually employed instead of a permanent magnet.

Another person to have constructed an electric motor is Moritz Hermann von Jacobi (1801–1874), whose speculations about the energy obtainable from motors were mentioned in Chapter 2. Von Jacobi described a practical motor in a paper that appeared in 1834, and he later constructed other motors. As mentioned in Chapter 2, he was at first under the delusion that there was no limit to the amount of energy that could be obtained from a

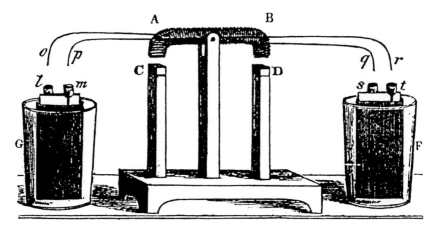

**Fig. 4.12** Joseph Henry's illustration of his reciprocating engine, published in the *American Journal of Science* (usually known as *Silliman's Journal*), **20**, 340–342 (1831). Two permanent magnets, C and D, are supported vertically, and pivoted above them is an electromagnet AB. Two independent solenoids are wound round AB, one of them ending at o and p, the other at q and r. On each side is an electric cell, the poles l, m, s, and t being brass thimbles containing mercury. The wiring of the electromagnet is arranged in such a way that when the wires o and p touch the mercury in the thimbles l and m, the current flows in such a direction as to cause the magnetic pole A to have the same polarity as the permanent pole C; there is repulsion, so that the electromagnet tilts over until the ends of the wires q and r are immersed in the thimbles s and t. The current then flows in the opposite direction; the polarity of the electromagnet is reversed and it tilts over in the opposite direction.

battery-operated motor, but later realized that he was wrong. He then related the energy produced in a motor to the amount of zinc consumed in the battery that operated it, something that Joule did more precisely. Some of von Jacobi's results were reported by him at British Association meetings.

The first person to patent an electric motor was John Davenport, a Vermont blacksmith. His patent was taken out in 1837, and his motor is still preserved in the Smithsonian Institution. Davenport was never able to make a commercial success of his motors, and he died in poverty. Fig. 4.14 shows motors installed in a locomotive by the Scotsman Robert Davidson.

## ELECTROMAGNETIC INDUCTION

In 1831 Michael Faraday made one of his most famous discoveries, electro-magnetic induction. It is often said that this discovery led to the invention of electric motors, but it is much more likely that they came directly from the work on electromagnets, as we have suggested in the previous section.

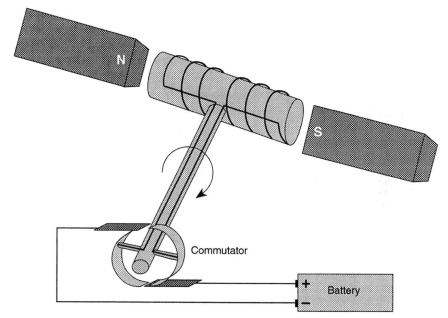

**Fig. 4.13** Schematic diagram of a simple type of electric motor.

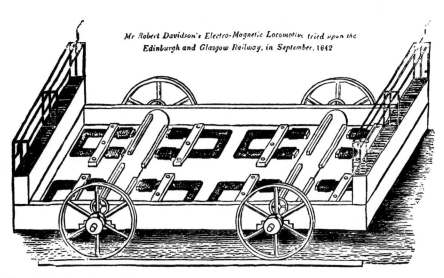

**Fig. 4.14** An electric locomotive, designed by Robert Davidson, which in 1842 performed an experimental run on the Edinburgh to Glasgow Railway. The motors consisted of wooden cylinders to which iron strips were fixed. The horseshoe magnets, to be seen on the sides of the axles, were magnetized alternately by use of commutators. Batteries, containing iron and zinc plates, are to be seen at both ends of the carriage. (Courtesy of the Science Museum, London.)

Faraday frequently experimented with the idea that a wire bearing a current might induce a current in a nearby wire. Until 29 August 1831, however, he always failed to find any effect. On that day he wound one side of a soft-iron ring with insulated wire, and arranged a secondary winding, connected to a galvanometer, around the other side (see Fig. 4.15). At first he thought that the experiment was again a failure. He noticed, however, that when he turned off the electric current in the primary coil, believing the experiment to have failed, the galvanometer revealed a sudden short flow in the secondary circuit. Closer investigation showed that a continuous current in the primary circuit had no effect; it was only when the current was started or stopped that there was an effect on the galvanometer. Joseph Henry later claimed that he had discovered electromagnetic induction at the same time as Faraday, and quite independently; there is no reason to doubt this, but Henry had not published his results.

Soon after his discovery of electromagnetic induction in 1831, Faraday demonstrated that if he pushed a magnet into a coil of wire, a transient current was produced (Fig. 4.16). A current was generated in the opposite direction when the magnet was withdrawn. No current passed when the magnet was stationary; to generate a current the magnet had to be moved in relation to the coil.

The important point is that electromagnetic induction only occurs if there is a change in an electric current, which means that there is a change in the magnetism. Oersted and Ampère had shown that a steady electric current produces a magnetic field, and Faraday and his contemporaries thought at first that if a wire were placed near to a magnet, an electric current would be generated. Attempts to demonstrate this were, however, unsuccessful, and it was first believed that success might be achieved if a stronger magnetic field

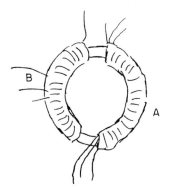

**Fig. 4.15** A diagram from Faraday's *Diary* illustrating his discovery of *electromagnetic induction* in 1831. When he caused a current to flow in coil A, called the primary coil, Faraday observed a short-lasting current in coil B (the secondary coil). As was his custom, Faraday at once published his report.

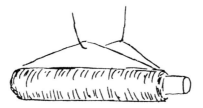

**Fig. 4.16** A diagram from Faraday's *Diary* of 1831, illustrating an experiment in which a magnet induces an electric current in a coil when a magnet is pushed into, or pulled out of, the coil.

were employed. Being wise after the event we can see today that such a result would be impossible. We are now aware—which Faraday was not at the time—of the necessity for energy to be conserved. If a stationary magnet (either a permanent magnet or an electromagnet) were to induce an electric current in a nearby wire we would be getting something for nothing—where would the energy come from? We think so much in terms of energy today that it is hard to realize that before the middle of the nineteenth century the word energy was hardly ever used, and that even the most competent scientific investigators were unaware of its great importance. In addition, even Faraday was only just beginning to realize that an electric current involves a movement of electric charge. Oersted had discovered that a steady current of electricity would affect a magnet; however, we now know that an electric current involves a moving electric charge, and that energy (from a battery, for example) is being used up in creating it. It had also been found by Arago and Sturgeon that an electric current (a moving electric charge) in a coiled wire would produce a magnet. Thus to produce an electric current we need some change in a magnetic field—this can be produced by moving a permanent magnet (which uses up energy), or by switching on or off the current in a solenoid. A stationary magnet cannot be expected to give rise to electro-magnetic induction, because there is no source of energy.

Faraday's 1821 discovery of electromagnetic rotation had shown that electrical energy could be converted into motion. In 1831 he succeeded in converting mechanical motion into electricity. He rotated a copper disc between the poles of a magnet, and found that a steady current flowed from the centre of the disc to its edge (see Fig. 4.17). This achievement encouraged Faraday to carry out the further researches, particularly those on electrolysis, which were to lead to the announcement in 1838 of his general theory of electric and magnetic fields.

We mentioned earlier that Joseph Henry also discovered electromagnetic induction, perhaps even before Faraday did, but that he failed to publish his result. Henry did discover self-induction, or auto-induction. He reported in 1832 that he wound a long wire into a coil round a magnet, and (without any use of a battery) completed the circuit. When he broke the circuit he

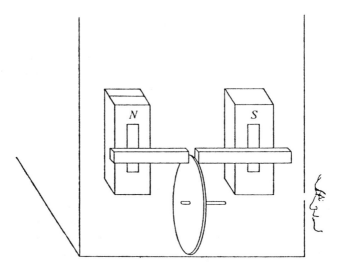

**Fig. 4.17**   Faraday's published diagram in 1831, showing his demonstration of the rotation of a copper disc.

observed a strong spark, showing that even though the coil was not initially carrying a current, a current was induced when the circuit was broken; this is self-induction. Henry also observed that if a current was passed through a coiled wire, and the current was broken, the spark obtained was much more vivid than if the wire was uncoiled; this he explained as due to the additional current self-induced in the coiled wire.

During Henry's 1837 visit to England, he on one occasion carried out some electrical experiments with Faraday and Wheatstone at the Royal Institution. Faraday and Wheatstone were having trouble making one of the experiments work, but Henry used his knowledge of self-induction to carry it out successfully. It is said that Faraday then showed his enthusiasm and high regard for Henry by jumping up and down, clapping, and shouting 'Hurrah for the Yankee experiment'. In Joseph Henry's honour the official unit of magnetic inductance is called the henry. (It might have been called the faraday, or farad, but both of these terms are used for other quantities.)

Before taking our leave of the eminent Joseph Henry we should make brief mention of his indirect influence on the invention of the telephone. One day at the Smithsonian Institution Henry was visited by a young man who was teaching the deaf and had what he thought to be a good idea for transmitting the sound of the voice over telegraph lines. He was, of course, Alexander Graham Bell (1847–1922). When he admitted to Henry that he did not have the technical knowledge to proceed, Henry's advice was positive: he told Bell that he had the germ of a great invention, and that he should acquire the necessary knowledge. Bell followed the advice, for which he often expressed gratitude, and exhibited his telephone in Philadelphia in 1876.

The telephone is outside the scope of the present book, but one detail deserves mention. When the great Lord Rayleigh was first shown a telephone he remarked 'It is certainly a wonderful instrument, although I suppose not likely to come to much practical use'. This was an uncharacteristic lapse on Rayleigh's part; his attitude to radio and flying was much more positive.

## FARADAY'S RESEARCHES ON ELECTROLYSIS

In 1832 Faraday began to devote much attention to the electrolysis of water, and the ideas he had formed about electricity led him to conclusions of the greatest importance. In 1833 he showed that electrolysis can be brought about by electricity produced in a variety of ways, such as from electrostatic generators, voltaic cells, and even electric fish. In particular, he showed that electrolysis can occur if an electric discharge is passed through a solution, without any wires being introduced into it. Experiments done by him in 1834 convinced him that if electrolysis is brought about by inserting wires into a solution, and then a current is passed, the effect cannot be explained in terms of action at a distance. He performed one experiment in which two solutions were separated from each other by a 70-foot string soaked in brine. Gases were evolved at the two wires, and Faraday thought it impossible that the inverse square law could explain an effect extending over such a distance. He carried out another experiment in which a solution was placed near a source of static electricity which produced an intense electric field. No electrolysis occurred, and Faraday concluded that it is necessary for a discharge to take place or for a current to flow. He also showed that the effects of electrolysis do not follow straight lines, which they would do if there were action at a distance. Also, Faraday argued, if a substance were attracted to a wire by the inverse square law, would it not remain bound to the wire rather than being released from it?

Faraday's conclusion that action at a distance does not apply led him to dislike the word pole that had come to be used in connection with electrolysis, since he felt that it implied a centre of force. In its place, at the suggestion of his friend and personal physician Whitlock Nicholl (1786–1838), he introduced the word electrode, from the Greek ελεκτρων (electron) and όδοζ (hodos), meaning 'way'. Nicholl also suggested the word electrolysis, from λυσιζ (lysis), meaning 'splitting', and also the word electrolyte, to mean something that undergoes electrolysis.

As Faraday proceeded with his investigations into electrolysis he came to view the process as follows. He considered that the electrical effect is transmitted from molecule to molecule, and that the affinities of the components of water, or of a dissolved salt, are weakened, so that each component would be able to leave its partner and jump to another close by.

For other help with words Faraday approached William Whewell of Trinity College, Cambridge, who had shown much interest in Faraday's research. In

1834 Faraday asked Whewell for words that would refer to the regions of solution present at the surfaces of the positive and negative electrodes. He
pointed out that if we imagine the north and south magnetic poles of the
earth to be due to an electric current passing round lines of latitude, the
current would travel like the sun from east to west. He said that he had considered the words 'eastode' and 'westode', but realized that 'these are words
which a scholar...could not suffer for a moment.' In his letter of reply dated
the next day Whewell suggested the words anode and cathode, and explained
why he favoured them. Anode comes from the Greek $\alpha\nu\alpha$ (ana), upwards,
and $\acute{o}\delta o\zeta$ (hodos), a way, and therefore suggests the rising of the sun, in the
east. Cathode comes from $\kappa\alpha\tau\alpha$ (kata), downwards, and is related to the
setting of the sun, in the west. Whewell also suggested three other words that
are used today, anion, cation, and ion, the latter being a general term for
either an anion or a cation. The Greek word $\iota o\nu$ (ion) is the neuter present
participle of a Greek word that means 'to go'; an ion is therefore something
that moves. Whewell explained to Faraday why we must say cathode and not
catode, but cation and not cathion; the word $\acute{o}\delta o\zeta$ has an aspirate h, indicated
by the accent over the o, whereas $\iota o\nu$ (ion) does not.

Faraday made good use of the words cation and anion, but commented to
Whewell that he thought he would find little use for the word ion. Now
that it is known—which Faraday did not know—that ions do exist in solution, the word has found a wide use. The words cathode and anode have
undergone a substantial and undesirable change of meaning since Faraday's
time. Faraday intended the words to apply to the electrode surfaces, or to
regions of the electrolyte close to the surface, in an electrolytic cell.
However, by a perhaps inevitable extension of meaning they soon were
applied to the electrodes themselves. This is not so bad, but a further extension of meaning has been most unfortunate. It is now common to apply the
words to the electrodes of a battery or other device that generates electricity.
This means that whereas in an electrolytic cell the anode is the positive electrode, in a battery it is the negative electrode. This confusing extension of
meaning was never intended by Faraday, who would certainly not have
approved of it. At the present time there is no need for the words anode and
cathode; the expressions 'positive electrode' and 'negative electrode' are
more helpful and are quite unambiguous.

In 1834 Faraday suggested two fundamental laws of electrolysis. It is often
thought that he discovered these two laws empirically, but this is not the case;
they were deduced by him on the basis of his ideas about how electricity
interacts with a solution, and were later confirmed by him. If electrolysis had
proceeded by action at a distance, as had been thought previously, the
amount of material deposited or evolved would have depended on the electric
potential and on the size of the vessel used. According to Faraday's ideas,
however, the amount would depend only on the product of the current and
the time, i.e., on the quantity of electricity that passes through the solution,

and Faraday confirmed that this is the case; this was his *first law of electrolysis*. And since, in Faraday's view, the electricity was concerned with weakening the affinities between the constituents of matter, the amount should depend on the equivalent weight of the substance evolved or deposited. This is his *second law of electrolysis*. Faraday never expressed the second law in quite this form, but his experiments showed the law to be obeyed.

It is somewhat ironic that Faraday should have deduced his laws of electrolysis on the basis of ideas that we now know to be not entirely correct. The effect of the current is not to bring about dissociation, as Faraday thought; instead the ions are already present and the electric field brings about their movement towards the electrodes, where they are neutralized. The relationship between the quantity of electricity and the amount of material deposited or evolved arises not because the electricity is concerned with dissociation, but because it is concerned with the neutralization of ions that are already present. This, however, did not matter as far as Faraday's conclusion was concerned.

Faraday's laws of electrolysis were of great theoretical significance, since they suggested that electricity itself is not continuous, and that fundamental particles of electricity are in some way associated with atoms. Faraday himself was always rather reticent in referring to atoms, but he did write, in one of his early papers on electrolysis:

> ...if we adopt the atomic theory or phraseology, then the atoms of bodies which are equivalent to each other in their ordinary chemical action, have equal quantities of electricity naturally associated with them.

At other times, however, Faraday tended to neglect the atomic theory, and to adhere to the idea that electricity is a fluid.

It was not for several decades that more definite statements were made about the relationship between electricity and matter. In the 1873 edition of his *A treatise on electricity and magnetism* Maxwell suggested the possibility of a particle of electricity—but apparently never really believed in it. In 1881 Helmholtz presented the Faraday Memorial Lecture to the Chemical Society and discussed the atomic nature of electricity in the following terms:

> Now the most startling result of Faraday's law is perhaps this. If we accept the hypothesis that the elementary substances are composed of atoms, we cannot avoid concluding that electricity also, positive as well as negative, is divided into portions, which behave like atoms of electricity.

In 1874 the Irish physicist George Johnstone Stoney (1826–1911) had also suggested the idea of a particle or atom of electricity, and in 1891 he suggested that the unit of negative electricity should be called the *electron*, which is what it has been called since. The important investigations in which the real existence of electrons was established will be described in Chapter 6.

Besides being of great theoretical significance, Faraday's work on electrolysis in the 1830s had important practical consequences. The electroplating industry soon became well established, particularly in Birmingham, England. Old Sheffield Plate, produced by heating and annealing sheet silver on to copper, was quickly superseded by the more convenient technique of silver plating. Faraday's work also led to the establishment of the electroforming industry, in which metal is electrically deposited on a mould which can be removed; this was used in the production of early gramophone (phonograph) records and in electrotyping. Faraday's work on electrolysis also led to new scientific instruments, such as coulometers used for the measurement of quantity of electricity.

Until 1834 Faraday's stature as a scientist had not been adequately recognized by the Royal Institution, by either a suitable appointment or a suitable salary. When in 1812 Davy resigned as Director and Professor at the Institution, his professorship of chemistry went to William Thomas Brande (1788–1866), who had little distinction in research but did a competent job of teaching chemistry, giving a lecture at nine o'clock every morning for over 40 years. Faraday was sensitive to his own inferior position and salary and, although he retained considerable loyalty to the Institution, gave some thought to alternative appointments. In 1827 he was offered the first professorship of chemistry at the newly-formed University College of the University of London, but declined, the appointment going to J F Daniell instead. In 1829 Faraday did accept a part-time professorship at the Royal Military Academy at Woolwich, and this significantly augmented his Royal Institution salary. His duties at the Academy involved giving 25 lectures a year, and he retained the appointment for about 20 years.

In 1834 Faraday's situation improved greatly, as he was appointed Fullerian Professor of Chemistry at the Royal Institution, a post that was endowed especially for him by John Fuller, a wealthy Member of Parliament. A jovial and eccentric man, Fuller had attended many lectures at the Royal Institution, and admitted that he had often slept through them; his endowment was, in his words, 'in gratitude for the peaceful hours thus snatched from an otherwise restless life'. In a lecture on Faraday given in 1981, Sir George (now Lord) Porter told the story about Fuller's endowment, and commented that he did not mind if anyone slept during his lecture—provided that a suitable endowment to the Royal Institution followed.

At the same time that Fuller endowed a professorship of chemistry, he also established a professorship of physiology at the Royal Institution, and Peter Mark Roget (1779–1869; Fig. 3.17) was appointed to it. Besides his important work on the persistence of vision (Fig. 3.18), mentioned in Chapter 3, Roget did useful work on electricity and magnetism.

## FARADAY'S WORK ON PROPERTIES OF MATTER

Over a period of years Faraday carried out investigations of the greatest importance on the behaviour of various substances in the neighbourhood of magnets and wires carrying currents. He would have said that he was working with magnetic and electric fields, but that point of view was not generally accepted for some time. Beginning in 1837 he was particularly interested in the possible influence of substances on the distribution of an electric field. The apparatus he used in one series of investigations is illustrated in Fig. 4.18;

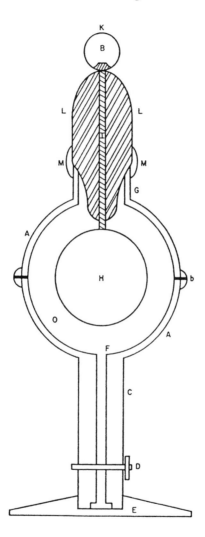

**Fig. 4.18**  The published diagram (1837) showing the device used by Faraday for demonstrating induction. A and H are two concentric metal spheres, separated by the space O. B is a metal sphere connected to the inner sphere H. Faraday compared the capacities of these condensers or capacitors to carry an electric charge. He found that the capacity varied greatly with the insulating material, which he called the *dielectric*, present in the space O. The factor by which the capacity was increased he called the specific inductive capacity; we now call it the *dielectric constant*.

he arranged for several identical such devices to be constructed. Each consisted of a pair of concentric spheres, the outer sphere (A in the diagram) being made as two hemispheres which could be separated, so that the outer sphere could be placed around the inner sphere H; in other words, he had constructed capacitors, widely used in modern electronic technology. He made careful quantitative experiments with condensers modified by replacing the air between the spheres, in the space O, with other insulating substances, and made the important observation that such condensers always held a greater share of charge than when air was present in the space. He characterized the behaviour of the insulator in terms of a property which he called the specific inductive capacity, but which is now always called the dielectric constant. (There is another quantity called the permittivity, which has units; the dielectric constant is just the ratio of the permittivity of a substance to that of a vacuum.)

The dielectric constant of a particular substance is the ratio of the capacities of two condensers to carry a charge: it is the ratio

$$\frac{\text{capacity when the condenser contains that substance}}{\text{capacity when the condenser contains a vacuum}}$$

In practice, the capacity when there is a vacuum is almost the same as the capacity when air or any other gas is present; Faraday at first could find no difference between different gases. Another convenient way of looking at the dielectric constant is to consider it from the point of view of the attraction and repulsion between two charged bodies. If the medium between two charges is first a vacuum, and then water is introduced, the force is reduced by a factor of about 80, and we say that the dielectric constant of water is 80. This result provides an important part of the explanation of why substances like common salt, when dissolved in water, exist as separated sodium ($Na^+$) and chloride ($Cl^-$) ions; the attraction between them, which holds them together in the crystal, is reduced by a factor of about 80 when water is added.

These experiments in which Faraday discovered the great importance of the dielectric (as he called the substance placed between the plates of a condenser) were of great significance in confirming his ideas about the electric field, and in leading others to accept his ideas. The attention of most investigators had previously been concentrated on the conducting materials, such as metals, on which charges were placed. The medium between two charged bodies was regarded as somewhat incidental—as merely something through which action at a distance could operate. Faraday's discovery that the medium could affect the capacity of a condenser, sometimes by a large factor, showed convincingly that the medium was of great importance.

Faraday also attempted to find support for his ideas by observing some physical effect in matter through which his lines of force were passing. His first discovery of such an effect was in 1845, when he made experiments with plane-polarized light, which we discussed briefly in Chapter 3 in connection with Edwin Land's invention of polaroid. When plane-polarized light is passed through certain substances the plane of polarization becomes twisted through an angle, and the angle of twist can easily be measured using an instrument called a polarimeter. Faraday used this property of the rotation of the plane of polarization in his experiments of 1845. Ordinary glass does not rotate the plane of polarization, but Faraday placed a special lead borate glass close to an electromagnet, as shown in Fig. 4.19. He then passed plane-polarized light through the glass, and found that the plane was rotated. This was the first observation of any effect of magnetism on light. The result suggested to Faraday that substances like glass, hitherto regarded as non-magnetic, were not entirely indifferent to a magnetic field.

Faraday then carried out many experiments in which substances, including gases, were placed between the poles of an especially powerful horseshoe electromagnet. He found that some substances, such as iron, tended to align themselves along the lines of force, and were attracted into the more intense parts of the electromagnetic field; he called such substances paramagnetic. In November 1845, however, he tried the experiment with a piece of heavy glass, and found that it tended to set itself perpendicular to the magnetic field and to move into regions of less intense field. Bismuth behaved in a similar way. Such substances he called diamagnetic. He explained the difference between paramagnetics and diamagnetics in terms of the way the magnetic field passed through them (Fig. 4.20).

All of these experiments convinced Faraday that the idea of action at a distance must be discarded, and that the results should instead be explained in terms of the fields of force that were established. His point of view was neatly summed up in Maxwell's *Treatise on electricity and magnetism* (1873):

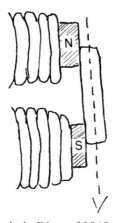

**Fig. 4.19** A drawing from Faraday's *Diary* of 1848 illustrating his discovery of the effect of a magnetic field on the plane of polarization of light passed through a piece of glass.

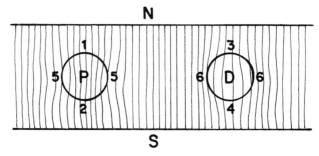

**Fig. 4.20**   Faraday's published diagram (1845) in which he gave his interpretation of the difference between paramagnetic (P) and diamagnetic (D) substances. In the former, the lines of force tend to be concentrated within the material; these are more lines of force in a unit volume of the paramagnetic material than in air or in a vacuum. With a diamagnetic material there are fewer lines of force.

> ....Faraday, in his mind's eye, saw lines of force traversing all space where the mathematicians saw centres of force acting at a distance: Faraday saw a medium where they saw nothing but distance: Faraday sought the seat of the phenomena in real actions going on in the medium, they were satisfied that they had found it in a power of action at a distance impressed on the electric fluids.

In the last paper he submitted for publication, in 1860, Faraday included gravity as involving a field of force—an idea that was somewhat ridiculed at the time, but was later realized to be correct. The way in which Maxwell brilliantly developed Faraday's theory along mathematical lines is explained in the next chapter.

## THE GENIUS OF MICHAEL FARADAY

Faraday worked unrelentingly for many years. He and his wife lived in an apartment at the Royal Institution, so that his laboratory was close at hand. His extensive notebooks show that he wasted no time, and worked long hours six days a week; his religion required Sundays to be days of rest, but he took no others. It was not uncommon for him to work in his laboratory from nine in the morning to eleven at night, with only short breaks. He made all his own extensive notes on his experimental results, and had no help in writing the four 450 articles he published, and the several books he wrote.

The only assistance Faraday had was from Charles Anderson (*c.*1791–1866), a retired army sergeant, who did technical work in the laboratories from about 1827 to his death in 1866. Anderson contributed nothing to the planning of the experiments, and for most of the time not a word passed between the men. Anderson did no more than follow Faraday's instructions, which he did meticulously and precisely. The story is told that on one occasion Faraday arrived early at his laboratory, and was surprised to find Anderson there before him. It emerged that Anderson had worked through the night, keeping the fire stoked; since Faraday had not told him to stop work the previous evening, he felt he had to continue. It was apparently not on a Monday morning that Faraday found Anderson at work.

Faraday avoided social events, and would have sympathized with Sherlock Holmes's comment that 'This looks like one of those unwelcome social invitations which call on a man either to be bored or to lie'. Faraday, however, did not lie or become bored; he gracefully declined to attend purely social events. But it should not be thought that Faraday was antisocial. He particularly enjoyed the company of young people, and was always happy to attend children's parties. On at least one occasion Faraday entertained young friends by riding his velocipede round the lecture theatre of the Royal Institution. No one can read Faraday's *The chemical history of a candle* without realizing how much joy he himself derived from giving his many lectures to young audiences, and how entertained and instructed his listeners must have been.

Faraday had a large number of friends, many of them scientific investigators, and he was always generous in his assistance to them. When Mary Somerville (1780–1872) wrote her second great book, *On the connexion of the physical sciences* (1834), Faraday went to much trouble on her behalf, reading drafts of the book with great care and making numerous suggestions to improve the wording.

In 1839, when not yet 50, Faraday suffered a nervous breakdown from which he never fully recovered. It seems possible that the illness was brought about by mercury poisoning as much as by overwork. It is likely that mercury had been spilt in the laboratories, so that the atmosphere would have been contaminated with the vapour, not then recognized as presenting a medical hazard. Faraday did not take leave from the Institution, but devoted himself more to administration, and spent less time in his laboratory. It is sometimes said that he rested for a period, but his scientific output during that period was still remarkable.

Within a few years his condition had improved sufficiently for him to be able to resume experiments, but he never showed his former originality and brilliance. After the mid-1850s his memory had become erratic, and he found it hard to concentrate on his researches or his lectures. He was able to give lectures to juvenile audiences at Christmas until 1861, in particular his famous *The chemical history of a candle*.

In 1862 Faraday resigned his position at the Royal Institution, and retired to a house at Hampton Court Palace provided to him by Queen Victoria; it is near the west entrance to the palace grounds, and is still called Faraday House. He died on 25 August 1867. Consideration was given to burying him in Westminster Abbey, or at least raising a monument to him there, but it was decided that such actions would be inappropriate as he would not have

been willing to worship at the Abbey in his lifetime. He was buried in
Highgate cemetery in North London. Karl Marx was later buried there also,
but his grave is much more impressive, and is visited much more often, than
that of Michael Faraday.

## FARADAY'S SCIENTIFIC LEGACY

Some idea of Michael Faraday's scientific eminence is gained by a speculation
as to how many Nobel Prizes he would have received if there had been such
prizes in his time. The first Nobel Prizes were awarded in 1901, and by now
we have a good idea of the criteria that are used in awarding them: we can
therefore make a reasonable guess as to how Faraday would have fared if the
Prizes had been awarded earlier. So far no one has won more than two Nobel
Prizes, but it seems likely that Faraday would have won six. He would surely
have got a Nobel Prize for chemistry for his discovery of benzene (1825), the
prototype of a vast number of organic compounds. It is hard to see how he
could have been overlooked for Nobel Prizes for physics for his discovery of
electromagnetic rotation (1821), his discovery of electromagnetic induction
(1830), his laws of electrolysis (1834), his work on dielectrics (1837), and his
discovery of paramagnetism and diamagnetism (1845).

It is hard to think of anyone else who would have deserved so many Nobel
Prizes. Newton might have won two, for his mechanics and his optics, and
possibly a third (shared with Leibnitz) for his calculus. Maxwell would probably have won two, for his distribution of molecular speeds (Chapter 2) and
his theory of electromagnetic radiation (Chapter 5). Einstein did win one
Prize, for his work on the quantum theory (Chapter 8), but surely should
have had another one for his work on relativity. The suggestion that Faraday
deserved six prizes puts him in a class by himself.

Faraday's researches into electricity and magnetism fall clearly into the cat-
egory of pure science or basic research. They were motivated mainly by his
curiosity about the nature of electricity and magnetism, and only to a small
extent by his desire to make practical use of electricity. His discoveries nevertheless had far-reaching practical implications.

It is important not to exaggerate these, as is often done. It is sometimes
said, for example, that the electric motor owes its origin to Faraday's work,
but this is not the case. As discussed earlier, the work of William Sturgeon
and Joseph Henry on electromagnets, done before Faraday discovered
electromagnetic induction, had more to do with the development of electric
motors.

It is, on the other hand, reasonable to claim that it was Faraday's work on
electromagnetic induction (1831) that led to the *dynamo*, or rotary electric
generator. The principle of such devices is that coils of wire are moved relative to the field of a magnet, so that an electric current is produced; a genera-

tor, in fact, is an electric motor operated in reverse. The first to construct an electric generator seems to have been the instrument maker Hyppolyte Pixii (1808–1835), who exhibited one to the Académie des Sciences in Paris in 1832. A permanent horseshoe magnet was rotated relative to a wire coil, and in later devices coils were rotated relative to magnets. Soon permanent magnets were replaced by electromagnets activated by batteries. In 1866 C F Varley (1823–1883) discovered and patented the principle of self-excitation, in which the electromagnets are activated by the electricity produced by the generator itself. By that time it had been realized that electromagnets having soft-iron cores possessed enough residual magnetism to provide the magnetic field necessary to initiate the output from a generator. At about the same time Werner von Siemens (1816–1892) demonstrated similar devices to the Berlin Academy of Sciences, and it was he who invented the word dynamo.

It is correct to credit Faraday with producing the prototype of the electrical transformer. If in his experiments of 1831 (Fig. 4.15) he had continuously reversed the direction of the current in coil A, he would have observed an alternating current in coil B; that is, the current in coil B would have continually changed its direction. Today most of the electric power that is transmitted from one place to another is sent not as a steady, or direct, current, but as an alternating current. In North America alternating current usually changes its direction 60 times a second, which we call 60 cycles a second, or 60 hertz; in Britain and elsewhere the alternating current is at 50 cycles a second. Such alternating current is directly produced by rotary generators, and is readily converted from one voltage to another by the use of a transformer based on Faraday's iron ring (Fig. 4.15). If coil A, called the primary of the transformer, has fewer turns of wire than the secondary coil (coil B), an alternating current at low voltage is converted into an alternating current at higher voltage; we then have what is called a step-up transformer. A step-down transformer is one in which an alternating current at higher voltage is converted into one at lower voltage, by having more turns in the primary coil than in the secondary.

The transmission of power over long distances, by the use of the familiar pylons that are conspicuous in the countryside in many parts of the world, or by underground cables, is based on the use of transformers. The higher the voltage the further the electricity can be sent. The procedure is therefore to use a step-up transformer to raise the voltage at the transmitting end, and then to step the voltage down at the receiving end. For safety the voltage used by the public should not be more than about 200 volts, but much higher voltages are used in transmission lines.

The history of how power transmission has developed is interesting and instructive. The first transmission station in the world which sold electricity to the public was opened in 1882 by Thomas Edison at Pearl Street in New York City. It transmitted direct current, which Edison at the time favoured over alternating current, as did many other experts, including William

Thomson (later to be Lord Kelvin). The matter was at first controversial, and powerful arguments were raised in favour of alternating current as well as direct current. In the end alternating current prevailed, since in spite of some disadvantages it was easier to change its voltage by transformers. The great advantages of using high voltages for long-distance transmission were beginning to be recognized in the early 1880s.

In 1883, at Grenoble, Switzerland, alternating current at 3000 volts was transmitted a distance of 14 kilometres, and in Paris in 1886 the distance was raised to 56 kilometres by the use of 6000 volt alternating current. A particularly important advance was made in 1889 with the opening of the Deptford Power Station on the Thames estuary. This station, which was the prototype of all subsequent power stations, was the brainchild of Sebastian Ziani de Ferranti (1864–1930), who had been born in Liverpool, England, of Italian extraction. His education was limited (it included some evening classes at University College, London), but at an early age he carried out experiments on dynamos and transformers, and devised schemes for the large-scale generation and distribution of electricity at high voltages. As a result, at the age of twenty-three he was appointed chief electrician to the London Electric Supply Corporation. His choice of Deptford as the site of the power station was supported by the facts that land was cheap and that coal could easily be supplied by ship from the north of England. The generators produced alternating current at 86 hertz and transformers raised the voltage to 10 000 volts, which was distributed to substations in the heart of London some 12 kilometres away. No cables available at the time would support such a voltage, and Ferranti designed a new type of cable, in six-metre lengths, in which copper conducting wires were separated by waxed-paper insulators. Ferranti took out a large number of patents, and the firm of Ferranti Ltd which he established in 1905 is still in the forefront of electrical and electronic engineering.

Subsequently higher and higher voltages were used to transmit alternating current over greater distances. By the end of the 19th century 60 000 volts were being used, and today voltages of several hundred thousand are common.

# FIVE

# James Clerk Maxwell and radio transmission

We saw in the last chapter that the basic research work done by Michael Faraday on electricity and magnetism led directly to the large-scale generation and distribution of electric power. Similarly, the researches carried out by James Clerk Maxwell led directly to radio transmission and all its refinements, such as television and radar. Maxwell's influence was wide, since his unique way of looking at scientific problems led to a radical change in the whole theoretical structure of the physical sciences. Albert Einstein remarked that Maxwell's theory of electromagnetic radiation was 'the most fruitful that physics has experienced since the time of Newton'. Einstein's own great contributions to science followed to some extent Maxwell's lead.

James Clerk Maxwell (Fig. 5.1) was born on 13 June 1831 at 14 India Street, Edinburgh. His father, John Clerk Maxwell, had been born John Clerk, but added the name Maxwell to satisfy some legal difficulties relating to an inheritance. John Clerk Maxwell had been trained as a lawyer, but was interested in technical matters, and was a Fellow of the Royal Society of Edinburgh. When James was eight years old his mother died at the age of 48 of abdominal cancer, the disease of which James himself was to die, in 1879 and at the same age. James inherited a strong intellectual background, and also a comfortable income which allowed him to be independent of any paid appointment.

James's father was a Scottish Presbyterian and his mother, whose ancestry was English rather than Scottish, was a member of the Episcopal Church of Scotland. Throughout his life James, like Michael Faraday, retained a strong Christian faith, and could quote extensively from the Bible. From an early age he was imbued with a great modesty and lack of pretentiousness, which remained with him all his life.

John Maxwell was laird of Glenlair, a position that James inherited on his father's death. Much of James's boyhood, and some periods of his adult life, were spent at the manor house at Glenlair, which is near Corsack in County Galloway, in the south west of Scotland. The countryside surrounding Glenlair is as beautiful as anywhere in the world. In that rustic neighbourhood he acquired such a broad Scottish brogue that he was often understood only with difficulty even by fellow Scots, particularly those from Edinburgh. At the age of ten he was sent to Edinburgh Academy, where on account of his accent and his homespun ways he was somewhat ridiculed by his fellow

**Fig. 5.1**   James Clerk Maxwell (1831–1879), at Cambridge in 1855, holding the colour top he used in many of his earlier experiments on colour vision.

schoolboys, who inappropriately nicknamed him 'Daftie'. When aged 15 and still at school Maxwell published a paper, on a geometrical problem, in the *Proceedings of the Royal Society of Edinburgh*. It had received the approval of James David Forbes (1809–1868), professor of natural philosophy at the University of Edinburgh and also a distinguished glaciologist, and Forbes communicated the paper to the journal.

In 1847, at the age of 16, Maxwell became a student at the University of Edinburgh. Here he came particularly under the influence of Forbes, who was a man of strong personality. Forbes was an excellent lecturer who took great pains with his lecturing techniques, even to the extent of taking elocution lessons from the famous actress Sarah Siddons (1755–1831). Maxwell was also greatly influenced by the great philosopher and educationalist Sir William Hamilton (1788–1856). Forbes and Hamilton, by the way, were bitter enemies, holding completely different and uncompromising views on politics, education, and almost everything else.

During his undergraduate years in Edinburgh Maxwell did a good deal of research, in Forbes's laboratory, particularly on colour theory; Fig. 5.1 shows him holding a colour top, which we mentioned in Chapter 3. He also began to think about the implications of Faraday's great work on electricity and magnetism.

Maxwell took his Edinburgh degree in 1850 and then became an undergraduate at Cambridge. He was first attached to St Peter's College, now called Peterhouse, but decided that the chances of a Fellowship would be greater at Trinity College. After a term at St Peter's he therefore transferred to Trinity, where he came under the influence of William Hopkins (1793–1866). As a married man Hopkins could not become a College fellow (according the rules at the time), but he became a private tutor, and was remarkably effective in helping students to get good degrees, being called the 'wrangler-maker'; G G Stokes, William Thomson (the future Lord Kelvin), P G Tait, and many others had profited from his tutoring. At Cambridge Maxwell was also influenced by the Revd William Whewell (1794–1866), the Master of Trinity, who did distinguished work in science as well as in other fields, and by George Gabriel Stokes (1819–1903), who was Lucasian professor of mathematics from 1849 until his death.

While still an undergraduate at Cambridge, Maxwell continued his work in colour theory, which we discussed in Chapter 3. He also began to write a paper on Faraday's lines of force. In 1854 he graduated from Cambridge as Second Wrangler in the Mathematical Tripos. In 1855 he was elected a Fellow of Trinity, a position he retained even after he left Cambridge, but had to relinquish in 1858 on his marriage.

Unlike some scientists, who work on various topics in sequence, with little overlap between them, Maxwell tended to work on several topics simultaneously over a considerable number of years (Fig. 5.2). He worked on colour vision from 1849, when he was a student at Edinburgh, until 1871, when he went to Cambridge as Cavendish Professor. His work on electromagnetic theory began shortly after his graduation from Cambridge in 1854 and continued until his death in 1879. His work on the kinetic theory of gases was done for the most part between 1859 and 1879. There was often an interval of several years between his papers on the same subject. Twelve years elapsed between his two most important papers on kinetic theory (in 1867 and 1879), and six years between his first and second papers on electromagnetism (in 1855 and 1861).

Maxwell's method of carrying out research was unusual. During his career he considered a number of broad themes, and his interests covered an unusually wide range. On each theme he carried out the appropriate experimental and theoretical research, and in some cases perfected his ideas gradually, often publishing a series of papers on the topic. What was remarkable about his approach was that he often changed his outlook considerably on a given topic. Each of his major papers on electromagnetic theory written between 1855 and 1868 presented a complete account of the subject, but each viewed it in a different way, presenting different aspects of the problem. This variety led Sir James Jeans to remark that Maxwell's writings led one into a kind of

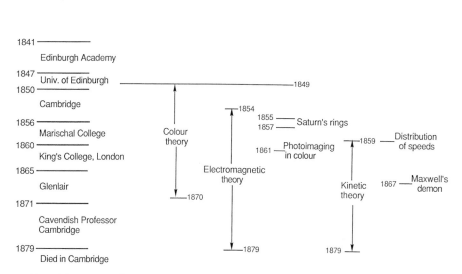

**Fig. 5.2**    Maxwell's research projects, shown in relation to the events of his life.

enchanted fairyland—one never knew what to expect next. Maxwell himself remarked that 'there is no more powerful method for introducing knowledge into the mind as that of presenting it in as many different ways as we can'. Those of us who teach science recognize how valuable that advice is.

Maxwell's method of doing research differed greatly from that of Michael Faraday. Faraday worked hard and steadily at collecting experimental results, which he recorded daily in his diary, and his ideas gradually became modified as his results accumulated. Faraday was something of a compulsive worker, spending long hours in his laboratory and refusing to deviate from his rigorous programme, even declining to see important visitors until he was ready for them. Maxwell, on the other hand, never gave the impression of working particularly hard, often devoting only a few hours a day to research, and entering freely into social activities. No doubt while engaging on these activities Maxwell was at the same time pondering the broad scientific problems with which he was concerned, such as electromagnetic theory.

Although he was prone to making careless mistakes, which were easily corrected, Maxwell's great strength was that he was at once able to think correctly about any problem he attacked, so that he wasted little time.

We outlined Maxwell's work on colour theory and colour vision in Chapter 3. It was begun while he was a student at Edinburgh, and was continued, along with other research, for about twenty years. Some of it was done while he held his first appointment, at Marischal (pronounced 'Marshal') College, Aberdeen, and it was essentially completed during the five years (1860–1865) in which he held his second appointment at King's College, London. To some extent Maxwell's work on colour involved extending the work of others, so that it was not as highly original as some of the rest of his work. At the same time there is no doubt that Maxwell was one of the great pioneers of the theory of colour, and of colour physiology, being the first to put the subject on a quantitative basis.

## MARISCHAL COLLEGE, ABERDEEN (1856–1860)

In 1856, at the age of 25, Maxwell was appointed professor of natural philosophy at Aberdeen's Marischal College and University, as it was then called (it was administratively distinct from King's College, Aberdeen, another degree-granting institution). The academic year was short, from early November to early April, to accommodate the many students who worked on farms. Maxwell's teaching load was heavy, involving 15 hours of lecturing a week besides laboratory duties. Maxwell never enjoyed lecturing, and was not a success at it. His Scottish brogue was strong, his diction was poor, and he had something of a stutter when he was nervous; as a result he could not hold the attention of his students. He also had an unfortunate tendency, throughout his career, to make mathematical mistakes, being particularly careless (one biographer has said 'cavalier') about the use of plus and minus signs. He was even apt to express his ideas in a confused way. In a man who wrote with such great clarity this is most surprising; even though he was very conscientious in preparing his lectures he became muddled when he started to speak. He complained that the Aberdeen students did not understand his jokes, which was perhaps partly because they did not hear them properly. He always had a very amusing style of speaking and writing, but it was not always appreciated, as his was a paradoxical mixture of boyish and sophisticated humour, sometimes requiring for its understanding an intimate knowledge of Latin and Greek.

Maxwell had a strong sense of humour, and his correspondence with his friends was always amusing. He took great pleasure in writing humorous verse, such as a wonderful parody of Robert Burns's 'Coming through the rye', in which he summarizes his ideas about the science of mechanics. The title is 'Rigid Body Sings', and the first verse gives us the flavour of it:

'Gin a body meet a body
    Flyin' through the air.
Gin a body hit a body,
    Will it fly? and where?
Ilka impact has its measure,
    Ne'er a ane hae I,
Yet a' the lads they measure me,
    Or, at least, they try.'

In 1858 Maxwell married Katherine Mary Dewar, the daughter of the Principal of Marischal College. She was taller than he (he was only five feet four), and seven years older. The union was childless, but there is no doubt that their married life was one of great mutual devotion. Katharine, however, seems to have been disliked by her husband's relations and friends, of whom there were many, as he was very popular. None of them ever had anything nice to say about her, and some were very critical; J J Thomson's wife called her a 'difficult woman'. She seems to have been dour and humourless, and devotees of old films will remember a superb example of the type, played by Jean Cadell, in *Whisky Galore*. The Maxwells appear to have resided at the Principal's house after their marriage.

Katherine Maxwell did not much appreciate her husband's scientific career, preferring to perform the social duties of the wife of the laird of Glenlair, which they visited as often as possible. She was, on the other hand, of some help to her husband in his experiments on colour vision and on the viscosities of gases. In letters to his friends Maxwell referred to her as his 'better 1/2', but said little else about her. He was certainly an exemplary husband, taking care of her when she was ill, as she did of him when he was ill. His letters to his friends were always full of wit, but his letters to her showed none of that side of him, which she presumably did not appreciate. Instead they were full of moral advice and quotations from the Scriptures.

## SATURN'S RINGS AND THE KINETIC THEORY OF GASES

A good deal of Maxwell's time at Marischal College was taken up by his work on the rings of Saturn, and this led to his important work on the kinetic theory of gases. It seems convenient to consider this work here, even though it was completed at his next appointment, at King's College, London.

In 1855 Cambridge University announced that the subject assigned for its 1857 Adams Prize was a theoretical study of Saturn's rings (Fig. 5.3), with special reference to two possibilities: that the rings were solids, and that they were fluids (liquids or gases). The Adams Prize had been instituted in 1848 for an essay on a specified topic in celestial mechanics; the examiners for the prize expected the essays to be based on substantial pieces of original research, worthy of publication in a scientific journal. The prize could be awarded annually, but no essay had ever been submitted until Maxwell, shortly after his arrival in Aberdeen, met the 16 December 1856 deadline for the 1857 award. His essay, which won the Prize, was the only one submitted.

At the time, astronomers had observed three concentric rings about Saturn, all in the same plane. It was known that at least some regions of the rings must be quite thin, since in some areas the planet behind can be seen plainly. Maxwell began his work at Cambridge and continued it after taking up his appointment at Marischal College. He carried out a careful theoretical calcu-

**Fig. 5.3** Saturn and its rings, as it appears in a recent photograph from a *Voyager* spacecraft. Maxwell concluded, from theoretical studies based on the known properties of solids and liquids, that the rings could not be completely solid or liquid, but must be made up of solid particles.

lation, and concluded that the rings could not be solid or liquid, since the mechanical forces acting upon rings of such immense size would break them up. He suggested that instead the rings must be composed of a vast number of individual solid particles rotating in separate concentric orbits at different speeds. His final article on the subject, 'On the stability of the motion of Saturn's rings', published in the *Proceedings of the Royal Society of Edinburgh* in 1859, ran to 90 pages and is a monumental, meticulous, and lucid analysis of the problem.

Later studies, including observations from *Voyager* spacecraft in the latter years of the twentieth century, confirmed Maxwell's conclusions. The fact that the rings are composed of particles is supported by observations of stars seen through portions of the rings. Spectroscopic studies have shown that the particles are composed of impure ice, or at least are ice-covered. Radar observations making use of the Doppler effect have confirmed the range of speeds predicted by Maxwell. It appears that the particles have diameters from a few centimetres to about a hundred metres.

The kinetic theory of gases is concerned with the understanding of the properties of gases in terms of the behaviour of the individual molecules. Two simple relationships apply to gases to a good approximation; at a given temperature the product of the pressure and the volume is a constant (Boyle's law), and at a given pressure the volume is proportional to the absolute (Kelvin) temperature (Gay–Lussac's law). In 1738 the German mathematician Daniel Bernoulli (1700–1782) had suggested that the pressure of a gas is due to the bombardment of the molecules on the surface of the vessel (Fig. 5.4), but this idea, undoubtedly correct, was not taken seriously for many years.

It was, in fact, not until over a hundred years later that Bernoulli's suggestion was developed into a mathematical theory. This was done in 1846 by John James Waterston (1811–1883), but unfortunately his paper was rejected for publication, and did not see the light of day until 1892. The first published work on the problem was by the German physicist Rudolph Clausius, whom we met in Chapter 2 in connection with the second law of thermodynamics. In papers published in 1857 and 1858 Clausius derived the fundamental relationship between the pressure–volume product for a gas, and the number of molecules, their mass, and their mean speed.

The kind of mathematics used in Maxwell's treatment of Saturn's rings was directly applicable to the kinetic theory of gases. Early in 1859, when still at Aberdeen, he saw a translation of one of Clausius's papers, and he set out to develop the theory. At the 1859 meeting in Aberdeen of the British Association for the Advancement of Science (Fig. 5.5) he presented a theory of the viscosity of gases on the basis of kinetic theory. Viscosity is a property which is related to the difficulty with which a substance flows. When we pour oil we find that it flows more slowly than water, and we say that oil has a higher viscosity than water. Glass is an interesting substance, since it is really a liquid, but it has such a high viscosity that it behaves like a solid. Gases have much lower viscosities than liquids. When Maxwell got interested in their viscosities, he found that hardly any data had been obtained. His mathematical treatment of the properties of gases in terms of the movement of their molecules led him to conclude that their viscosities are independent of pressure, and that they increase approximately with the square root of the absolute

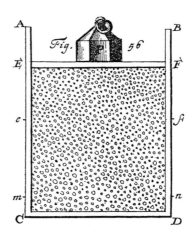

**Fig. 5.4**  Daniel Bernoulli's diagram (1738) depicting his ideas about the kinetic theory of gases; the pressure of a gas is due to the bombardment of the molecules on the surface of the container.

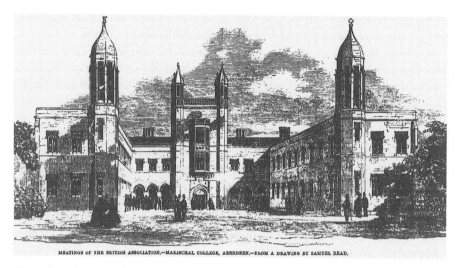

MEETINGS OF THE BRITISH ASSOCIATION.—MARISCHAL COLLEGE, ABERDEEN.—FROM A DRAWING BY SAMUEL READ.

**Fig. 5.5** Marischal College, Aberdeen, as depicted in the *Illustrated London News* at the time of 1859 meeting of the British Association for the Advancement of Science. Maxwell's lecture room is believed to have been on the upper floor behind the tower at the left of the drawing. It may have been in that lecture room that he announced his distribution law for molecular speeds.

temperature. He thought it rather peculiar that the viscosity should not depend at all on the pressure, and decided that experimental results were needed.

In the attic of his house in Kensington, and with the help of his wife Katherine, Maxwell made many experimental measurements of gas viscosities, to confirm the conclusions he had drawn about the effects of pressure and temperature. Many of the experiments were made at temperatures between 51°F (10.6°C) and 74°F (23.3°C), and it appears that these temperatures were brought about by simply changing the temperature of the attic room; this was arranged by Katherine, who organized the appropriate stoking of the fire. Some work was also done at 185°F (85°C), and this temperature was achieved by a suitably directed current of steam. The results of this investigation were communicated to the Royal Society in Maxwell's Bakerian Lecture entitled 'On the viscosity and internal friction of air and other gases'; the paper was published in the *Philosophical Transactions* in 1866.

At the British Association meeting in Aberdeen Maxwell also announced a theory, for which he has become famous, of the distribution of molecular speeds. It had already been recognized that in a gas, as well as in a liquid, some molecules are moving rapidly and others more slowly. Clausius was well aware of this, and in his papers he had explained, in terms of the variation of molecular speeds, why liquids become cooler when evaporation occurs. The more rapidly moving molecules are more likely to leave the surface of a liquid, so that the molecules left behind after evaporation have on the average less energy; the temperature of the liquid is therefore lower.

In 1860 Maxwell published a paper, based on his presentation at the Aberdeen meeting, in which he derived an equation for the distribution of molecular speeds in a gas. His derivation was based on a mathematical treatment of statistical probability given ten years earlier by John Herschel—whom we met in Chapter 3 in connection with photographic techniques. Maxwell realized that it would be hopelessly complicated to try to deal with the individual motions of a large number of molecules, but that probability theory provides a short cut to a solution. Probability theory deals, for example, with the problem of the number of ways in which a certain number of balls can be distributed among a certain number of boxes. With characteristic insight, Maxwell realized that molecular distributions could be dealt with in the same way. He considered how molecular speeds would distribute themselves amongst a large number of molecules.

Maxwell's first derivation of the distribution law, although leading to the right mathematical expression, was not entirely satisfactory, and he had to grapple with the problem for some years before finding something with which he was happy. In 1867, when he was in 'retirement' at Glenlair, he published a much improved version of his kinetic theory, including a better derivation of his distribution law. Maxwell's electromagnetic theory, to be considered later, is often regarded as his most important achievement, since its consequences have been so widespread. His theory of the distribution of molecular speeds is perhaps just as important. Particularly in the hands of Ludwig Boltzmann, whom we met in Chapter 3, it led to the theory of the distribution of molecular energies, and to the important branch of science known as statistical mechanics.

Later in 1867 Maxwell made another contribution, the Maxwell demon, which we discussed in Chapter 2 in connection with the interpretation of the second law of thermodynamics.

Although he made important contributions to kinetic theory, especially by his distribution law, Maxwell was never convinced of its validity. In the same paper that he gave at the Aberdeen meeting in 1859, Maxwell also presented an 'equipartition theorem' for molecular energies. In its original form, this theorem stated that the average energy of a molecule resulting from its translation (movement from one place to another) would be the same as its energy of rotation. This theorem is perfectly correct within the framework of Newtonian mechanics, but when it was realized that quantum mechanics (Chapter 8) applies, the theorem had to be modified considerably. Since some of the experimental results on the specific heats of gases were clearly inconsistent with the equipartition of energy, Maxwell always had serious doubt about the validity of kinetic theory. There is an irony here, as Maxwell had done so much to establish the theory, which over 20 years after his death could be modified by quantum theory to give a satisfactory interpretation of all the experimental results on specific heats.

We should now jump back in time a few years. In 1860 it was decided to combine Marischal College with another college at Aberdeen, King's College. It had become apparent that a city as small as Aberdeen could not sustain two separate degree-granting institutions; a Royal Commission recommended their fusion into one university. As a result there was need for only one professor of natural philosophy. In view of Maxwell's poor performance as a lecturer it is not surprising that it was the King's professor, David Thomson, nicknamed 'Craftie', who was retained. Daftie does not seen to have been unduly upset at being defeated by Craftie; he had, after all, ample means, and he had not enjoyed the Aberdeen interlude. It might have been thought that being the son-in-law of the Principal of the College would give Maxwell an advantage, but the opposite may have been the case. It has been suggested that Katherine Maxwell exerted some influence in having her husband removed, believing that he would retire to Glenlair, but if so she was frustrated in her intention.

On losing his position at Aberdeen, Maxwell applied for the newly vacant Edinburgh chair of natural philosophy, but the appointment went to his close friend Peter Guthrie Tait (1831–1901), with whom he had been at school and at Cambridge. Two men can hardly have been more different than Maxwell and Tait. Maxwell was short and slight, with a diffident and nervous manner; Tait was over six feet tall, strongly built, looked like a prizefighter, and had an overpowering personality; he was a great golfer, and published two papers on the ballistics of the golf ball. Whereas Maxwell was a poor lecturer, Tait was superbly good. Tait later collaborated with Thomson (Kelvin) on a *Treatise on natural philosophy* (i.e., physics), which became a scientific classic; it was commonly referred to as T and T', and since the Archbishops of York and Canterbury also happened to be named Thomson and Tait, Maxwell referred to his two scientist friends as the 'Archepiscopal Pair'. Throughout their careers Maxwell and Tait kept up a lively and amusing scientific correspondence, in which Maxwell often addressed Tait as O T', and signed himself $dp/dt$; this was a rather contrived private joke between them, Tait having published a paper in which the equation $dp/dt = JCM$ appeared, the latter being Maxwell's initials.

Maxwell then applied for the vacant chair of physics and astronomy at King's College, London, and there he was successful; he was still under 30 when he took up the appointment in 1860. Before doing so he was struck down with smallpox at Glenlair, and for some days was close to death. His wife ministered to him conscientiously, reading to him from the Bible, and he recovered completely. Maxwell later insisted that she had saved his life.

## KING'S COLLEGE, LONDON (1860–1865)

King's College, an Anglican foundation which opened its doors in 1831, was housed in a building on the Strand, next to Somerset House. It had already achieved some distinction for scientific research. John Frederick Daniell, whom we met in Chapter 4 in connection with his battery, had been professor of chemistry there until 1845, and had done distinguished work in electrochemistry and spectroscopy, some of it in collaboration with William Allen Miller (1817–1870) who succeeded him. Maxwell and Miller had many scientific interests in common; as we saw in Chapter 3, Miller was particularly interested in astronomical photography. Maxwell also saw much of Michael Faraday, who was 40 years his senior, and became very friendly with him.

The facilities for research at King's College were almost non-existent; Daniell and Miller had done their experimental research in a lumber room under the lecture theatre. Maxwell carried out his experimental work at his house at 8 Palace Gardens, Kensington. It still stands, now numbered 16, a substantial house with four main floors in addition to a basement and attic with windows. The house was chosen because of its proximity to Hyde Park. Maxwell's life in London was a leisurely one—in marked contrast to that of Faraday. He lectured and carried out research only in the mornings. In the afternoons he went horseback riding in the Park with his wife, while evenings were devoted to social activities. In spite of this, the London period was probably the most original and fruitful of his career. His instinct for thinking correctly on any physical subject allowed him to accomplish a great deal with little effort. He was elected a Fellow of the Royal Society in 1861.

It was at the beginning of his appointment in London that he carried out his famous demonstration of a coloured photoimage, projected on a screen, of a striped ribbon; we described this in Chapter 3. It emerged much later that there was a curious anomaly about the demonstration. The photographic emulsions available at the time were sensitive only to the blue end of the spectrum; they were only slightly sensitive to green, and not at all to red. How then was it possible to produce an image that did show the greens and reds? The answer was provided in 1961, just 100 years after the demonstration, by Ralph M Evans of the Eastman Kodak Company. The greens, which show up only faintly, can be explained by the slight sensitivity of the emulsions to green. The reds, however, should not have shown up at all, and yet they did. By reproducing the experiment under the original conditions, and using copies of the original transparent positives, which are still at the Cavendish Laboratory, Evans was able to show that the red dye used in the ribbon also reflected a good deal of ultraviolet light, to which the emulsion was sensitive. As a result, the red stripes on the ribbon produced a good image not because they were red but because of the ultraviolet light they reflected. This is a rare example of where a piece of scientific research was accompanied by good luck as well as good science.

Throughout the years they lived in London, Maxwell and more particularly his wife felt a longing to live at Glenlair, to which they paid frequent visits. In 1865, aged 34, he tendered his resignation to King's College, with the intention of spending his remaining years in retirement at Glenlair (Fig. 5.6). He

**Fig. 5.6**    The manor house at Glenlair, as it appeared after Maxwell extended it in the late 1860s. His father had planned something more grandiose. Unfortunately, the building was badly damaged by fire in 1929, and has not been restored; little more than the outer walls remain. It was at this house that Maxwell wrote most of his *Treatise on electricity and magnetism*.

especially wanted to have the manor house enlarged; his father had designed the house, but had not been able to complete it to his satisfaction.

## INTERLUDE AT GLENLAIR (1865–1871)

Almost as soon as Maxwell returned to Glenlair, in the spring of 1865, he scraped his head on a tree branch while riding a horse, and contracted erysipelas, which at the time was often fatal. Again his wife Kathleen nursed him back to health after an illness of about three weeks. He then devoted himself conscientiously to the traditional duties of a Scottish laird. In his house he conducted morning and evening prayers, which the servants were expected to attend. He visited his tenants and cared for them in times of illness, reading the Scriptures to them 'where such ministrations were welcome'. On Sundays he worshipped at the neighbouring churches at Parton and at Corsock, where he became a church elder. In 1867, when the construction at the house was being carried out, he and his wife made their only trip abroad, to France and Italy. A studio photograph of Maxwell and his wife, probably taken during the time they were at Glenlair, is shown in Fig. 5.7.

Maxwell did much more than carry out social duties at Glenlair. He had retained the lease of the Kensington house, and visited London a number of

**Fig. 5.7** Maxwell, his wife, and their West Highland terrier. This is a studio photograph in which they are posed in front of a painted backdrop which simulates a Highland scene.

times, particularly to attend meetings of the Royal Society. During the Glenlair period Maxwell's scientific correspondence was particularly heavy. On 11 December 1867 he wrote to Tait the famous letter in which he suggested what came to be called 'Maxwell's demon', which we considered in Chapter 2.

At Glenlair Maxwell also wrote a book *The theory of heat*, which was published in 1870 and gave a particularly lucid account of thermodynamics; in it he explained in a simple manner many ideas that had been put forward obscurely by Clausius, Boltzmann, and Gibbs. The book remained in print for many years, eventually running to 11 editions. Although Maxwell had a considerable interest in thermodynamics he wrote no major paper on the subject, but many original contributions appeared in *The theory of heat*.

During the Glenlair period Maxwell wrote several scientific papers, and most of his *Treatise on electricity and magnetism*, which was to be published in 1873.

# EARLY IDEAS ON ELECTROMAGNETIC RADIATION (1854–1864)

This seems a convenient place to begin to consider what was Maxwell's greatest contribution to science, his theory of electricity, magnetism, and light. This work began when he was a student at Cambridge, and continued until his death.

Much of Faraday's great work on electricity and magnetism, which we discussed in Chapter 4, had been completed by the time Maxwell became a student at Cambridge in 1850. Maxwell, who was one of the few to realize its importance, always insisted that he did nothing more than express Faraday's ideas in mathematical form, but he was being unduly modest. Producing the mathematical equations was far from a routine task that any highly skilled mathematician could have carried out; it also involved clarifying and modifying the basic concepts. Maxwell's first paper, 'On Faraday's lines of force', which appeared in 1856 when he was 25 and still at Cambridge, was important for showing mathematically that Faraday's ideas gave a valid alternative to Ampère's treatment based on central forces.

Maxwell's second paper on the subject, 'On physical lines of force', appeared in four parts in 1861–62, and was concerned mainly with devising a model for the ether which would account for the stresses associated with Faraday's lines of force. This ether had a rather complicated structure, consisting of spinning vortices, some of them electrical and some magnetic. The model was such that a changing magnetic field gave rise to an electric field, and that a changing electric field produced a magnetic field. This model is today mainly of historical interest, in showing how Maxwell's ideas developed, as he discarded the model in his final version of his electromagnetic theory, basing it entirely on the mathematical analogy.

## THE WAVE THEORY OF LIGHT

At this stage it is necessary for us to make an important digression to discuss developments that had occurred earlier regarding the nature of light. Newton had interpreted the result of his experiments in terms of the theory that light consists of a stream of particles, or corpuscles. In 1665 Newton's arch-enemy Robert Hooke (1635–1703) had put forward a wave theory of light, but he did not develop the idea to any extent. According to Hooke, the vibrations occurring are in the same direction as the light path, whereas the modern theory is that they occur at right angles to the light path (Fig. 3.8). A little later the great Dutch investigator Christiaan Huygens (1629–1695) brilliantly developed Hooke's wave theory and showed how it could explain the reflection and refraction of light.

Newton gave careful consideration to both corpuscular and wave theories, and did not discard either of them. Indeed, he tended to favour ideas that were partly wave and partly corpuscular, which is rather remarkable, as that is the modern view, which will be discussed in Chapter 8. Newton's main difficulty with the wave theory is that he could not see how it could explain the fact that light travels in straight lines; he thought that when light encountered an obstruction the waves would be sent out in different directions. There is indeed some spreading of this kind, but because the waves are so small (the wavelengths are very short, as will be discussed) the spreading is slight. Newton changed his views from time to time, but on the whole he favoured the corpusular theory. Because of his great prestige, this was the theory generally accepted until the beginning of the nineteenth century.

We mentioned in Chapter 3 the important work of Thomas Young on colour perception. Now we must discuss the wave theory of light that Young proposed early in the nineteenth century, and which the French physicist Fresnel developed, as it plays a vital role in Maxwell's theory of electromagnetic radiation.

In an important experimental investigation, Young discovered the interference of light. He passed a beam of white light through two pin holes in a screen, with another screen placed beyond the first (Fig. 5.8). Where the rays overlapped on the second screen, he saw a series of coloured bands. If instead of white light he used light of one colour, the bands were instead alternately light and dark. These bands obtained with light of one colour are explained as follows. The bands result from the interference with each other of the waves from the two pin-hole sources. If the distances travelled by the two waves is the same, the crests of the waves will be superimposed, and the intensity of

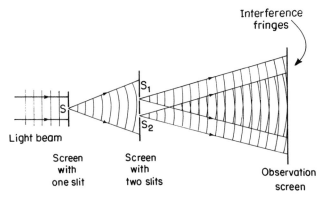

**Fig. 5.8**    A schematic representation of Young's interference experiment. The light waves spreading from the slit S reach the two slits $S_1$ and $S_2$. The light waves coming from these two slits give rise to interference fringes on the observation screen. If the light is monochromatic (involving only a narrow range of wavelengths) the observation screen shows light and dark bands. Otherwise coloured bands are seen.

the light will be doubled. If, on the other hand, one wave has half a wavelength further to travel than the other, the crest of one wave will coincide with the trough of the other, and darkness will result. Fig. 5.9 shows the shape of a light wave, and shows how two waves can annihilate one another.

When white light is used, the different colours interfere at different positions, and the result is coloured bands.

From the dimensions of his apparatus and the breadth of the bands Young was able to calculate the wavelengths of the different coloured rays. He found that the lengths are exceedingly small compared with the lengths of ordinary obstacles in the path of a beam of light. The light at one end of the visible spectrum is red, and Young found that red light has a wavelength of about 700 thousand-millionths of a metre ($7 \times 10^{-7}$ m, or 700 nanometres). At the other end of the visible spectrum there is violet radiation, which Young found to have a wavelength of about 400 nanometres, or roughly half that of red radiation.

The fact that the wavelengths are so small compared with ordinary objects overcomes a difficulty that Newton found with the wave theory of light. He thought that if a wave arrived at the edge of an object it would be diffracted and would not continue along the same straight line. Because of the shortness of the wavelengths, however, the extent of the diffraction is slight. The diffraction of light can be detected only in specially contrived circumstances.

Young did not pursue the problem much further, as he had so many other interests. It was left to Augustin Jean Fresnel (1788–1827) to develop the wave theory mathematically to a high degree of perfection. One of the most important changes he made was to regard the vibrations as transverse (at right angles) to the direction of the rays. It was mentioned earlier that in Robert Hooke's theory the vibrations were considered to be in the direction of the rays. Newton had argued that the polarization of light was incompatible with Hooke's ideas. Newton was aware of some of the polarization effects that we discussed in Chapter 3; some of them had been demonstrated by Christiaan Huygens with crystals of Iceland spar. Newton saw that a

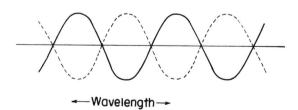

←—Wavelength—→

Fig. 5.9   A representation of wave motion. The wavelength of the light is the distance between neighbouring crests or troughs. The dashed curve shows another wave such that its troughs coincide with the crests of the first wave. The two waves will annihilate one another. The frequency is the time taken for the complete vibration to occur. The product of the frequency and the wavelength is the speed of light.

polarized beam must somehow have become changed at right angles to the direction of the beam, and this did not seem to be possible according to Hooke's type of wave theory.

Fresnel realized that Newton's difficulty was removed if the vibrations occurred at right angles to the direction of the beam. In the case of ordinary light, there would be vibration in all directions at right angles to the beam. When light passed through certain substances like Iceland spar, the resulting beam would have the vibrations only in one direction, for example up and down. The polarized beam could then pass through another crystal oriented in the same way, but not through one turned through 90°.

It follows from the wave theory of light that the wavelength multiplied by the frequency (the number of complete vibrations that occur in unit of time) is equal to the speed with which the light travels:

$$\text{wavelength} \times \text{frequency} = \text{speed of light}$$

The measurement of the speed of light is therefore a matter of great importance. Aside from allowing the frequency to be calculated from the wavelength, the speed allows one to distinguish between the corpuscular theory favoured by Newton and the wave theory favoured by Huygens, Young, and others. If the corpuscular theory applied, the speed would be greater in a medium such as water than it is in air. According to the wave theory, however, the speed will be less in water. It is therefore important to determine the speed in water as well as in air.

The first estimate of the speed of light in air was made in 1675 by the Danish astronomer Olaus Roemer (1644–1710). He determined his value from observations of the satellites of the planet Jupiter. The time of their revolution about the planet was known accurately, and the distances of the planet from the earth at various times were known. Roemer realized that if light did not travel instantaneously, but at a finite rate, the eclipses of the satellites should appear to take place earlier when the planet was nearer to the earth than when it was further away. From the expected and observed times he calculated a value for the speed of light. We now know that it was too low by about 40 per cent, but the important thing was that it was certain that light does not travel with infinite speed, which had until then been considered possible. Roemer's value was improved somewhat in 1726 by the British astronomer James Bradley (1693–1762). Further progress had to wait until non-astronomical techniques were developed more than a century later.

Reliable measurements of the speed of light were made by Hippolyte Fizeau and Léon Foucault, whom we met in Chapter 3 in connection with their early work on photography. They worked together from 1844 on the speed of light, and devised a satisfactory esperimental system, but a few years later dissolved their partnership over a personal dispute. They continued their work independently, and met with success almost simultaneously in May

1850. Both men paid special attention to calculating whether light travels more rapidly in air or in water, since this provided a way of deciding between the corpuscular and wave theories.

Fizeau and Foucault also devoted much effort to measuring the absolute speed of light in air, and this Fizeau succeeded in doing before Foucault. Fizeau's apparatus is illustrated in Fig. 5.10. The observer O was at Fizeau's father's house in Suresnes, a Paris suburb, while the other observer O' was at the top of a hill about 8 kilometres away in Montmartre, another Paris suburb. The source of light A is reflected by a half-silvered mirror, and the observer O adjusted his telescope so that he could see the beam through the half-silvered mirror D. The observer O' could see the beam after it had travelled from the mirror B to the mirror D and back again. As the large toothed wheel C was rotated faster and faster, the light reappeared and then disappeared. The time required for the light to travel the total distance could be calculated from the speeds at which the light became visible and invisible to the observer O. Fizeau's first value was 315 000 kilometres per second, which is rather high compared with the modern value of 299 792 kilometres per second. The experimental method, however, was reliable, and was used later in more precise determinations.

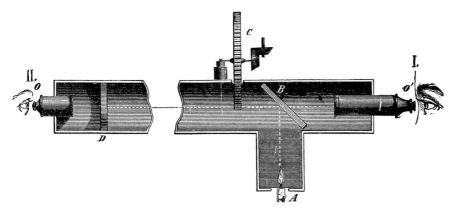

**Fig. 5.10** Fizeau's diagram of his apparatus to measure the speed of light in air. The light from the source A was reflected by a half-silvered mirror B, and the observer O adjusted his telescope so that he could see the beam through the half-silvered mirror D. The observer O', some 8 kilometres away, could also see the beam after it had travelled from the mirror B to the mirror D and back again; the time for the double trip was about 27 millionths of a second. The large toothed wheel C was spun rapidly round its axis. When the wheel was moving sufficiently rapidly the reflected rays struck the teeth of the wheel, and became invisible to the observer O' stationed behind the wheel. As the wheel was rotated faster and faster, the light reappeared and then disappeared. The time required for the light to travel the total distance could be calculated from the speeds at which the light became visible and invisible to the observer.

Foucault and Fizeau used similar apparatus for comparing the speed of light in air and water. Foucault's diagram of his apparatus is shown in Fig. 5.11. A beam of light is focused by a lens and reflected by a plane mirror to a focus on another mirror. The silvered mirror allows the observer to see both the source of light and its reflection. If the mirror is stationary the beam returns along the same path. If the mirror is rotated rapidly at a known speed, it has turned through a small angle during the time it takes for the light to travel from m to M and back again. It therefore continues along a different path towards the position a′. In the lower part of the diagram the light is shown passing through a tube containing water, and the image after reflection is at a″. Since the image a″ is deflected more than the image a′ it follows that light travels faster in air than in water. This was announced by Foucault on 30 April 1850, only a few days before the result was obtained by Fizeau.

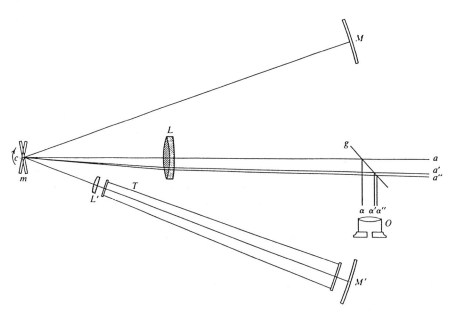

**Fig. 5.11** Foucault's diagram of his apparatus to compare the speed of light in water with that in air; similar apparatus was used by Fizeau. A beam of light from a is focused by a lens L and reflected by a plane mirror m to focus on a mirror M. The silvered mirror g allows the observer at O to see both the source of light and its reflection. If the mirror m is stationary the beam returns along the same path. If the mirror is rotated rapidly at a known speed, it has turned through a small angle during the time it takes for the light to travel from m to M and back again. It therefore continues along a different path towards the position a′. In the lower part of the diagram the light is shown passing through a tube T containing water, and the image after reflection is at a″.

# THE SPEED OF AN ELECTROMAGNETIC WAVE

Some time in the miidle of the nineteenth century it was realized that important insight can be obtained by comparing results expressed in the two sets of
units, *electrostatic* and *electromagnetic*, that were being used at the time. The
electrostatic units are particularly appropriate to static electricity, and relate
the force of attraction and repulsion between two charged bodies to the
quantities of electricity that they held. The electromagnetic system of units,
on the other hand, relates force to electric currents. The quantity of electricity
residing on a wire at a given time depends on the speed with which the
current travels along the wire. It turns out that the ratio of the value of an
electrical quantity expressed in electromagnetic units to that expressed in
electrostatic units is the speed with which the current travels.

In 1856–1857 the German physicists Wilhelm Edouard Weber (1804–
1891) and Rudolph Herrmann Arndt Kohlrausch (1809–1858) made careful
measurements of an electric charge, first in one set of units and then in the
other. When they divided one value by the other they obtained a velocity that
was very similar to the speed of light. They concluded that a current travels
along a wire with the speed of light.

In 1861 the British Association for the Advancement of Science set up a
committee under William Thomson's (Kelvin's) chairmanship to establish a
set of electrical and magnetic standards. Maxwell was an active member of
that committee, and besides attending meetings he carried out important
experimental investigations in connection with the objectives of the committee. His experiments were carried out using equipment made available to him
by John Peter Gassiot (1797–1877), a wealthy wine merchant and amateur
scientist who had already done some remarkable experiments using vast
numbers of electric cells. His conclusion from the experiments comparing
electrical quantities expressed in the two sets of units was that the speed of an
electric current was close to 300 000 km s$^{-1}$, which is the speed of light.

An alternative procedure, used by Maxwell and others, was to compare the
electrical quantity now called the permittivity of a vacuum, $\varepsilon_o$, with the magnetic quantity now called the permeability, $\eta_o$. Maxwell's mathematical treatment of Faraday's lines of force led to the conclusion that the speed $v$ of an
advancing electromagnetic field was given by

$$v = (\eta_o\, \varepsilon_o)^{1/2}$$

Measurements made by various physicists of $\eta_o$ and $\varepsilon_o$ also led to the finding
that $v$ obtained in this way was the speed of light. These results convinced
Maxwell that light is an electromagnetic wave; in his own words, emphasized
in italics in his 1861–62 papers:

> *Light consists in the transverse undulations of the same medium which is the cause of
> electric and magnetic oscillations.*

In other words, the same electromagnetic theory applies to light, to an electric field, and to a magnetic field. The field produced by an electric current has also a magnetic component, and the field produced by a magnet has also an electric component; light also has electric and magnetic components.

An important aspect of this work was that it led to a decision between the two alternative theories of electricity that were held at the time. One theory was that there were two electrical fluids, positive and negative, which moved in opposite directions when a current flowed. Use of the method of comparison of units led to the conclusion that if there were two fluids they would each flow with half the speed of light. The evidence thus supports the theory (which Benjamin Franklin among others had advocated) that only one type of electricity flows along a wire when a current passes. Today we know that a flow of electrons is involved.

## Maxwell's later electromagnetic theory (1864–1873)

Maxwell's final major paper on the subject was entitled 'A dynamical theory of the electromagnetic field'; it appeared in 1864. With uncharacteristic immodesty, but with perfect justification, he described this paper in a letter as 'great guns'. In this paper he ignored the rather elaborate and artificial model he had proposed for the ether, and concentrated on the propagation of electromagnetic waves through space. The position he took, and this is accepted today, is that the mathematical treatment remains valid without any assumptions about the nature of the medium through which the waves travel. In other words, he threw out the bathwater but carefully preserved the baby!

In doing so Maxwell made an important break with scientific tradition. Previously it had been felt necessary to base a scientific theory on a model that could be clearly visualized. Maxwell's theory, on the other hand, represented the truth not in terms of a model, but as a mathematical analogue. Maxwell's friend and colleague William Thomson (Kelvin) always insisted on a mechanical model; in his own words,

> I never satisfy myself unless I can make a mechanical model of a thing. If I can make a mechanical model I can understand it.

As a result, Kelvin never really understood Maxwell's theory, even though he had himself made important contributions to interpreting Faraday's lines of force mathematically. Similarly, he never understood Clausius's concept of entropy—another concept that cannot be understood in terms of a model— even though he had been one of the first to appreciate the second law of thermodynamics.

Maxwell's theory can be expressed in terms of a few equations which have been referred to as simple. They are indeed simple in form, but understanding them involves a considerable background knowledge of electrical and magnetic theory, and of vectors. Maxwell himself invented the names 'curl', 'grad' and 'div grad' for mathematical operators that appear in his equations.

Maxwell's famous book *Treatise on electricity and magnetism* was published by the Clarendon Press, Oxford, in 1873 when Maxwell was back at Cambridge as its first Cavendish professor. In many ways the book is something of a surprise. It is certainly not to be recommended to a student wanting to learn about the theory; the 1864 paper is much easier to follow. The book raises a number of aspects which Maxwell himself had not been able to clarify. It consists of four parts: Electrostatics, Electrokinematics, Magnetism, and Electromagnetism. The author recommended that these four parts should be read concurrently, but does not explain how this remarkable feat should be performed; presumably he meant that one should read a little of the first part, then a little of the second, and so on. In spite of its obscurities, the book was a great inspiration to many physicists, including Albert Einstein.

A significant feature of the book is that the word ether is mentioned only once, and that the model for the ether elaborated in his 1861–62 paper is referred to only incidentally. This does not mean that Maxwell had abandoned his belief in the existence of an ether; in his article on 'Ether' in the famous ninth edition of the *Encyclopaedia Britannica* (1875) he expressed very clearly his belief in the existence of an ether. He considered, however, that his theory of electromagnetic radiation was valid whether or not the ether exists, or what its nature is.

Maxwell also did some important theoretical work on governors, the feedback devices used on engines to provide automatic control. His work was inspired by the fact that in carrying out their work on electrical units for the British Association, Maxwell and his colleagues had used a speed governor to ensure the uniform rotation of a resistance coil. It somewhat resembled the governors used by James Watt on his steam engines; weights attached to a shaft were displaced by centrifugal force and activated a control valve. In 1868 Maxwell published a mathematical treatment of the subject, and his paper, 'On governors' in the *Proceedings of the Royal Society*, is regarded as founding the field of control theory, or *cybernetics*. The latter word, derived from the Greek word for steersman, was first used by the mathematician Norbert Wiener (1894–1964), who in 1948 published an important book with that title. Today the subject has been broadened to include control and communication in living systems and in computers.

# First Cavendish professor (1871–1879)

Until the 1870s, Cambridge University lagged considerably behind Oxford in its facilities for scientific research. Since 1683, Oxford science professors had made good use of the Ashmolean Museum for their research, and Oxford's substantial Science Museum was completed in 1858. In the meantime Cambridge scientists had been forced to do experimental work in their College rooms or homes. Stokes, for example, had done his research on fluorescence in a small room behind the kitchen of his house.

In 1870 William Cavendish, the Seventh Duke of Devonshire (1808–1891), who was the Chancellor of Cambridge University, offered to finance the construction of a building for experimental research, and to provide furnishings and equipment. The Duke had himself been a distinguished scholar at Cambridge, having gained high honours in the Classical Tripos and also having been Senior Wrangler in the Mathematical Tripos; he was perhaps the only hereditary peer to have gained such distinction. In 1871 the University created a chair of experimental philosophy, the holder to be the director of the new laboratory. The post was first offered to William Thomson (Kelvin), who had established a great reputation as professor of physics at the University of Glasgow. However, he was unwilling to leave Glasgow, where he remained until his death in 1907. The post was then offered (at Thomson's suggestion) to the distinguished German physiologist and physicist Hermann von Helmholtz, whom we met in Chapter 2; he had just accepted the chair at Berlin and also declined. As third choice, Maxwell was urged to stand for the position, which he did reluctantly, and he was appointed. He had just reached the age of 40.

It is noteworthy that the three prime candidates for the most coveted chair of physics in an English university should have been two Scotsmen and a German. Scotland, with only a fifth of the population of England, was far ahead of England in the teaching of science, as was Germany. For over a century, for example, large classes at Edinburgh and Glasgow Universities had been taught by physical scientists like Joseph Black and William Thomson (Kelvin). It is hard to think of Englishmen who would have been suitable for the Cambridge post. Faraday had died in 1867, and as he had worked alone he left no successors. John William Strutt (the future Lord Rayleigh; 1842–1919) was only 29 and had not yet made his reputation; eight years later he was to succeed Maxwell as Cavendish professor. John Tyndall (1820–1893)—an Irishman—was distinguished enough but his agnosticism would have made him unacceptable at Cambridge.

Although Maxwell (Fig. 5.12) was the third choice, he was undoubtedly a more original scientist than Thomson or Helmholtz. He was not the best choice as professor; Thomson and Helmholtz were both inspiring men and excellent lecturers, and would have attracted large classes. Maxwell's classes were pitifully small, averaging two or three students. His lecturing had not improved; he still rambled and made many mistakes at the blackboard. To some extent he made up for these deficiencies by attracting a number of excellent research students. His contributions to the design of the Cavendish Laboratory, which was completed in 1874 and not properly furnished until

**Fig. 5.12**  A photograph of Maxwell taken during his tenure of the Cavendish professorship at Cambridge.

1877, were effective, and he personally arranged for the construction of much of the apparatus to be used in the practical classes, paying for some of it out of his own pocket. The laboratory was first called the Devonshire Laboratory, but Maxwell suggested the change to the Cavendish Laboratory, so that it would jointly honour the Duke of Devonshire and his kinsman the brilliant scientific investigator Henry Cavendish (1731–1810).

Besides teaching and directing research at the laboratory, Maxwell accomplished a great deal during his eight years at the Cavendish. He prepared for publication, by the Oxford University Press in 1873, his *Treatise on electricity and magnetism*. He refereed papers for journals, examined students, and wrote a number of articles on a variety of topics, some for the *Encyclopaedia Britannica*. Some of these are particularly noteworthy, and worth reading today. His article on 'Capillary action', which first appeared in the ninth edition of the *Britannica* in 1875, was reprinted in the 13th edition in 1926, with no changes but a few additions in brackets made by Lord Rayleigh; it is still a valuable and lucid account of the basic theory. The same is true of his article on 'Diffusion', which also appeared in the ninth edition. His article 'Atom', although inevitably now much out of date, gives the historian a remarkably clear idea of the state of knowledge of the subject at that time.

One of the last pieces of research to be done by Maxwell was on the radiometer, a rather puzzling device that had been invented by William Crookes. We will defer discussing this work until the next chapter, as it is related to gases at low pressures and to the work of J J Thomson.

In the spring of 1879, Maxwell became ill. He returned to Glenlair where his condition did not improve, and in October he was told that he had abdominal cancer and could not live more than a month or so. He returned to Cambridge and although in great pain continued research almost to the day of his death, on 5 November at the age of 48.

Maxwell received few honours in his lifetime. He did receive an honorary Doctor of Laws degree from the University of Edinburgh in 1872, and an honorary doctorate from Oxford in 1876. There was little publicity at the time of his death, and it seems likely that his friends let it be known that because of his modesty and unpretentiousness any public recognition would be inappropriate. His funeral took place at St James the Less Church in Cambridge, and he was buried in the churchyard of the tiny Parton Kirk, where he had worshipped. For many years his grave was hard to find. At the entrance to the churchyard a brass plate installed in 1989 describes him with apt simplicity as 'A Good Man, Full of Humour and Wisdom'.

## Maxwell's scientific legacy

Maxwell's impact on the world of science and technology was far-reaching, and had many aspects. Perhaps his most important influence was to make scientists realize that a scientific theory need not correspond to an easily visualized mechanical model. The same lesson had been inherent in Clausius's concept of entropy, but not many had learnt it; Maxwell himself was one of the first to do so, and he explained it clearly in his book *The theory of heat*. His own theory of electromagnetic radiation was a uniquely important example of a theory that did not relate to a mechanical model.

Einstein said that the theory of electromagnetic radiation had inspired his theory of relativity, which has had such far-reaching consequences. In particular, Einstein's insight into the relationship between mass and energy led directly to the possibility of producing energy by causing nuclear processes to occur.

Maxwell's work on thermodynamics was also of lasting importance, and transformed the whole of physics and chemistry. In particular, his theory of the distribution of molecular speeds made possible, in Boltzmann's hands, the science of statistical mechanics, which has applications to some of the most useful technical devices, such as lasers.

## Experimental confirmation of Maxwell's theory of radiation

For several years after Maxwell proposed his theory of electromagnetic radiation a number of physicists devoted attention to gaining further understanding of it, to improving the way it was formulated mathematically, and to looking for ways of proving it experimentally. Maxwell's papers and his book were by no means easy to understand, and several great physicists, including Kelvin and Helmholtz, never properly grasped the theory. A small group of physicists, however, who were dubbed the Maxwellians, gained a solid understanding of it, and were able to help others to do so.

The most important of the Maxwellians, all of about the same age, were Oliver Heaviside (1850–1925), George Francis Fitzgerald (1851–1901), and Oliver Lodge (1851–1940). The most unusual and eccentric of them was Heaviside, who was a nephew of Charles Wheatstone. He had little education but somehow managed to become a highly accomplished mathematician. He first worked as a telegrapher in Newcastle-upon-Tyne, but being extremely shy and reclusive he retired in 1874 at the age of 24 (perhaps a record for early retirement). He was already quite deaf, and his hearing steadily deteriorated. For the rest of his life he led a frugal existence, at first in Kentish Town, North London and after 1889 in a seaside cottage (called by his colleagues the 'inexhaustible cavity') near Paignton, Devon; he had some financial support from his brother and later from friends and a government grant. Although he disliked meeting people he carried out an extensive correspondence, particularly with his fellow Maxwellians. His letters are remarkable for being heavily interspersed with quaintly worded irrelevancies and outspoken comments. In 1889, although impoverished, he published at his own expense a book on *Electromagnetic waves*, which is remarkable for its incomprehensibility. As he admitted in a letter to Hertz, it could be understood only by people who had read some of his previous publications, in which he had introduced a completely new and eccentric terminology; this had caused him to be called the 'interminable terminologist'. His name is still remembered for his introduction in 1902 of what is often called the Heaviside layer, or the Heaviside–Kennelly layer; also called the ionosphere, it is an ionized layer in the atmosphere capable of reflecting radio waves.

George Fitzgerald was the nephew of Johnstone Stoney, the inventor of the word electron (Chapter 4). He was educated at Trinity College, Dublin, where he was professor of natural and experimental philosophy (physics) from 1881 until his early death. He made important contributions to electromagnetic theory and in other fields. His name is particularly remembered today for the Fitzgerald–Lorentz contraction, a theoretical idea which played an important part in leading Albert Einstein to his theory of relativity.

Oliver Lodge (Fig. 5.13) belonged to a robust and prolific family. His grandfather the Revd Oliver Lodge had 25 children. He himself had 12, and he lived to the age of 89. He studied at the Royal College of Science in London and at University College, London. He heard Maxwell lecture in 1883, and was one of the few not bored but inspired; he bought Maxwell's book but was able to understand it only after a long struggle. From 1881 Lodge was professor of physics at University College, Liverpool (the precursor

**Fig. 5.13**   Sir Oliver Joseph Lodge (1851–1940). He made important contributions to physics, particularly to the understanding of electromagnetic radiation. In 1894 he was the first to transmit and receive a radio signal, and he made other important discoveries relating to radio.

of the University of Liverpool), and from 1900 until his retirement in 1919 he was Principal of the new University of Birmingham. He had a deep understanding of physics, and a sound judgement, so that he was constantly being called upon to resolve controversial issues. He also wrote many articles explaining science to the public. The radio talks on science that he gave in his later years led to his wide public reputation. He was knighted in 1902. Some idea of the popular esteem in which he was held is given by the fact that the vast Queen Victoria Monument in Liverpool, unveiled in 1906 (after Lodge had left Liverpool), has at its base an easily recognized sculpture of Sir Oliver Lodge, with a student beside him, representing Education. Lodge was also active in the field of psychical research, in common with Sir Arthur Conan Doyle and the scientist Sir William Crookes.

Heaviside, Fitzgerald, and Lodge were in regular communication on scientific matters for many years. Fitzgerald and Lodge were close friends and met often as well as writing to each other. The reclusive Heaviside communicated to the others often in writing, but it appears that he came face to face with Fitzgerald and Lodge only once, in 1889, when each of them separately visited his lodgings in Kentish Town to discuss Hertz's work. One wonders if Heaviside's escape to Devon later in the year was so that people could not easily drop in on him.

Oliver Lodge made many contributions both to the understanding of radiation and to the practice of radio transmission. In 1887 he carried out an experiment of great simplicity, using two Leyden jars, in which he produced an electromagnetic wave and demonstrated its transmission from one place to another. Lodge published this work in a paper entitled 'On the theory of lightning conductors', as it formed part of a wider investigation. When the paper appeared in print it contained a note added in proof, in which Lodge said that he had just seen a paper by Heinrich Hertz which described a similar but more detailed investigation along the same lines.

As early as 1879, the year of Maxwell's death, Helmholtz had recommended his student Heinrich Rudolph Hertz (1857–1894) to investigate electromagnetic radiation, and its possible transmission. At first, and for several years, Hertz could not think of any feasible way of proceeding, but was finally successful in 1887. He produced a spark in an electrical oscillator (the transmitter), and detected it by means of a coil (the receiver) a short distance away. One important aspect of his experiment is that Hertz showed that it took a finite time for the wave to travel from the transmitter to the receiver. Those who believed in action at a distance were therefore shown to be wrong, since according to them the transmission would have been instantaneous.

Heinrich Hertz (Fig. 5.14) was a remarkable man who accomplished much in his short life of 36 years. He was one-quarter Jewish, born in Hamburg into a prosperous and cultured family. During his early years he found that he had a special interest in science and technology, and at first prepared for an engineering career. Later he began to favour pure science and entered the University of Munich in 1877. In the following year he moved to Berlin, where Helmholtz, the professor of physics, exerted a profound influence on

**Fig. 5.14**   Heinrich Rudolph Hertz (1857–1894), who was the first, in 1887, to confirm Maxwell's theory of electromagnetic radiation, by transmitting radiation over a short distance and confirming that it travelled with the speed of light.

him, employing him as a salaried assistant from 1880 to 1883. During this period Hertz worked on a variety of subjects, including the evaporation of liquids, the properties of dielectrics (which Faraday had discovered), and on electric discharges.

In 1883 Hertz moved to Kiel, and in 1885 to the University of Karlsruhe. Here he found suitable apparatus to carry out the experiments on radio transmission that were to make him world famous. Fig. 5.15 shows a slightly simplified version of his own published diagram of the apparatus he set up and used in the darkened lecture hall of the University of Karlsruhe. The source of the electromagnetic radiation was the spark gap, 5 centimetres in length, between two brass knobs, each of which was attached to large zinc plates. A parabolic metal reflector focused the radiation on the receiver which consisted of two brass knobs connected by a circle or square of copper wire. Hertz found that when a spark passed across the spark gap of the transmitter he could see a spark at the receiver.

During the next few years he performed more detailed experiments which demonstrated that electric and light waves behave in a similar manner. By passing the waves through large prisms of pitch, he showed that they were diffracted in the same way as light waves. He polarized the radio waves by passing them through a grating of parallel wires, and he diffracted them by the use of a screen with a hole in it. He moved the receiver to various positions and reflected the rays from the walls of the room. He used large concave mirrors to focus the rays, and found that they would cast shadows of objects

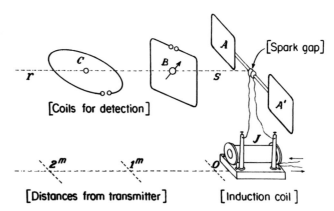

**Fig. 5.15**   Hertz's diagram, slightly modified to simplify it, of the apparatus he used to test Maxwell's theory of electromagnetic radiation, and to calculate the speed of propagation of an electromagnetic wave. Lettering in square brackets has been added. By means of an induction coil connected to the copper plates A and A' a spark was produced at the spark gap between two brass spheres. The waves were detected by means of square (B) or circular (C) loops of wire in which there were spark gaps. In later experiments Hertz showed that the electromagnetic radiation is reflected and diffracted in the same way as light.

placed in their path. From the results of these experiments he was able to calculate the wavelengths of the waves that were transmitted; they were about one metre. He could also calculate the speed of the waves, which was the same as the speed of light.

Hertz was just 31 years of age when he first obtained his confirmation of the Maxwell theory, and he at once became famous in scientific circles. From then on no scientist could reasonably deny the validity of the theory, or of Faraday's unconventional ideas on which the theory was based.

During the remaining few years of Hertz's life he thought and wrote much about Maxwell's theory, and was able to clarify many of the fundamental ideas. As soon as his paper appeared he received letters of congratulation from Lodge, Fitzgerald, and Heaviside, and he entered into an extensive correspondence with the three men. By this time they had a much deeper knowledge than Hertz of Maxwell's theory, and were able to give him much help. Lodge, whose own work had been rather eclipsed by that of Hertz, never showed any jealousy. He always admitted that Hertz's demonstration was much more convincing than his own, as Hertz had generated a continuous electromagnetic wave, while he had produced one of only short life.

One important contribution made by Lodge in his correspondence with Hertz was to convince him that he really had produced an electromagnetic wave. Not being sufficiently conversant with Maxwell's theory, Hertz had considered that he was working on the spreading of an electromagnetic force, and had not realized that the force was spread as a wave. When in 1892 Hertz collected his papers on the subject into a book he gave it the title *Arbeitung des Elektrischen Kraft*. An English translation appeared in the following year and at Lodge's suggestion it was given the title *Electric waves, being researches on the propagation of electrical action with finite velocity through space*.

That Hertz's experiments confirmed Maxwell's theory is emphasized in the following passage:

> Since the year 1861 science has been in possession of a theory which Maxwell constructed upon Faraday's views.... But as long as Maxwell's theory depended solely on the probability of its results, and not on the certainty of its hypothesis, it could not completely displace the theories which were opposed to it.... In this connection we can best characterise the object and result of our experiments by saying: The object of these experiments was to test the fundamental hypothesis of the Faraday–Maxwell theory, and the result of the experiment is to confirm the fundamental hypothesis of the theory.

It is interesting that when Hertz was asked if his experiments could be used as a basis of radiotelegraphy, he replied that it was impossible, as one would have to construct reflectors of unreasonably large size. He thought that the waves had to be focused in order for them to travel any distance, and did not realize that the invention of more sensitive receivers would render focusing unnecessary. He also pointed out that the waves travelled in straight lines, while the earth was round; he did not know about reflection from the ionosphere, which Heaviside suggested in 1902.

In 1890 Hertz was awarded the Rumford Medal of the Royal Society, and made his first and last visit to Britain. Most of his time was spent in London, where he was entertained at many lunches and dinners, and was conducted to many places of interest. He was never left alone, and in a letter to his wife mildly complained that he would have liked so much to wander around a little by himself. He also visited Cambridge where he was entertained by J J Thomson.

From about 1888 Hertz had suffered from severe head pains, which physicians could not diagnose but was probably a malignant bone condition. He had several operations, which brought only temporary relief, and much of his work was done while he was in considerable discomfort and was very depressed. He died of blood poisoning on New Year's Day 1894, in his 37th year. It is rather sad to report that he was perhaps the only famous scientist who had his biography written by his mother. This biography is of great interest, since reproduced in it are a number of letters written by Hertz to members of his family, and many passages from Hertz's diary. The book is thus partly a biography and partly an autobiography, and is a rich source of information about him.

On Hertz's death Oliver Lodge at once wrote a letter of condolence to his widow, and enquired about the possibility of borrowing some of Hertz's papers and equipment for use at a memorial lecture which Lodge was to give at the Royal Institution on 1 June 1894. He did receive from her a number of papers, but the equipment could not be made available (later it was retrieved and is now in the Deutsches Museum in Munich). Lodge's lecture, later published, gave an admirable account of Hertz's great contributions, and characteristically Lodge minimized his own contributions in the same field.

In honour of Hertz the decision was made to use his name as the standard of frequency; 1 hertz (1 Hz) is the same as 1 reciprocal second (1 s$^{-1}$). The unit Hz must be used for frequency only in the sense of cycles per second; the unit s$^{-1}$ used in any other sense must not be written as Hz (e.g., the unit metre per second, m s$^{-1}$, for speed must not be written as m Hz).

During the Nazi regime a curious position was taken by the German authorities with regard to Hertz's achievements. Previously he had been remembered with great honour, in Germany as well as everywhere else. The Nazis, however, could not contain their frustration at the fact that he was one-quarter Jewish. Particularly concerned was Philipp Lenard, winner of the 1905 Nobel Prize for physics, who was pathologically anti-Semitic. In his early days Lenard had been a research assistant to Hertz, had a high regard for his work, and even edited the published volume of his papers. Most people in his position would have felt that it was impossible to denounce

Hertz, but Lenard managed to do so in a way that must evince a grudging admiration for its effrontery. In 1943 Lenard published a statement that Hertz had somehow been able to suppress his Semitism until 1888, when he did his great work, but that Semitism emerged later and that then his work on Maxwell's theory took a turn for the worse. There is a delightful irony here; we know now from correspondence that has come to light that Hertz's change of attitude to the theory was in fact due to his correspondence with Heaviside, Fitzgerald, and Lodge, who were as 'Aryan' as could be.

On account of their Jewish blood, Hertz's widow and their two daughters felt it wise to leave Germany in 1936, and they lived in England to the end of their lives. This had the advantage for Britain that some of his papers were donated to the Science Museum in London. Hertz's nephew Gustav Hertz (1887–1975), who shared the 1925 Nobel Prize for physics with James Franck, was obliged by the Nazis to vacate his professorship in Berlin in 1935, and then went to the Siemens Research Laboratory; after a period in the USSR he returned to East Germany as director of the Institüt für Physik in Leipzig.

## THE ELECTROMAGNETIC SPECTRUM

When Maxwell and Hertz did their work, not much more than the visible spectrum was known. We saw in Chapter 3 that in 1800 William Herschel discovered, on account of its heating effect, that there was some invisible radiation beyond the red, and this was known as the infrared. In the following year Johann Wilhelm Ritter discovered ultraviolet radiation, from the fact that it blackened silver chloride. There was, therefore, before Hertz did his experiments, knowledge of just a little more than the visible spectrum. The frequency range was hardly more than a factor of two.

Even Hertz himself did not realize right away that what he has done was to extend greatly the electromagnetic spectrum; it was Oliver Lodge who first pointed this out. Fig. 5.16 shows a modern version of the electromagnetic spectrum. It includes several radiations unknown to Hertz; for example, X-rays were discovered in 1895 by the German physicist Wilhelm Konrad Röntgen (1845–1923). Gamma rays, which are emitted by radioactive substances, were discovered in 1900 by the French physicist Paul Ulrich Villard (1860–1934).

The figure shows the wavelengths and frequencies associated with the various types of radiation. Visible light covers only a small fraction of the spectrum known today. Red light has a wavelength of about 700 nanometres ($7 \times 10^{-7}$ m). The frequency corresponding to it is about 400 million million Hertz (400 THz). Violet radiation has a wavelength of about 400 nanometres, or roughly half that of red radiation; the frequency is roughly twice that of red radiation.

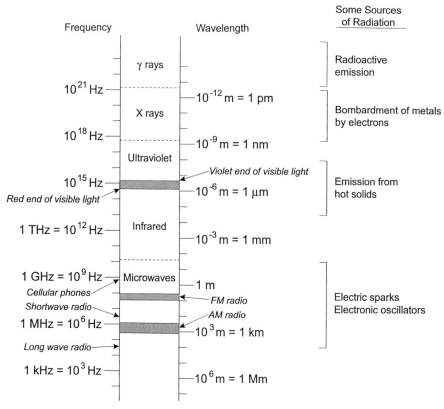

**Fig. 5.16** The electromagnetic spectrum, showing the types of radiation known today, and some of the ways in which different kinds of radiation are produced. The following prefixes are commonly used with wavelengths and frequencies:

tera, T = 1 million million ($10^{12}$)
giga, G = 100 million ($10^9$)
mega, M = 1 million ($10^6$)
kilo, k = 1000 ($10^3$)
milli, m = one thousandth ($10^{-3}$)
micro, $\mu$ = one millionth ($10^{-6}$)
nano, n = one thousand millionth ($10^{-9}$)
pico, p = one million millionth ($10^{-12}$)

What is remarkable about Hertz's contribution was that he had discovered radiation having wavelengths of a few metres, which is about a million ($10^6$) times greater than previously observed. The frequencies are therefore a million times smaller.

We should note a few important general characteristics of different types of radiation. High-frequency radiation (at the top of Fig. 5.16) is high-energy radiation, which penetrates matter more easily, and brings about chemical change more readily. For example, exposure of the skin to ultraviolet radiation, X-rays, or gamma rays, can have undesirable effects—the more so the greater the energy. At the other end of the spectrum, low-energy rays such as radio waves cause no ill effects when they pass through our bodies, which as a result of radio and television broadcasting they are doing all the time.

## OLIVER LODGE AND THE INVENTION OF RADIO

Maxwell wrote his first paper on electromagnetic radiation in 1861, but it was another 50 years before the first commercially profitable radio signal was transmitted. The Marconi Company, the first successful radio company, paid its first dividends in 1911. This delay emphasizes that much had to be done to turn the pure science into a practical commercial enterprise. Even after Hertz had transmitted the first signal more than 20 years had to elapse before radio became profitable.

There is wide confusion among the public, and even among modern scientist and engineers, as to who invented radio. Almost everyone will say that it was Marconi, and perhaps will add that he made the first radio transmission in 1896. However, two years before that (when Marconi was only 20 and knew nothing about radio) Oliver Lodge had given a public demonstration of radio to a large audience in Oxford. Five years previously he had made demonstrations of the same kind; he had written about them and they were seen by others. All of Marconi's work in fact was a repetition of what Lodge and others had already done. Marconi's contribution to radio, to be outlined later, was considerable, but his innovations were commercial rather than scientific.

This is a little like that involving the invention of the telegraph, which we discussed in Chapter 4. Many people think that Samuel Morse invented it, but in fact all his ideas came from Joseph Henry. The difference is that while there is ample evidence that Morse was given much help by Henry, Marconi always denied that he knew anything about what Lodge had done—although there is evidence to the contrary.

Lodge's first demonstration of radio transmission was at a Royal Institution lecture on 8 March 1889. This lecture attracted a large audience, and Lodge emphasized the great importance of what he called syntony, by which he meant the appropriate tuning of the receiver to receive the signal. Lodge gave a second major lecture to the Institution of Electrical Engineers on 25 April 1889, and this also attracted much attention. Two features of this demonstration are of particular interest. It was noticed that the waves transmitted by Lodge were being picked up by the telephone system in the building. Also, the lecture was dramatically interrupted by the caretaker of the building, who rushed in white-faced saying that there was sparking between the gas and water pipes in the basement. Lodge, of course, was delighted to hear that his waves were being picked up at that distance; after reassuring the caretaker that there was nothing to worry about he invited Sir William Thomson (Kelvin) who was in the audience to accompany him to the basement and see the phenomenon.

One result of these demonstrations was that Lodge noticed that a couple of little metal knobs, initially in light contact with each other, cohered (became stuck together) when a tiny spark was produced by a radio wave. Previously the same sort of effect had been discovered at the Catholic University in Paris by the physicist Edouard Branly (1844–1940). Branly found by accident that if metal filings were loosely packed in a glass tube, and sparks were produced in the vicinity, the electrical resistance of the tube was greatly reduced. When the tube was shaken its resistance resumed its original value. Several people had suggested that this effect might be used to detect radio waves. Lodge, in particular, decided to experiment with tubes of that kind, to which he gave the name coherers, and he make improvements to their effectiveness.

Lodge's most famous demonstration of radio transmission was made on 14 August 1894 at a meeting in Oxford of the British Association for the Advancement of Science. His transmitter, consisting of an electrical oscillator activated by an induction coil, was set up in the Clarendon (physics) Laboratory. The receiver consisted of an iron-filings coherer connected to a sensitive galvanometer and also to a 'Morse inker' which recorded the signals on paper tape. The demonstration was made at the Oxford Museum (Fig. 5.17), where Lodge gave the lecture. (As we will see in Chapter 7, this building was already famous as the site of the confrontation between Bishop Wilberforce and Thomas Huxley on Darwin's theory of evolution). The transmitter and receiver were 180 feet apart and there were two stone walls between them.

Lodge's assistant in the Clarendon Laboratory tapped out Morse code letters, and these were clearly observed by the audience in the lecture theatre. The large audience included many distinguished people, some of whom are well known in the history of radio. Present were Lord Rayleigh, Ambrose Fleming (later to invent the diode, Chapter 6), and Campbell Swinton, an electrical engineer of great distinction; he immediately repeated the experiments in his own office. An important foreign visitor was the great Austrian physicist Ludwig Boltzmann, already famous for his work on the distribution of energy (Chapter 2). Boltzmann was a sensitive man, and after the lecture he took Lodge to task for having referred to Hertz as 'no ordinary German'. The remark seems inoffensive—after all, no ordinary person of any nationality would have accomplished what Hertz did—but Boltzmann apparently thought it insular and patronizing. Lodge at once apologized.

One would have thought that this 1894 demonstration would have made it evident that Marconi did not invent radio in 1896. Lodge had made his demonstration at an important meeting, before many distinguished people, at a university which is not entirely unknown. The demonstration was reported in *Nature*, in *The Electrician*, a journal widely read by engineers, and in the *Proceedings of the Royal Institution*.

**Fig. 5.17**   The Oxford University Museum, from a steel engraving by J. H. Le Keux. This engraving was first published in the *Oxford Almanac* for 1860. The museum is famous for several important events. In 1894 it was the scene of the first public demonstration of radio anywhere in the world. Earlier, in 1860, it was the scene of the famous confrontation between Bishop Wilberforce and Thomas Huxley, on the subject of Darwin's theory of evolution (Chapter 7).

Lodge did not at once take out a patent on radio, and this was perhaps a tactical error on his part, as it led to much trouble. His initial attitude was that he was a scientist and not an engineer, and he felt that publication of his work was enough. His attitude changed later when he realized that others were making use of his ideas without giving him the credit. He took out his first patent in 1897, after Marconi had taken out his patent but before Marconi's system had been revealed. Lodge's position was later completely vindicated in the British and US courts, but in spite of this, modern accounts of radio often overlook Lodge's contributions and his priority. Even Abraham Pais, in his excellent *Inward bound* which gets nearly everything exactly right, mentions Marconi as transmitting signals without wires in 1896; Lodge is mentioned three times, but only rather incidentally.

## COMMERCIAL RADIO TRANSMISSION

Guglielmo Marconi (1874–1937) undoubtedly made important contributions to making radio commercially possible, but he did little if anything that was

scientifically or technically new. He was born in Bologna of a well-to-do
Italian father and a Scots–Irish mother (née Annie Jameson) who was 'well-
connected' in the sense that she was related extensively to the families that
made Jameson Irish Whiskey and Haig and Ballantine Scotch Whisky. He
showed little early promise, failing the examinations for entrance to the
Italian Navy and the University of Bologna.

In 1894, the year of Lodge's demonstration in Oxford, Marconi became
fascinated by Hertz's investigations on the transmission of radio waves, and
after obtaining minimal technical training he repeated some of Hertz's experiments. He was always secretive about the exact nature of his early radio
experiments, but it appears that he used a coherer that was not essentially
different from that used by Branly and later by Lodge. He also earthed
(grounded) his transmitter and receiver, and used antennas connected to both
transmitter and receiver. After achieving some success in transmitting radio
signals over a distance of one or two kilometres, he approached the Italian
government for financial support, which was denied. In February 1896 he
moved to England, where he received considerable moral support from
William Henry Preece, who in 1888 had become engineer-in-chief of the
British Post Office (which in Britain was in charge of telegraphy as well as the
mail service). Preece was a powerful figure in electrical engineering circles,
and was later knighted for his services. However, he had little knowledge of
mathematics or physics; his understanding of electricity was limited to the
simplest laws relating to direct current, and he seemed to be unaware of self-
induction, which as we saw in Chapter 4 was discovered in about 1830 by
Joseph Henry. Worse still, Preece was openly contemptuous of mathematics
and electrical theory, regarding himself as a practical man who could get on
better without such frills.

For many years Preece had tangled with the Maxwellians. Heaviside was
particularly contempuous of him; being such a recluse Heaviside probably
never met Preece, but in published papers and in his letters to Lodge and
Fitzgerald he expressed himself in powerful and colourful language. He even
invented a number of offensive appelations for Preece, such as 'man of brass'
(apparently a play on British Association, or British Ass), 'the bouncer',
'Mr Priggs', and 'the unscientific speculator'. There had also been bitter
confrontations between Lodge and Preece at scientific meetings. After one of
them Lord Rayleigh commented that Preece seemed to use the word mathematician as a term of abuse.

In siding with Marconi Preece was no doubt getting his revenge on the
physicists, particularly on Lodge. Preece did have the virtue of being enthusiastic about the possible development of radio, but with his limited scientific
knowledge he would have found it embarrassing to work with such an
eminent scientist as Lodge. Marconi, whose understanding of physics was if
possible inferior to that of Preece, and who was much less experienced on the
technical side, was a much more satisfactory ally. Also, Marconi with his

wealthy relatives would require no financial aid from the Post Office; the same could not be said of Lodge.

At the time there was no general enthusiasm for the development of radio. Many people thought that there was no need for it, since the land masses of the earth were by this time covered by telegraph lines, and telegraph cables were under all the oceans. The telegraph and telephone were being used only for communication, with no plans for extending it to public entertainment, which would have been technically possible. Radio was first considered only as a substitute for the telegraph, with no thought of its use in entertainment, which is its major use today. What tipped the scales in favour of the development of radio was that the telegraph had one serious limitation: cables could not connect ships to shore. The fact that radio was developed principally in Britain is undoubtedly because Britain was then the largest maritime nation in the world, half of the world's ships being registered in Britain. It was partly for this reason that Marconi had gone to Britain. Besides getting support from Preece he got useful cooperation from the Royal Navy, especially from Henry Jackson, who later rose to high rank in the Navy.

Marconi took out a British patent, his first patent, in 1896, the year of his arrival in England. The details of it were kept secret, and Lodge and other scientists were very concerned about it. As far as they were able to discover, Marconi's system was essentially the same as that demonstrated by Lodge in Oxford two years previously. By British law a patent cannot be taken out on anything that has been publicly demonstrated, so that Marconi's patent was almost certainly irregular. Marconi dealt with criticisms by insisting that his system did not use Hertzian or electromagnetic waves at all, but 'Marconiwaves'. These, he said, have different characteristics; Herzian waves spread out from their source, but Marconiwaves could be directed from any point to any other on earth, and would pass through any mountains that were in the way. The suggestion that Marconiwaves were any different from Hertzian waves is, of course, absolute nonsense.

In July 1897, with the financial help of his mother's family and friends, Marconi formed the Wireless Telegraph and Signal Company, which was usually known as the Marconi Company. He succeeded in transmitting a radio signal across the English Channel in 1899. By this time Marconi was using wavelengths that were a hundred or a thousand times longer than the 4 to 5 metres used by Hertz. Soon afterwards, radio transmitters were installed in lighthouses in various parts of England.

After obtaining his most important patent in 1900, his 'four sevens' patent (it was patent number 7777), Marconi was able to make rapid progress. One good result did come from Marconi's ignorance of the nature of electromagnetic waves and his belief that his Marconiwaves behaved in a special way. The physicists knew that the waves had to travel in straight lines, and concluded that transmission over long distances would be impossible because of the curvature of the earth. Marconi was perfectly prepared to try the 'impossible', and it worked. On 12 December 1901 his assistant sent Morse signals across the Atlantic Ocean, from Cornwall, England, to St John's, Newfoundland, where Marconi himself received them. This unexpected behaviour led the physicists to think again. In 1902 Heaviside, and independently the American physicist Arthur Edwin Kennelly (1861–1939), postulated the existence of an ionized layer in the upper atmosphere which would reflect the waves and allow them to travel great distances over the earth's surface.

Marconi shared the 1909 Nobel Prize for physics for his work, and this created something of a sensation among the leading physicists of the time. Nobel Prize decisions are usually reasonable, but it is hard to justify this selection. Not only had Marconi made no innovations in physics and few if any in technology: he understood very little physics. His achievements were entirely in commercialization. He later developed short-wave radio equipment, and developed a world-wide radio telegraph network for the British government. Marconi's allegiance was always with his native country of Italy, to which he finally returned, and his later activities were on behalf of the Italian government. In 1929 he was created a Marchese and an Italian Senator. He later became a strong supporter of Mussolini, and his death two years before the outbreak of World War II spared him the embarrassment of becoming an enemy of Britain which had given him so much support. When he died in 1937 he received the unique tribute of a two-minute silence by all radio stations throughout the world.

Marconi's commercial successes naturally caused annoyance to Oliver Lodge, as he knew that Marconi was stealing all his ideas. To give himself some protection he took out a British patent in 1897; it included a special tuning system, which he called syntony, as well as his other innovations. In the following year he took out a corresponding US patent. In 1911 Lodge and Alexander Muirhead, a highly competent electrical engineer, set up the Lodge–Muirhead Syndicate, which began to manufacture and market radio equipment. This led to legal action between them and the Marconi Company. In the end there was an out-of court settlement; it was acknowledged that Marconi had used Lodge's techniques, and Lodge and Muirhead received a substantial amount of cash, in return for which they undertook not to continue to manufacture radio equipment. This, of course, was what Lodge wanted; he had not wished to go into business, but objected to others profiting from his work. An interesting sequel to all the controversies was that in 1943, three years after Lodge's death and six years after Marconi's, the US Supreme Court ruled that the only valid patent held by the Marconi Company was Lodge's US tuning patent of 1898, which they had acquired from Lodge by the cash settlement of 1911. On scientific and technical matters, therefore, Lodge was completely vindicated.

In Chapter 4 we saw that a murder helped to make telegraphy popular in Britain, and the same is true of radio. In 1910 Dr Hawley Harvey Crippen murdered his wife in London by poisoning her with hyoscine, and buried her body beneath the cellar of his house. When the remains were discovered, Crippen and his young mistress Ethel Le Neve were on a ship, the *Montrose*, sailing across the Atlantic. She was inadequately disguised as a boy; the two were travelling as a Mr Robinson and his son, and had already attracted the attention of other passengers on account of their odd behaviour towards one another. The ship's captain became convinced of their true identity when he saw that Le Neve's suit did not fit her, and had to be adjusted with safely pins; also, he noticed that she 'squeezed Crippen's hand immoderately'. He sent a Marconigram, as a cable was then called, to the British authorities, and received his instructions in reply. The captain reported that Crippen greatly enjoyed watching Marconigrams being sent out and received, and remarked 'What a wonderful invention it is'.

When the ship reached Father Point, the customs station for Quebec, it was boarded by Inspector Dew and a sergeant from Scotland Yard; they had travelled by a faster ship, and Dew was disguised as a pilot. After extradition proceedings in Quebec the two were returned to London, where they were tried at the Old Bailey. Crippen was found guilty and hanged. Le Neve was tried separately for being an accessory after the fact, and was found not guilty. The trial had a sequel that made legal history. Crippen had left his estate to Le Neve, but it was denied her. His estate had been in his wife's name, and since he murdered her he could not profit from it; he had therefore nothing to leave.

Credit for the first transmission of speech is due to Reginald Aubrey Fessenden (1866–1932) who was born and educated in Canada. He applied for a teaching position at McGill University in Montreal, but was turned down, and as a result did most of his work in the United States. (Canada, in the nineteenth century, had a rather bad record for turning down good people in favour of scientific nonentities. In 1851 Thomas Huxley was turned down by the University of Toronto in spite of having been strongly recommended by Charles Darwin. Later Toronto also turned down the eminent physicist John Tyndall, in spite of his strong support from Michael Faraday!)

After working with Thomas Edison, Fessenden held academic appointments in the US, and worked mainly on radio communication. He invented the technique of amplitude modulation (AM), and on Christmas Eve, 1906, he spoke by radio from Brant Rock, Massachusetts, to a banana fleet of the United Fruit Company. The ships had purchased their radio equipment from Fessenden, but had not been told that there was anything special about it. The operators of the ship-to-shore telegraph systems had been told to expect a special broadcast on Christmas Eve, and had assumed that they would receive the usual dots and dashes of the Morse code. Instead they heard to their amazement a short voice announcement, followed by an Edison–Bell recording of Handel's Largo. Some carols were then sung, and then Fessenden himself scraped out Oh Holy Night on his violin. He sang out the last verse, and then said 'If anyone hears me, please write to me at Brant Rock.'

Fessenden was thus, among other things, the first disc jockey. Earlier Edison had told him that speech over the radio was as likely as a man jumping over the moon. In 1900, Fessenden had sent a voice message to an assistant one mile away, saying 'Is it snowing where you are, Mr Thiessen?' During his career he took out over 500 patents, and many of his ideas were adopted without his consent during the First World War. In 1928 the US Radio Trust paid Fessenden $2.5 million in compensation.

Radio was later transformed by the introduction of electronic techniques, a subject that is touched on in the next chapter.

Since radio communication was made possible by Maxwell's theory of radiation, it is appropriate that an important radiotelescope has been named after him. The James Clerk Maxwell Telescope (Fig. 5.18), situated on the top of a mountain in Hawaii, is operated by the Royal Observatory of Edinburgh and was opened in 1987. It is concerned with radiation in the microwave region of the spectrum, which is between the infrared and the commercial radio regions. By means of this telescope, information is gathered from regions of outer space vastly more distant than the planet Saturn, which had been the subject of Maxwell's early research.

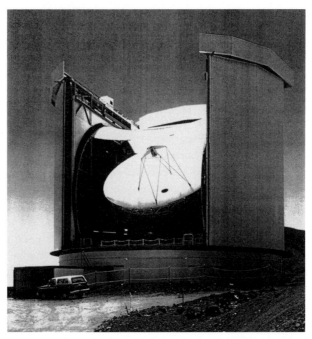

**Fig. 5.18**   The James Clerk Maxwell telescope, which is on the top of a mountain in Hawaii. Opened by the Duke of Edinburgh in 1987, it is the largest telescope in the world designed to explore outer space by the use of radiation in the region of the spectrum between the infrared and radio waves.

# SIX

# *J J Thomson and the electronic age*

The discovery of the electron, a negatively charged elementary particle that is much lighter than any atom, had a great impact on both science and technology. Electrons are important constituents of atoms, and their arrangement within atoms controls how atoms combine together to form molecules. Electrons are important in many other ways. The conduction of electricity through many substances turns out to be a flow of electrons, and the whole electrical industry was greatly affected when this was realized. A new branch of electrical engineering, known as electronics, has grown up since the first decade of the twentieth century; it is concerned specifically with the control of the movement of electrons. In particular, the radio industry was transformed by the development of devices such as diodes and triodes which involve the flow of electrons. The adjective electronic was in common use by the 1920s, and at the same time the noun electronics began to be used. In April 1930 the word gained wide recognition by the founding, by the McGraw-Hill Publishing Company of New York, of the periodical *Electronics.*

Two lines of evidence led to the realization that electrons exist. One line sprang from Faraday's laws of electrolysis, which we discussed in Chapter 4. Another, more direct, came from investigations of electric discharges in gases. Perhaps the most important of these investigations were those of J J Thomson and his associates at Cambridge University. Their work provided the first measure of the charge and the mass of an electron, and had a powerful effect on the progress of both pure science and technology. Before considering their work we must say something about the earlier investigations on which it was based.

## ELECTRICITY AND MATTER

As we saw in Chapter 4, Faraday's laws of electrolysis were important in suggesting that electricity is not continuous but consists of particles. Faraday himself was always a little lukewarm about accepting the idea of atoms and molecules. To him, experiments were all-important, and he retained a scepticism about theories. Similarly, Maxwell speculated about the possibility of a particle of electricity, but never with much enthusiasm.

The most specific idea about a particle of electricity was advanced in 1874 by George Johnstone Stoney (1826–1911) at a meeting of the British Association. Stoney, who was born in County Dublin, Ireland, was at the time Secretary of Queen's University, Dublin. He deduced from the value of the Faraday constant and other data that the ratio of the charge of the particle to the mass of a hydrogen atom is about 100 000 coulombs per gram ($10^5$ C $g^{-1}$). We will see later, in connection with J J Thomson's work on the ratio of the charge to the mass of an electron, that Stoney's result was of great importance. Later, as we saw in Chapter 4, Helmholtz in 1881 presented in his Faraday Memorial Lecture another clear statement of the idea of electrical particles.

In 1888 Johnstone Stoney made a particularly interesting and original contribution—but since it was ahead of its time people did not then pay much attention to it. He suggested that the negative particles move in orbits within the atoms. He tried to develop this idea into detailed explanations of the spectra of atoms and molecules, but in the absence of quantum theory (Chapter 8) these were inevitably in disagreement with experimental findings. This is partly why his idea did not receive much attention, although we now know that it was quite correct. It was Stoney who first suggested, in 1891, that the unit of negative electricity should be called an electron.

Stoney's idea that spectra have an electrical origin did later begin to be accepted. Thus at the 1893 meeting of the British Association Oliver Lodge (1851–1940) expressed the opinion that 'radiation is due to the motion of electrified parts of molecules, not to the molecules as a whole.'

## MUCH ADO ABOUT NOTHING

The lines of reasoning outlined so far, based on the laws of electrolysis, did not lead to the isolation of the electron and the measurements of its properties. Those important advances came by a different route, by the study of electric discharges in gases. Electric discharges are familiar to everyone today in the form of fluorescent lamps, many varieties of which are used in public places for advertising. The production of a discharge depends considerably on vacuum technology, which is the attempt to produce a vacuum in a vessel. It is impossible to pump absolutely all the air out of a vessel and create a perfect vacuum; what we do is to produce as low a pressure as we can. To produce a discharge at all the pressure must be low, and for some modern techniques the pressure must be very low indeed.

For many years there had been interest in the vacuum—or 'nothingness' as T S Eliot called it in his verse-play *The elder statesman*. At about the beginning of the seventeenth century pumps began to be used for the draining of mines, and it was noticed that it was impossible to pump water from a depth of more than about ten metres (34 feet). Galileo was called upon to explain this, and he gave the incorrect explanation that the reason for the height limit was that a longer column of water would break up under its own weight. One of Galileo's students, Evangelista Torricelli (1608–1647), had the idea of experimenting with mercury instead of water; the density of mercury is about fourteen times that of water. In 1643 Torricelli performed an experiment in Florence which is often repeated today. He filled with mercury a tube, about a metre long and sealed at one end. With his finger over the open end he upturned the tube into a dish of mercury. The level of the mercury fell, leaving a void above it, and in this way he created a vacuum. This apparatus was the first barometer, and Torricelli noticed that the level of the mercury varied slightly from day to day, and correctly explained this as the result of changes in the pressure of the atmosphere. His hypothesis was that the column of mercury was supported by the weight of the air.

In 1646 news of this experiment reached the young Frenchman Blaise Pascal (1623–1662) who was working in the inland revenue department in Rouen. He at once began an important series of investigations in which he constructed barometers of water, mercury, and even red wine. He also persuaded his brother-in-law to take a Torricellian barometer to the top of the Puy de Dôme to test whether the level of the mercury fell with increasing height. The experiment was successful, and was repeated by taking the barometer to various other high places. Barometers then became used as altimeters for measuring the heights of mountains.

These experiments soon became well known, and there was great interest in making pumps that would remove air from a vessel and produce something like a vacuum. As far as is known the first successful pump was constructed in 1654 by Otto von Guericke (1602–1686), who was an engineer and mayor of Magdeburg. He arranged for the construction of two heavy metal hemispheres fitted with greased gaskets and then joined together to make a sphere. Fig. 6.1 shows one of the several engravings that were made to depict a demonstration that was performed in the presence of the German emperor Ferdinand III. As much air as possible was pumped out of the sphere, and then two teams of eight horses were given the task of separating them. They were unable to do so until air had been allowed to enter the sphere. It must be said that there is something rather fishy as well as horsy about this demonstration, because if we estimate the diameter of the sphere and the force required to separate the two hemispheres, we conclude that the horses should have had little difficulty. Perhaps, to impress the emperor, they had been trained to pretend that they were pulling harder than they were.

In 1658 Robert Hooke (1635–1703) constructed at Robert Boyle's request an air pump that was a great improvement over the one used by von Guericke. At that time Robert Boyle (1627–1691) was working at a house he rented on High Street, Oxford, and was conducting experiments on a variety of scientific problems, including the properties of gases. In 1653 Hooke became an undergraduate at Christ Church, Oxford, and he was hired by

**Fig. 6.1** An engraving showing the famous experiment of Otto von Guericke, carried out in Magdeburg in 1654.

Boyle in about 1655 as a research assistant, holding that position until about 1662. It was with the aid of this pump that Boyle and Hooke confirmed the relationship that came to be called Boyle's law. According to this law, at a given temperature the pressure of a gas multiplied by its volume is a constant; if we double the pressure, we halve the volume. Later Hooke designed an improved air pump for Boyle, which Boyle used after he had transferred his laboratory to London in 1668.

Hooke was an experimentalist of great skill, and he made many original contributions to science. His name would be better known today if it had not been for Newton, who was intensely jealous of him and did all he could to discredit him. Hooke was a distinguished microscopist, and his book *Micrographia*, published in 1665, contained remarkable drawings of his microscopical observations. For many years, from 1661 to 1703, Hooke was curator of experiments to the Royal Society, and he had the onerous task of performing an original experiment or demonstration at each of the Society's weekly meetings. Many of these involved pumping the air out of vessels to approach a vacuum. This had become a matter of great interest not only to Hooke but to other Fellows of the Society. However, it incurred the irritation of King Charles II, the Founder and Patron of the Society who showed great interest in its work, although he made no financial contribution to it, the Fellows having to support it by their dues (which too often were in arrears). King Charles could not understand how pumping air out of a bottle so that there was nothing inside it could possibly be a useful thing to do, and he expostulated

These gentlemen spend their days debating nothing.

His Majesty did not realize that important conclusions regarding gases could be obtained only by the use of vacuum pumps. Nor could he have known of the remarkable future advances that would be made using vacuum techniques, including the discovery of the electron and of other charged particles.

For two centuries or so there were few improvements to air pumps. Mechanical pumps not much better than those designed by Hooke continued to be used, and the pressures were never very low. Today we often express pressure in torr, named in honour of Torricelli; 1 torr is one millimetre of mercury, and the pressure of the atmosphere at sea level is never far from 760 torr. Pumps of the kind made by Hooke probably gave pressures not much lower than about 10 torr.

An important advance was made in 1855 by Heinrich Geissler (1814–1879), who was a manufacturer of scientific instruments in Bonn. Geissler took advantage of Torricelli's famous experiment and designed a mercury pump which had no moving mechanical parts. A column of mercury was moved up and down, the vacuum above the mercury being used to suck the air out of a vessel step by step; after some time the pressure in the vessel was close to that above the mercury. By the use of such pumps he was able to reach pressures that were lower than could previously be achieved. Their effectiveness was limited only by the vapour pressure of mercury, which is about a thousandth $(10^{-3})$ of a torr. A little later, in 1862, August Töpler (1836–1912) designed a similar but somewhat more convenient pump, and Töpler pumps continue to be used at the present day.

Subsequently there have been great advances in vacuum technology. By avoiding liquids such as mercury and oil, which contaminate the vessel with their vapours, it is now possible to reach pressures as low as a million millionth $(10^{-12})$ of a torr.

## ELECTRIC DISCHARGE TUBES

An electric discharge tube (Fig. 6.2) consists of a tube into which two electrodes are sealed. Air is pumped out of the tube until the pressure is as low as possible. If the electrodes are connected to an electrical source of sufficiently high voltage, a luminescent discharge is observed. The colour of the discharge can be changed by admitting various gases at low pressures, as is done in many advertising displays. Electric discharges were investigated for several

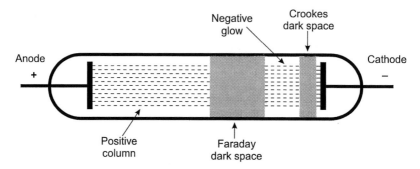

**Fig. 6.2**   An electric discharge in an evacuated tube. The light coming from such a tube varies from place to place, and various regions have been named.

decades before it was realized that there is a flow of electrons from one electrode to the other. At first discharges were thought to involve only light, but it was soon realized that there is more besides.

Among the early studies of discharges, carried out with mechanical pumps, were those of Humphry Davy in 1822 and of Michael Faraday in 1838. Faraday found that the glowing column of gas in the tube is separated from the negative electrode by a dark space, now known as the Faraday dark space (Fig. 6.2). Although Faraday could not explain these results he correctly predicted that experiments of this kind 'will have a far greater effect on the philosophy of electrical science than we at present imagine'. In 1852 William Grove described in some detail the influence of the pressure of the gas on the nature of the discharge. Further advances, however, could be made only when lower pressures could be obtained.

Mention was made of the fact that in 1855 and 1862 Geissler and Töpler introduced their mercury pumps and obtained much lower pressures and more satisfactory discharges; discharge tubes are still often called *Geissler tubes*. In 1858 Geissler's friend Julius Plücker (1801–1868), who was professor of mathematics at the University of Bonn, made a detailed study of the discharges in a Geissler tube. Plücker's most important contributions were in analytical and projective geometry, and his unlikely but effective invasion of experimental physics was no doubt due to Geissler's influence. His most important contribution was to report that the discharge shifted its position when an electromagnet was brought near to it. He also examined the effects of pressure on the discharges. Plücker's papers are far from lucid, but make entertaining reading.

Important contributions were also made by Johann Wilhelm Hittorf (1824–1914), who had been Plücker's student and was professor of physics at the University of Münster. Earlier he had made important contributions to the understanding of the electrolysis of solutions, and in 1869 he studied electrical discharges in gases. He made the important observation that if an object is placed in front of the negative electrode it casts a shadow, proving that the rays travel in straight lines. Hittorf referred to these rays as Glimmerstrahlen (glow rays).

Similar investigations were begun in 1876 by Eugen Goldstein (1850–1930) in a laboratory at the Potsdam observatory. Goldstein began to use the term cathode rays for the rays emitted by the negative electrode, which is called the cathode. In 1886, during discharge-tube experiments in which the cathode was perforated, he observed faint streamers of light behind the cathode, and found that these were rays travelling in the opposite direction. He referred to them as Kanalstrahlen (canal rays), and in 1898 Wilhelm Wien (1864–1928) showed that they too were deflected by a magnetic field. We shall see that these rays, consisting of positively-charged particles, later played a prominent part in the investigations of the structures of atoms.

In the 1870s William Crookes (1832–1919), one of the most colourful figures in the history of science, began a more detailed investigation of electric discharges. He confirmed Hittorf's discovery that the beam emitted by the cathode in a discharge tube is deflected in a magnetic field (Fig. 6.3), and concluded that the rays consist of particles each carrying a negative electric charge. However, many scientists were unable to confirm this result, which became a matter of controversy. It later became clear that the people who had failed to observe the deflection by a magnet had not obtained sufficiently low pressures in their vessels.

## CROOKES'S RADIOMETER

At this point it is convenient to digress a little to consider some work that Crookes had done just before his discharge tube work. In 1873, in the course of investigating the atomic weight of thallium, an element he had discovered spectroscopically a few years earlier, Crookes was led to construct what

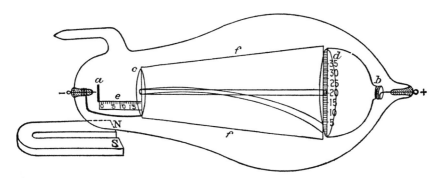

**Fig. 6.3**   Crookes's diagram of the cathode-ray tube he used to show that the beam is deflected in a magnetic field.

became called a 'light-mill' or radiometer, which later became a popular toy. This is an eye-catching device that window-shoppers often notice at the stores of opticians and jewellers. A radiometer consists of an evacuated bulb inside which four vanes are mounted on spokes (see Fig. 6.4(a)). Each vane is silvered on one side and blackened on the other. The vanes spin around in the direction that suggests that there is a greater force on the blackened sides than on the silvered sides.

The device at once attracted attention, not only from scientists but from the general public. Queen Victoria was amused; she tended to be fascinated by scientific 'toys' such as the radiometer and the kaleidoscope, which Sir David Brewster (1781–1868) had invented in 1816. In 1876 she invited some scientists to come to Buckingham Palace and explain it to her. One of them was Maxwell, then Cavendish Professor in Cambridge, and his letter to his uncle describing the visit is worth quoting as it provides a good example of his boyish humour:

> ...I was sent for to London, to be ready to explain to the Queen why Otto von Guericke [of 'Magdeburg hemispheres' fame] devoted himself to the discovery of nothing, and to show her the two hemispheres in which he kept it, and the picture of the 16 horses who could not separate the hemispheres, and how after 200 years W. Crookes has come much nearer to nothing and has sealed it up in a glass globe for public inspection. Her Majesty however let us off very easily and did not make much ado about nothing, as she had much heavy work cut out for her all the rest of the day....

When Crookes first announced his radiometer most scientists accepted his explanation that the effect was the result of the pressure of light. Maxwell refereed Crookes's paper, and accepted that explanation, although he commented that his own theory of electromagnetic radiation predicted a light pressure much smaller than seemed to be involved in the radiometer; Maxwell was always willing to admit that he might be wrong. Ideas about light pressure were much discussed at the time, and the fact that the tails of comets always point away from the sun had been explained as due to the light pressure (the modern explanation is in terms of the 'solar wind', due to the movement of charged particles). There is a story that during a discussion of comets in Maxwell's house the word 'tail' was used so often that Maxwell's terrier created much amusement by running around in circles chasing his tail. When Maxwell noticed that the dog was exhausting himself by this running around he controlled the situation by applying his extensive knowledge of mechanics. The way he did this was explained in 1931 by his biographer and former assistant William Garnett in an article in *Nature* commemorating the centenary of Maxwell's birth:

> Maxwell's way of dealing with the case was, by a movement of the hand, to induce the dog to revolve in the opposite direction, and after a few turns to reverse him again, and to continue these reversals, reducing the number of revolutions for each,

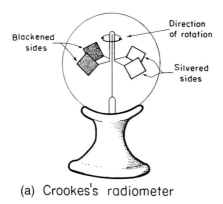

(a) Crookes's radiometer

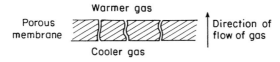

(b) Thermal transpiration

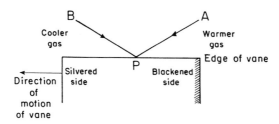

(c) Explanation of the radiometer

**Fig. 6.4** The radiometer designed by William Crookes. (a) Diagram of the instrument. (b) Thermal transpiration, which was first explained by Osborne Reynolds in 1879. (c) The explanation of the movement of the vanes of the radiometer, in terms of the force on the edges.

until like a balance wheel on a hair spring with the maintaining power withdrawn, by slow decaying oscillations the body came to rest.

However, it soon became evident that light pressure could not explain the spinning in the radiometer. For one thing, the light ought to produce a greater force on the silver side, which reflects it, than on the blackened side, which absorbs more of it. In fact the vanes move in the opposite direction.

Another explanation therefore had to be looked for. Unquestionably the blackened side will become warmer than the silvered side, because of absorption of light, and the gas near the blackened side will be warmer than that near the silvered side. This led to the second explanation of the movement: because the molecules striking the blackened side are on the average moving faster than those striking the silvered side there will be a greater force on the blackened side. This is a plausible explanation, and indeed is the one originally accepted by Maxwell; it is still often given to unsuspecting students at the present day, and it is presumably the one Maxwell gave to Queen Victoria. Maxwell's detailed study of the problem, however, showed to his surprise that the explanation is incorrect. His analysis showed that when a solid is present in a gas in which there is a variation of temperature, a steady state is soon reached in which there is a steady flow of heat. As far as the surfaces of the vanes are concerned, therefore, there is no net force.

The solution to the problem was provided by Osborne Reynolds (1842–1912), professor of engineering at Owen's College, Manchester, and still remembered for the *Reynolds number*, a dimensionless number related to the motion of a body in a fluid, and to the possibility of turbulent flow. Early in 1879 Reynolds submitted a paper to the Royal Society in which he considered what he called thermal transpiration, and also discussed the theory of the radiometer. By thermal transpiration Reynolds meant the flow of gas through porous plates caused by a temperature difference on the two sides of the plates (Fig. 6.4(b)). If the gas is initially at the same pressure on the two sides, there is a flow of gas from the colder to the hotter side, resulting in a higher pressure on the hotter side. The explanation is in terms of oblique reflections on the surfaces (Fig. 6.4(c)). A gas molecule moving in the direction AP will on the average move faster than one moving in the direction BP, and because surfaces are never completely smooth there will be a net force from right to left on the solid. If, as in a transpiration experiment, the solid cannot move, the gas will instead move from left to right, i.e., from the cooler to the warmer gas.

To explain the radiometer, therefore, one must focus attention not on the faces of the vanes, but on their edges. As seen from Fig. 6.4(c), the net movement of the vane is away from the warmer gas and towards the cooler gas. The behaviour is just as if there were a greater force on the blackened side of the vane (which as Maxwell showed is not the case), but the explanation must be in terms of what happens not at the sides of the vanes but at their edges.

Maxwell refereed Reynolds's paper, and so became aware of Reynolds's suggestion. Maxwell at once made a detailed mathematical analysis of the problem, and submitted his paper, 'On stresses in rarified gases arising from inequalities of temperature', for publication in the *Philosophical Transactions*; it appeared in 1879, shortly before his death. The paper gave due credit to Reynolds's suggestion that the effect is at the edges of the vanes, but criticized Reynolds's mathematical treatment. Reynolds's paper had not yet appeared (it was published in 1881), and Reynolds was somewhat incensed at the fact that Maxwell's paper had not only appeared first, but had criticized his unpublished work. Reynolds wanted his protest to be published by the Royal Society, but after Maxwell's death this was thought to be inappropriate.

Maxwell's 1879 paper, and the later paper of Reynolds, were important pioneering contributions to the theory of rarified gases, a field that was to be greatly extended in the present century.

## THE EDISON EFFECT

An important observation was made in 1880 by the great American inventor Thomas Alva Edison (1847–1931; Fig. 6.5) in connection with his work on electric light bulbs having carbon filaments. He was investigating why the bulbs had a short lifetime, which was generally attributed to the ejection of carbon from the filament. Edison sealed a wire into a bulb (Fig. 6.6), and found that if the wire was connected to the positive end of a battery, a current would flow, which meant that a current was flowing through the vacuum in the bulb. No flow occurred if the side wire was connected to the negative end of a battery. This discovery attracted great interest, and it was referred to as the Edison effect. No explanation could be given for the effect for some years; it was then realized that there was a flow of electrons from the filament to the side wire, as will be further discussed later. In 1884 Edison took out a patent on a voltage indicator that was based on the effect, and this might be said to be the first patent in electronics.

**Fig. 6.5**   Thomas Alva Edison (1847–1931).

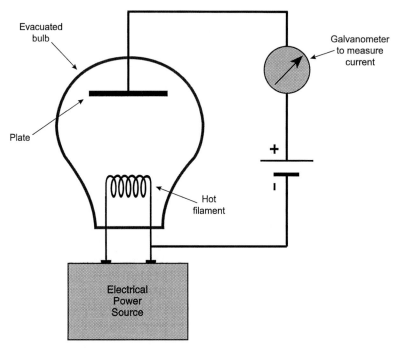

**Fig. 6.6** Illustration of the Edison effect. If the plate is positively charged, current flows between the plate and the hot filament, but not otherwise. The effect was discovered by Edison in 1883. It was later realized that the current flow is a flow of electrons from the hot filament to the electrode, and that the device can be used as a 'valve' to rectify an alternating current. The diode, invented by J A Fleming in 1904, is essentially this device.

## JOSEPH JOHN THOMSON AND THE ELECTRON

Beginning in about 1890 Joseph John Thomson (1856–1940; Fig. 6.7), the Cavendish professor in Cambridge, carried out a series of detailed investigations on cathode rays. His work, and other work done in his laboratory, completely transformed the whole of physical science. He identified in a clear way the particle occurring in the discharge (the electron) and measured its charge and mass.

J J Thomson's life and career were of the greatest interest and distinction. He was born near Manchester, and it was only by accident that he became a physicist at all. He intended to be an engineer, and his father had arranged for him to be apprenticed to a leading engineer. His father's death, and resulting lack of money to pay the premium for the apprenticeship, made it necessary for him to look towards another career. He had been studying engineering at Owen's College, Manchester (an excellent college, particularly for science, which later developed into the University of Manchester), and

**Fig. 6.7** Joseph John Thomson (1856–1940), from a portrait by Arthur Hacker at the Cavendish Laboratory, Cambridge.

one of its professors encouraged Thomson to continue his studies and then take a scholarship to Trinity College, Cambridge.

He was successful in winning the scholarship, and in 1876 went to Cambridge, where he studied for the Mathematical Tripos. The course of study for this included much theoretical physics as well as pure mathematics, but he did not acquire any experience in experimental physics. Because of this—and rather paradoxically, as Thomson's fame is as an experimental physicist—he did no experimental work in the Cavendish Laboratory. Even more remarkably, he apparently never met Maxwell, who was its director for three of the four years that Thomson was an undergraduate, and whose work later inspired his own research.

Thomson took the Tripos examinations in 1880, and was Second Wrangler (the First Wrangler was Joseph Larmor, who also became a distinguished physicist and a Cambridge professor). Thomson was elected a Fellow of Trinity College in the following year, and remained at the College for the rest of his life, serving as its Master for many years.

His first research efforts were competent but not of much distinction, and certainly not of lasting value. They were mainly mathematical in character, and were a development of some of Maxwell's ideas. Previously Maxwell and William Thomson (Kelvin) had toyed with an idea that seems rather strange today, the 'vortex atom'. A vortex is a rotating ring of fluid, such as a smoke ring or a whirlpool. The idea was that atoms were vortices present in the 'ether', the hypothetical medium that was then supposed to pervade all space.

Thomson took Maxwell's ideas about the vortex atom a little further, and in 1882 his work won the Adams Prize—which Maxwell had won for his work on Saturn's rings in 1855–57. Thomson then tried to apply the vortex theory to some problems of chemical constitution—a matter in which Thomson, unlike most physicists at the time, was particularly interested.

The vortex theory, which several physicists had taken seriously, died a natural death when more plausible theories came along. In his Baltimore Lectures, given at Johns Hopkins University in 1884, Kelvin jokingly remarked that the most interesting thing about the theory of molecular vortices was its name. Later, after Einstein proposed his theory of relativity, it was no longer necessary to postulate an ether, so that the vortex atom died along with the ether.

In his early research years in Cambridge J J Thomson also carried out, not very successfully, some experimental work on electricity. It is a rather curious paradox that although Thomson is remembered much more for his experimental work than for his theoretical work, he was actually a poor experimentalist, being particularly clumsy with his hands. The important experimental work for which he is famous was done for the most part by his assistants, and it is said that when he approached their apparatus they prayed silently that he would not touch it, as almost inevitably he would do it some damage. Thomson did have a genius for designing apparatus and finding out what was wrong if it did not work properly. Since in addition he was enthusiastic, and treated his assistants with great consideration and geniality, he was an ideal research director. He was also a superb lecturer, always able to present his material in an interesting way at a level suitable to his audience.

In 1884 the Cavendish professorship of Experimental Physics fell vacant. We will recall from Chapter 5 that Maxwell was the first Cavendish professor. On his untimely death in 1879, the appointment went to John William Strutt, third Baron Rayleigh (1842–1919). Although a scientist of great distinction Rayleigh was a reluctant professor, preferring to carry out his researches at his family seat at Terling in Essex, and he resigned the Cavendish professorship in 1884. It is a rather curious fact, which we noted in Chapter 5, that although the chair was of such great prestige, it had at first been difficult to persuade people to accept it. William Thomson (Kelvin) and Helmholtz turned it down in 1870, and then Maxwell accepted it only reluctantly. Rayleigh accepted it reluctantly in 1879 and then stayed for only five years.

When the chair became vacant in 1884 there was, for a change, much competition for it. Several distinguished physicists were interested. One was Osborne Reynolds, of Owen's College, whose work on the Crookes's radiometer we met earlier in this chapter. Another was Joseph Larmor, who had been ahead of Thomson in the Tripos examinations. A third was the brilliant Irish theoretical physicist George Francis FitzGerald (1851–1901), whom we met in Chapter 5 and who was later to do important work (the 'FitzGerald contraction') that paved the way for Einstein's theory of relativity. At first sight all of these men would seem to have been more promising than Thomson. The electors for the appointment, however, thought otherwise, and Thomson, aged only 28, secured the professorship—much to the annoyance of some of his competitors. Being wise after the event we can see that an excellent choice was made, but the reasons for it are not obvious. The electors cannot have been much impressed by the research Thomson had done, but perhaps his enthusiasm and excellent lecturing persuaded them that his future career would be a distinguished one.

After being appointed professor Thomson chose for his research the gas discharges, mainly because Maxwell had recommended their study. At the time there was still considerable controversy as to the nature of cathode rays. Many investigators regarded them as a form of electromagnetic radiation, and some of the evidence seemed to support that idea: Hertz and others had shown that the rays could pass through thin metal foils, which seemed unlikely if they were beams of particles. Another argument in favour of electromagnetic radiation, and against charged particles, was that some of the initial attempts to deflect them in an electric field had proved unsuccessful. Thomson correctly suspected that the reason for this failure was that the tubes had not been evacuated sufficiently, and that the residual gas in the tubes was preventing the deflection. In his own words (in a rather long sentence, but nevertheless surprisingly clear):

> We must remember that the cathode rays, when they pass through a gas, make it a conductor, so that the gas acting like a conductor screens off the electric force from the charged particle, and when the plates are immersed in the gas, and a definite potential difference [is] established between the plates, the conductivity of the gas close to the cathode rays is probably enormously greater than the average conductivity of the gas between the plates, and the potential gradient on the cathode rays is therefore very small compared with the average potential gradient.

He therefore realized that it was necessary to obtain a better vacuum so as to reduce the electrical screening effect by the gas molecules. At the time the techniques for producing high vacua were nothing like as efficient as they are today, but by evacuating his apparatus for a number of days Thomson achieved sufficiently low pressures to be able to measure the deflections in an electric as well as a magnetic field. He first accomplished this satisfactorily in 1897, and confirmed that the particles carry a negative charge, which is consistent with the fact that they leave the cathode (the negative electrode).

The way was now open for Thomson's famous experiments in which he measured the ratio of the negative charge to the mass of the cathode-ray particles, which he always called by the old-fashioned name of 'corpuscles'; this word had often been applied by Boyle, Newton, and other early investigators to atoms and molecules. Thomson balanced the effects of electric and magnetic fields, finding a value for the charge/mass ratio of approximately two hundred million ($2 \times 10^8$) coulombs per gram ($C\ g^{-1}$). This result was of great significance since, as we have seen at the beginning of this chapter, in 1874 Johnstone Stoney had deduced from the value of the Faraday constant and other data that the corresponding ratio relating to a hydrogen ion is about one hundred thousand ($10^5$) $C\ g^{-1}$. The two values differ by a factor of about 2000, which can be explained in several ways, the simplest of which are: (1) that the masses of the particles are the same, but the charge on a cathode-ray particle is 2000 times that on a hydrogen ion, or (2) that the charges are the same (but opposite in sign) but the mass of the cathode-ray particle is 2000 times smaller than the mass of the $H^+$ ion. At the time there was not much strong evidence that would allow a decision, but preliminary measurements were beginning to suggest that the charges were the same. Thomson rather boldly accepted this possibility, and concluded that the mass of his 'corpuscle' (electron) was 2000 times smaller than the mass of the hydrogen ion. As a result Thomson, at a Friday Evening Discourse at the Royal Institution on 30 April 1897, made the momentous announcement of the existence of charged particles considerably lighter than any atom. As it turned out, he had made the right choice; soon afterwards other investigators at the Cavendish Laboratory confirmed that the charge on the 'corpuscle' was the same as that on the hydrogen ion, and therefore that its mass was about 2000 times smaller than that of the hydrogen atom.

One important consequence of Thomson's work was that it established the real existence of the electron, and therefore the real existence of atoms. Even in the early years of the twentieth century there were still a few sceptics who considered that atoms were no more than a convenient fiction (although almost everyone else had believed in them for at least a hundred years). Until 1909, for example, the distinguished physical chemist Wilhelm Ostwald maintained that there was no real evidence for the existence of atoms. In the edition of his textbook of physical chemistry that appeared in 1909, however, Ostwald finally admitted that he was convinced, and that Thomson's work had played an important part in persuading him.

## FURTHER WORK AT THE CAVENDISH LABORATORY

The first investigations on the charge of the electron were reported in 1897 by John Sealy Edward Townsend (1868–1957), a graduate of Trinity College, Dublin, who in 1895 had become a research student under Thomson. The technique he used was to cause the electrons to become attached to water droplets and to observe their rate of fall under gravity so that their radii could be estimated; from the total charge and the number of droplets the charge on each droplet could be calculated. The value obtained was consistent with estimates that had previously been made, on the basis of electrolysis results and other information, of the charges carried by some ions in solution. Townsend's value for the unit of charge was $1.57 \times 10^{-19}$ C; the accepted modern value is $1.602 \times 10^{-19}$ C, so that Townsend was in error by only 2%.

Townsend's later career was rather disappointing. In 1900 he left Cambridge to become Wykeham professor of physics at Oxford; this was a second chair in physics, intended for a person whose research was in electricity. Townsend was less than successful as a professor, carrying out little research and refusing to take any notice of the many important advances in science, such as the quantum theory. At the time there was no age limit for a professor, and the Oxford authorities were only able to persuade Townsend to retire in 1942 after he was prosecuted for disrupting a course for servicemen. Another professor of physics in Oxford, Robert Bellamy Clifton (1836–1921), had been an equal disappointment. He was professor of experimental philosophy (physics) from 1865 to 1921 and, as we saw on p. 54, he concentrated on teaching and did no research. Because of the long and unproductive tenures of Clifton and Townsend (both of whom had shown much promise at Cambridge), Oxford physics did not flourish until much later than that at Cambridge, which had been so fortunate with its early professors—despite their reluctance to serve.

Townsend's procedure of estimating the charge of electrons by attaching them to water droplets led to more detailed work of a similar kind by the American physicist Robert Andrews Millikan (1868–1953) at the University of Chicago. In 1906 Millikan attached electrons to drops of water, and then in 1911 used oil drops to reduce evaporation, in each case measuring their rate of movement under gravity and in an electrostatic field. The value he obtained for the charge on the electron is close to the one accepted today. Millikan's work was subject to controversy, in that it was alleged that he ignored results that were inconsistent with his idea of what the electronic charge should be. In particular, the Austrian physicist Paul Ehrenfest (1880–1933) was strongly of the opinion that particles existed which had charges that were smaller than that on the electron, and it may be that some of Millikan's results supported that point of view; Millikan, however, denied that he had obtained any such results. Later work has confirmed that particles with fractional charges do exist, so that Millikan may have missed something of importance.

It was mentioned earlier that another type of ray was found in discharge tubes in which there was a perforated cathode. These are the canal rays, which travel in the opposite direction and are positively charged particles. These rays were carefully investigated by Thomson during the course of his investigations of the cathode rays, and in 1899 he commented (again in a rather long sentence):

> The results of these experiments show a remarkable difference between the properties of positive and negative electrification, for the positive, instead of being associated with a constant mass $\approx 1/1000$ of that of the hydrogen atom, is found to

be always connected with a mass which is of the same order as that of an ordinary molecule, and which, moreover, varies with the nature of the gas in which the electrification is found.

The first thorough investigation of the positive rays was made by Thomson in 1907, and the apparatus he used is illustrated in Fig. 6.8.

Thomson was well aware of the potential usefulness of the positive ray machine that he had devised. In 1913 his book *Rays of positive electricity and their application to chemical analysis* was published, and he remarked that one of the main reasons for writing the book was to induce chemists to try this method of analysis. He added:

> I feel sure that there are many problems in chemistry which could be solved with far greater ease by this than by any other method. The method is surprisingly sensitive—more so even than the method of spectrum analysis, requiring an infinitesimal amount of material, and does not require this to be specially purified.

This prediction has been amply justified, in that in modern laboratories mass spectrometry is a widely used technique for analysis in research and in technology.

Aside from the simple instrument described in Thomson's 1907 paper, the first mass spectrometers were built independently and at almost the same time by Dempster and by Aston. Arthur Jeffrey Dempster (1886–1950) was born and educated in Canada but did most of his work at the University of Chicago, where he was professor of physics. Francis William Aston (1877–1945) began his career as a chemist and later worked under Thomson at Cambridge, where he developed a mass spectrometer that was a great improvement over the simple instrument used by Thomson. Dempster's first paper on his mass spectrometer appeared in 1918, Aston's in 1919, and both men continued to make modifications for a number of years. They both succeeded, by somewhat different techniques, in improving the focusing of the beams, in this way increasing the accuracy of their instruments. Aston subsequently made a number of important discoveries about isotopes and the

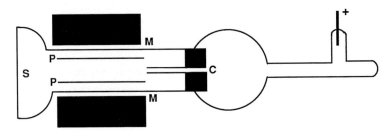

**Fig. 6.8**  Apparatus used by J J Thomson in 1910 to investigate positive rays ('canal rays'). C is the negative electrode (cathode). The beam is deflected by the poles MM of an electromagnet, and by the parallel plates PP which were connected to a source of electricity. This simple apparatus was the prototype of the mass spectrometers which are used widely today to measure the masses of molecules.

nature of atomic nuclei, and he was awarded the 1922 Nobel Prize for chemistry. Dempster discovered the isotope uranium-235 which has played an essential part in connection with nuclear energy (Chapter 8).

J J Thomson continued to direct the research at the Cavendish laboratory for many years. Many members of the laboratory did work of great distinction, and several received Nobel Prizes. He himself received the 1905 Nobel Prize in physics, and was knighted in 1908; in 1912 he was made a member of the Order of Merit, an exclusive British honour limited to 24 members. He was made Master of Trinity in 1918, and guided the College with his usual good humour and common sense until shortly before his death in 1940 at the age of 84.

Thomson's appearance was unmistakable, and he was a familiar figure in Cambridge for many years. On visits to Cambridge in the 1930s I remember seeing him several times. He was famous for his absentmindedness, especially when he was engrossed in research, and many stories were told about him. One that I remember from the 1930s relates to the fact that in his laboratory he customarily wore an old pair of baggy grey flannel trousers. One morning after he had left the Master's Lodge at Trinity for his laboratory, Lady Thomson observed these trousers in the bedroom. Fearing the worst, she telephoned the Porter's Lodge: 'Did you see the Master pass by this morning?'. When the reply was in the affirmative she asked 'And did he look all right?'. Again the reply was in the affirmative, and she sighed her relief. It emerged that the Master had uncharacteristically bought a new pair.

According to another story I heard many years ago, Thomson and the great physiologist Lord Adrian, a later Master of Trinity, were seen walking together in Cambridge. Thomson was overheard saying to Adrian 'Do you realize, Adrian, that when the word intellectual is used today, it doesn't seem to apply to people like you and me?'

Professor Max Born (1882–1972), who after distinguished work in Germany on quantum theory was for many years professor at the University of Edinburgh, told a nice story that brings out Thomson's absorption in his research. Born had worked with Thomson for a period in 1906, and visited him again 15 years later. He was taken to Thomson, who was bent over some apparatus, and was announced. Thomson lifted his head briefly and said 'How do you do. Now look here, this is the spectrum of.....' Born's comment was 'We were in the midst of the realm of research, forgetting the chasm of years, war and after-war, which lay between this rencontre and the days of our first acquaintance.'

## JOHN AMBROSE FLEMING AND THE DIODE

The work of J J Thomson and his colleagues in the Cavendish Laboratory soon led to the vast field of electronics, which plays such an important part in the world today. Not surprisingly, one of the pioneering roles in the field was by a man who had received some of his education at the Cavendish Laboratory.

He was John Ambrose Fleming (1849–1945) who was born in Devon but educated in London. He was of a mechanical turn of mind and from his early years had intended to become an engineer. Like J J Thomson, however, he found that he could not afford the premium required to become apprenticed to an engineer, so that he had to approach engineering by another route. In any case, he fully realized that to excel in engineering he needed a sound knowledge of basic science. He therefore studied mathematics and physics for two years, from 1867 to 1869, at University College, London. He was unable to complete his studies there for financial reasons, and for a time earned his living in several rather unproductive ways. He completed his academic studies in the evenings, and took his BSc degree in 1870.

From 1871 for 18 months he taught at Rossall School, but then returned to London to study at the Royal School of Mines in South Kensington (it later became part of Imperial College). He also did some teaching there, and carried out research, mainly of a chemical nature on the reactions occurring in voltaic cells. When the Physical Society of London was founded in 1874, Fleming was invited to give the first paper at its first meeting. It is interesting to note that 65 years later, in 1939, he gave his last scientific paper to the same society, at the age of 90. He remained scientifically active almost until his death at the age of 95.

As funds were again in short supply Fleming taught at Cheltenham College, with the intention of later going to Cambridge to work under Clerk Maxwell, with whom he corresponded. Along with his teaching he prepared himself for Cambridge, and made a careful study of Maxwell's brilliant but difficult *Treatise on electricity and magnetism*. On arriving in Cambridge in 1877 he was assigned to work under William Garnett, one of Maxwell's assistants, and later to be Maxwell's co-biographer. We have seen in the last chapter that Maxwell's lectures were less than inspiring, and were usually attended by only two or three students. Fleming attended them conscientiously and took detailed notes, a task of some difficulty in the circumstances. Much later, in 1931, he presented his notes to the Cavendish Laboratory, where they can still be seen.

His research at the Cavendish was on electrical circuits and bridges, and because of its shape Maxwell called one of his devices Fleming's banjo. At that time Maxwell was studying the scientific papers of Henry Cavendish (1731–1810), a remarkable eccentric genius who had made many discoveries that were years ahead of their time but had never been published. At Maxwell's suggestion Fleming repeated some of Cavendish's experiments, and this work was of help in the preparation of Maxwell's book on Cavendish's unpublished experiments.

In 1881 Fleming became professor of physics and mathematics at the newly-formed University College at Nottingham. However, the scientific attractions of living in London were too strong, and he resigned after a year. For a time he served as adviser to the London Telephone Company and the Swan Lamp Factory. In 1885 he was appointed professor of electrical engineering at University College, London, a post he was to hold for 41 years. Although he was only 36 on his appointment, his career had been rich and varied. His great success in engineering was no doubt to a great extent due to his broad experience with technical problems and his sound knowledge of basic physics and mathematics.

The facilities available to Fleming at University College were extremely primitive; he himself said that at first he had no more than a blackboard and a piece of chalk. His laboratory was a room about twenty feet square containing apparatus most of which had been given or lent by friends. Fleming managed to overcome all these difficulties. He was an outstandingly good lecturer, and was particularly successful at presenting science to an audience of non-scientists. He was in great demand as a lecturer to young people, and gave four of the famous Christmas Lectures at the Royal Institution.

His research covered a wide range. He made important contributions to the development of alternating-current transformers, being particularly concerned with the distribution of electric power. He also did valuable work on the development of electric lamps.

The work for which Fleming is best remembered is his invention of the diode, modifications of which played an important part in radio transmission for many years. Throughout his career he experimented with radio transmission, and lectured on the subject. In 1884 he had visited Edison in the United States. Mention was made earlier in this chapter of the Edison effect (Fig. 6.6), and Edison may have demonstrated the effect to Fleming; at any rate the two discussed it. Beginning in 1889 Fleming repeated Edison's experiment, and for the next few years he cooperated with Marconi in much of his work. In particular, he helped Marconi to design the transmitter used to span the Atlantic in 1901.

At first it was assumed that the Edison effect was due to the escape of carbon particles from the hot filament, and that these carried a negative charge. As a result of Thomson's experiments, however, it was realized that instead there was a flow of electrons from the hot filament to the plate. This led Fleming in 1904 to the idea that the device, since it allowed current to pass only in one direction, could be used as a rectifier for alternating current; it would cut out the current flowing in one direction, and allow it to pass only in the other.

The importance of this work was that the tube could be used to receive radio signals. Until then there had been considerable difficulty in detecting radio waves. The meters that had been used for receiving telephone messages were too slow to detect the positive–negative cycling in a radio wave, and only indicated the average value—which is zero. As mentioned in Chapter 5, some success had been obtained with coherers, and these were being used by

Oliver Lodge and Marconi, but they were notoriously unreliable. Fleming had been successful in using his tube to detect radio waves, and early in November 1904 he wrote to Marconi 'I have been receiving signals on an aerial with nothing but a mirror galvanometer and my device (i.e., diode). He secured a patent for what he called his thermionic valve on 21 September 1905; others called it a Fleming valve, or a diode.

Although the diode was the forerunner of the many other thermionic tubes that played an important part in the development of radio, Fleming never received adequate credit for it. He was honoured for his many other achievements, receiving a knighthood in 1929.

## OTHER THERMIONIC TUBES

Part of the reason for Fleming's failure to receive credit for his diode was that in 1907 the American inventor Lee De Forest (1873–1961) patented a tube that eventually, after many changes made by others, led to great improvements. This was the triode, in which an additional electrode was inserted between the filament and the plate (Fig. 6.9). With appropriate circuitry, this extra electrode, called the grid, exerted control over the electron flow. It

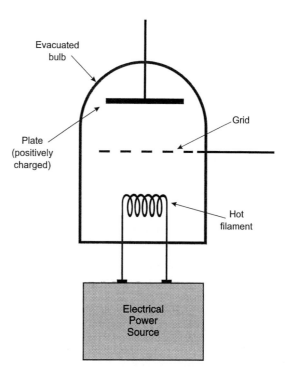

**Fig. 6.9**   The principle of the triode, invented by Lee De Forest.

appears, however, that De Forest, whose knowledge of science was much more limited than that of Fleming, had little idea of how his triode operated or of its potential advantages. In its original form, in fact, De Forest's triode was no more effective as a detector than Fleming's diode. He made broad claims about his triode that were technically unjustifiable. He claimed in his patent application that his triode was able to 'amplify feeble electric currents', but it could not do so. It was not for some years that other workers were able to use the triode as an amplifier, and with suitable circuitry it could be highly effective for this purpose. The essential feature of the operation of the triode as an amplifier is that a tiny change in the potential of the grid greatly affects the number of electrons that pass from the electron-emitting hot coil to the plate.

The situation became much confused by personal animosity between De Forest on the one hand and Fleming and Marconi on the other. Technical progress was delayed, and from time to time halted altogether, by extensive patent litigation. In 1912 De Forest and his associates were charged with fraud as a result of some of their claims; De Forest himself was exonerated, but two of his associates were jailed.

In the end, as a result of the work of many people, vacuum tubes (often called valves in Britain) came to have a vital role in radio. A surprisingly large variety of them came into use. The 1930 *RCA tube handbook* listed 59 types of tubes, but by 1949 over 10 000 types were listed, nearly half of them being suitable for receiving radio signals.

## ELECTRONIC RADIO TRANSMISSION

There are many complexities in radio technology, and the present account can do no more than outline the basic principles. A radio transmitter consists essentially of an antenna, an electronic oscillator, and a modulator. The oscillator consists of electronic circuits, formerly involving vacuum tubes, now replaced by transistors. The function of the oscillator is to set up oscillating currents, which 'drive' the antenna, setting up an alternating current that is characteristic of the oscillator. It is a consequence of Maxwell's electromagnetic theory that whenever electric charges are accelerated (which is the case when there is an alternating current in the antenna), a radio wave is produced which radiates away from the antenna with the speed of light.

If information is to be transmitted by means of this radio wave, it must be modified in some way. This is the function of the modulator, which may include a microphone. One way for the modulator to operate is by causing the signal from the microphone to modulate the energy of the emitted radio wave, which means that it modifies the amplitude of the waves. This technique, still used today, is referred to as amplitude modulation, or AM.

However, various extraneous effects also give rise to amplitude modulation, with the result that AM radio suffers from interference from miscellaneous sounds referred to as 'static'. Various ways were devised to reduce static, and one of the most effective of these was to use a different type of modulation, known as frequency modulation, or FM. The modulator in FM radio is designed in such a way that instead of moderating the energy or amplitude of the radio waves, it moderates their frequency. The great pioneer of FM radio was the American engineer Edwin Howard Armstrong (1890–1954). Although his technical achievements were of high quality and great ingenuity, Armstrong did not fare well as far as patents and business connections were concerned. In 1954, not having received the financial rewards due to him, he committed suicide. Some years after his death, and after extensive litigation, Armstrong's widow received a settlement which provided partial compensation for all his efforts.

A radio receiver consists essentially of an antenna, a detector and amplifier, and a loudspeaker. The radio waves sent out by the transmitter fall upon the antenna of the receiver, and set up oscillations in it. These oscillations produce small potential variations in the grid of a triode (or more complicated tube). Suitable electronic adjustments are made to allow detection of the desired signals. Particularly if a loudspeaker is used, there must be much more energy than is provided by the output of a single triode, and more tubes are used to produce this additional power.

## THE TRANSISTOR

Since the Second World War there have been remarkable improvements in electronics. These have been brought about largely by two inventions, the transistor in 1947 and the microprocessor in 1971. One important effect of these improvements has been to allow electronic devices to be much smaller than was previously possible. In addition they have greatly extended the versatility of electronic devices of all kinds. Today radio receivers that can be held in one hand can have a much wider range than the much larger receivers of pre-war days. The effect on computers has been particularly striking. During the Second World War there were few computers in existence, and they were bulky, involving thousands of vacuum tubes and occupying several large rooms. Today many students, at schools as well as universities, possess pocket-sized electronic calculators that are vastly more efficient that the early computers. Versatile desk-sized computers are now in many private homes.

Only a general account can be given of transistors, as to understand them fully requires a detailed knowledge of physics and of advanced technology. The basic ideas, however, are fairly simple. Transistors are based on solid substances known as semiconductors. There is a wide variation in the electrical conductances of substances, that is, in the ease with which substances allow electricity (i.e, a flow of electrons) to pass through them. Metals allow electric current to pass through them easily, and are said to have a high conductance or a low resistance (the resistance is the reciprocal of the conductance. Many other solids, hovever, have conductances that are smaller by many powers of ten and are said to be insulators.

The theory of electrical conductance is rather complicated, but the main ideas may be understood with reference to Fig. 6.10. According to the quantum theory, to be discussed more fully in Chapter 8, there are restrictions as to the energy levels into which electrons can go. In a metal (Fig. 6.10(a)) the electrons are normally in what is called the valence band. Immediately above the valence band there is a so-called conduction band into which it is possible for electrons to move. When they do move into the conduction band it is easy for them to move from one place to another. What happens when a metal wire is attached to the poles of a battery is that a few of the electrons in the metal can move easily into the conduction band, and they can then move freely along the wire. The electrical conductance is therefore high.

With an insulator (Fig. 6.10(b)) the situation is quite different. Again the electrons in the solid are confined to the valence band. Now, however, the conduction band is separated from the valence band by a wide gap. When an insulator is connected to the poles of a battery, the electrons in the valence band are unable to jump into the conduction band, because they have insufficient energy to do so. The current therefore cannot flow.

A semiconductor is in between (Fig. 6.10(c)). Now there is an energy gap between the valence band and the conduction band, but it is narrow. As a result, when we try to pass an electric current though the solid, a few electrons may be able to jump the gap and pass into the conduction band. There will therefore be some conductance, but less than that of a metal. If the temperature is raised it becomes easier for electrons to reach the conduction band; the conductance of a semiconductor therefore increases with temperature.

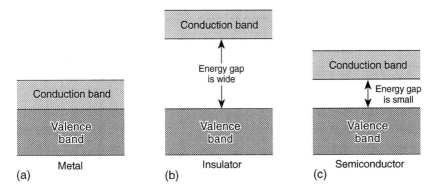

**Fig. 6.10**   Valence bands, energy gaps, and conduction bands in conductors, insulators, and semiconductors. Electrons in the valence band are held by atomic nuclei and do not move freely from one part of the surface to another. Electrons in the conduction band move much more freely.

If a pure substance is a semiconductor, it is known as an intrinsic semi-conductor. Crystals of pure germanium and silicon, for example, are intrinsic semiconductors. At absolute zero all of the electrons are in the valence band, but at room temperature a few electrons are in the conduction band. This means that there is a shortage of electrons in the valence band, and we say that there are positive holes in the valence band (Fig. 6.11(a)). When a current is passed through the solid, the electrons move in the conduction band, and positive holes move in the opposite direction in the valence band. We thus say that the electrical conduction is due both to the movement of electrons in one direction, and to the movement of positive holes in the other. (At first sight it may seem rather odd to talk about the movement of holes. A familiar analogy is provided by a bubble rising in a liquid. If an underwater swimmer exhales a bubble of air, we say that the bubble rises to the surface, and it is easy to understand what that means. In fact, of course, that is simply a convenient way of speaking; we should really have said that the water flows down beside the bubble, but that would be a more awkward way of expressing ourselves. Similarly, when we say that positive holes move in one direction, we are merely using a convenient way of saying that the surrounding electrons are moving in the opposite direction.)

In many cases it is possible to start with an insulator, in which the energy gap is wide, and add an impurity which narrows the energy gap and causes the material to act as a semiconductor. This procedure is known as doping, and the resulting semiconductor is known as an impurity semiconductor, or an extrinsic semiconductor.

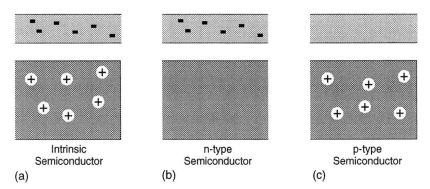

|               |               |               |
|---------------|---------------|---------------|
| Intrinsic     | n-type        | p-type        |
| Semiconductor | Semiconductor | Semiconductor |
| (a)           | (b)           | (c)           |

**Fig. 6.11**  Three types of semiconductors: (a) An intrinsic semiconductor. At temperatures above absolute zero there are some electrons in the conduction band, and an equivalent number of positive holes in the valence band. (b) An n-type semi-conductor. The added impurity has introduced electrons into the conduction band, but there are no positive holes in the valence band. (c) A p-type semiconductor. The added impurity removed electrons from the valence band, which we say contains posi-tive holes.

It is also possible to change the electrical properties of an intrinsic semi-conductor, such as germanium or silicon, by doping it. Suppose, for example, that we dope pure germanium with a small amount of phosphorus or arsenic. These atoms have an electron which they can supply, and the effect is that some additional electrons are added to the conduction band (they cannot enter the valence band, which is full). The conductance (Fig. 6.11(b)) is now due solely to the electrons in the conduction band. This type of semiconduc-tor is called an n-type semiconductor, the letter n standing for negative. It must be understood that although there are extra electrons in the conduction band, the entire material is electrically neutral.

Alternatively, an intrinsic semiconductor might be doped with something that withdraws electrons from the valence band. If, for example, silicon is doped with aluminum or boron, the valence band is left with a shortage of electrons. We describe this by saying that there are positive holes in the valence band (Fig. 6.11(c)). The conductance is now not due to the move-ment of electrons in the conduction band (as there are none there); it is due to the movement of positive holes in the valence band. What this really means is that, because of the holes in the valence band, the electrons that are there are able to move in the opposite direction. Semiconductors formed in this way are called p-type semiconductors, the p standing for positive.

The way in which an n-type and a p-type semiconductor can combine to allow current to pass in one direction but not in the other is illustrated in Fig. 6.12. The device could be constructed by doping one region of a tiny piece of silicon with phosphorus, in this way producing an n-type semiconductor.

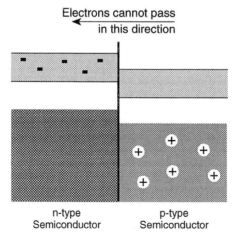

**Fig. 6.12**   A diode, formed by joining an n-type semiconductor to a p-type semi-conductor. Electrons entering from the right cannot cross the interface, because the electrons already present in the conduction band repel them. Electrons can, however, pass from left to right. The arrangement therefore acts as a diode (Fig. 6.6).

The neighbouring region could then be doped with aluminum, to give a p-type semiconductor. If the device is then inserted in an electrical circuit, electrons will be able to pass from the n-type end to the p-type end. They will be unable, however, to cross the interface towards the n-type semiconductor, because of the repulsion of the electrons already present in the conduction band. The device therefore acts as a 'valve'; for example, it rectifies an alternating current. It acts in the same way as the tube devised by Edison (Fig. 6.6) and later developed by Fleming into the diode. Transistors of this type are known as diodes.

The construction of a transistor triode involves the use of two n-type semiconductors separated by a p-type semiconductor, as illustrated in Fig. 6.13(b). The result is called an n-p-n transistor. It can act as an amplifier in the same way as a triode tube, as explained earlier (Figs 6.9 and 6.13(a)).

The account that has been given of transistors may make it appear that the matter is fairly simple, but nothing could be further from the truth. A great deal of difficult theoretical physics, experimental physics, and engineering was required to produce the first transistor. A considerable number of years elapsed before transistors could become commercially available. The transistor, in fact, provides us with one of the best examples of an invention that

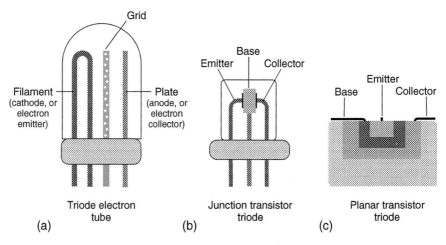

**Fig. 6.13** A diagram showing the relationship between a triode electron tube, an n-p-n junction transistor, and an n-p-n planar transistor. (a) A triode electron tube (compare Fig. 6.7). A positive charge on the grid allows electrons to flow from the filament to the plate. (b) A junction triode. On the left is an n-type semiconductor, which acts as an electron emitter, like the filament in the electron tube. On the right is another n-type semiconductor, which acts as an electron collector (like the plate in a). In between is a p-type semiconductor which acts as the grid, allowing electrons to pass only when it is positively charged. (c) A planar transistor, which is essentially the same as b, but which can be much smaller.

would not have been possible without a great deal of pure science together with ingenious technology.

The main work on transistors was done in the AT & T Bell research laboratories, and three men were principally involved. The first of them to carry out research on the electrical properties of semiconductors, before World War II, was Walter Houser Brattain (1902–1987), a graduate of the University of Minnesota. He was joined in 1936 by William Bradford Shockley (1910–1989), who was born in England but grew up in Palo Alto, California, and was a graduate of the California Institute of Technology. The third member of the group was John Bardeen (1908–1991), a graduate of Princeton University, who joined the Bell laboratories in 1945.

The first successful transistor was made on 23 December 1947 by Bardeen and Brattain. They used as a base a doped germanium semiconductor, and brought two pieces of gold foil into contact with it, separated by a very small distance. With this arrangement they were able to achieve an electrical amplification of a factor of about 50. The device they had constructed was rather clumsy in appearance and was first called a transresistor, a word that was soon shortened to transistor.

For the next few years Shockley carried out extensive experiments on the device, and made important contributions to the theory of its functioning. In 1951 he produced a highly effective n-p-n germanium transistor, of the type illustrated in Fig. 6.13(b). It was enclosed in a metal case about one centimetre in height, and was much more efficient than the point-contact device used by Brattain and Bardeen. In 1956 the three men shared the Nobel Prize for physics for their achievement. (Bardeen won a second Nobel Prize in 1972 for his work on the theory of superconductivity).

It took some time for transistors to be commercially feasible. At first a transistor cost ten times as much as a vacuum tube; this was partly due to manufacturing difficulties, and partly to the high price of germanium, which is more expensive than gold. In 1954 a transistor was made from silicon instead of germanium. Silicon, the main ingredient of sand, is the earth's second most abundant chemical element (oxygen is the most abundant), and is much less expensive.

Most modern electronic devices use silicon. The price of the transistor was greatly reduced not only by the change to silicon, but also by improved production techniques. In particular, the rather clumsy junction transistor triodes (Fig. 6.13(b)) were superseded by planar triodes (Fig. 6.13(c)), which were much smaller. It was found possible to grow large single crystals of very pure silicon. Improved methods of doping the silicon were also developed.

It has been seen that pure science, particularly the theory of semiconductivity, played a vital role in the development of the transistor. Later advances in the area of communication depended less on science than on highly ingenious technology. In about 1958 the first integrated circuit, or chip, was produced. This was an electronic circuit all of the components of which are formed on a single piece of semiconductor material, which today is nearly always silicon. A chip contains thousands of tiny transistors and other electronic components.

A chip containing all the components of a computer is referred to as a microprocessor, or as 'a computer on a chip'. Since about 1971 many such devices have appeared, the latest containing many millions of electronic components but occupying a very small space.

# SEVEN

# The Braggs and molecular architecture

William Henry Bragg (1862–1942) and his son William Lawrence Bragg (1890–1971) exerted a profound influence on the experimental study of the structures of molecules. In 1912, aged 22, Lawrence Bragg derived the key equation that is used in the X-ray analysis of crystal structures. From that early beginning he was concerned directly with the elucidation of the structures of many minerals. Later in his career he inspired many of the people who were successful in discovering the structures of proteins and other complicated substances that are important in living systems.

Structural work of this kind is of great practical importance, as can be shown by many examples. There are many pharmaceutical drugs, anaesthetics, and other materials that cannot easily be obtained naturally; they must therefore be synthesized, so that knowledge of their structure is essential.

There is another way in which structural work, particularly that done by the use of X-rays, has had enormous practical consequences. The discovery in 1953 of the double-helix structure of DNA, to be outlined later in this chapter, led to fundamental changes in our understanding of how, in humans, animals and plants, characteristics are transmitted from one generation to another. Already the discoveries have had a great impact on the breeding of animals and plants, and on the treatment of human diseases. The search for cures for many diseases has been greatly affected by this advance in scientific knowledge.

The first serious efforts to understand molecular structures were made in the middle of the nineteenth century. Previously, chemistry had been largely an empirical science; chemists prepared new chemical compounds, calculated how many of the various kinds of atoms each molecule of them contained, and discovered how the substances reacted with one another. From these studies, inferences could be drawn about the way in which the atoms were connected together in molecules. Many of the inferences proved to be correct, but uncertainties remained until X-ray methods and other physical methods could be used.

An important advance was made in 1852 by the English chemist Edward Frankland (1825–1899), who had received some of his training in chemistry in the laboratories of Robert Bunsen at the University of Marburg, where he obtained his doctorate. On 10 May 1852, in a paper to the Royal Society, Frankland drew attention to the fact that groups of elements tended to combine with a fixed number of other atoms. Nitrogen, phosphorus and arsenic, he pointed out, tend to join on to three atoms, as in the compounds $NH_3$, $NI_3$, $PH_3$, $PCl_3$, $AsH_3$, and $AsCl_3$.

He commented that

> no matter what the character of the uniting atoms may be, the combining power of the attracting element, if I may be allowed the term, is always satisfied by the same number of atoms.

Frankland realized that an atom could have two different valences; phosphorus, for example, can form both $PCl_3$ and $PCl_5$. In about 1865, the word valence or valency—derived from the word equivalence—began to be used for what Frankland had called the 'combining power'.


## STRUCTURES OF ORGANIC MOLECULES

The German chemist Friedrich August Kekulé (1829–1896) is usually given the main credit for advancing our ideas about the structures of organic molecules, which are the molecules containing carbon atoms. More of the credit, however, should go to Archibald Scott Couper (1831–1892), a Scotsman who went to work in Paris in 1856. In 1858 he published a paper which gave many more correct structures than were given by Kekulé at the same time. Unfortunately, he began to develop serious psychiatric problems, and after his return to Scotland spent time in a mental institution. As a result, Kekulé was able to claim much credit that should have gone to Couper.

Born in Darmstadt, Kekulé attended the University of Giessen where he studied under the great chemist Justus von Liebig (1803–1873). He later went to Paris and undertook further studies with the distinguished chemist Jean Baptiste Dumas (1800–1884). He spent a year, 1851–1852, at St Bartholomew's Hospital in London, and there he came under the influence of two English chemists who had done pioneering work; they were Alexander William Williamson (1824–1904) and William Odling (1829–1921). (Later, although a good chemist, Odling became a most unsatisfactory professor at Oxford, and greatly held back the progress of chemistry at that university. He considered that it was beneath the dignity of a professor to appear in a research laboratory. In those days professorships were for life—which in Odling's case was a long one—and for many years he resisted efforts to persuade him to retire; he remarked to a colleague that resignation was not one of his virtues. He did finally stand down in 1912, at the age of 83, after 40 undistinguished years in the chair.)

As a result of the work of Couper, Kekulé, and others, the structures of the compounds of carbon began to be understood. It was found that a carbon atom tends to combine with four atoms of elements such as hydrogen and chlorine, which have a valence of one. It also combines with two atoms of oxygen to form $CO_2$; the oxygen atom has a valence of two. In modern notation some simple compounds can be represented in two dimensions as

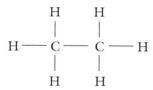

Four valence bonds emanate from each carbon atom, which has a valence of four, and one from each hydrogen or chlorine atom. The oxygen atom, on the other hand, has a valence of two; two bonds therefore connect it to a carbon atom.

There are many organic molecules in which carbon atoms are joined together. A simple example is ethane, in which there are two carbon atoms and six hydrogen atoms. One valence bond of each carbon atom is used to bind two carbon atoms together, while the other bonds are used to attach six hydrogen atoms:

Kekulé became professor of chemistry at the University of Ghent in 1858, and in 1867 he moved to Bonn. By introducing the idea that a chain of carbon atoms can be linked together; Kekulé was led to suggest a structure for the benzene molecule. This substance had been discovered in 1825 by Michael Faraday, who rather surprisingly had found it in oils rather than in the many aromatic spices and resins in which it is commonly present. Benzene is of particular importance in forming the basis for a vast number of other compounds of carbon. Because of their prevalence in living systems, compounds containing carbon have come to be called organic compounds, and compounds which contain structures like benzene are called aromatic compounds because they often have a characteristic smell.

The structure of benzene presented chemists with quite a difficult problem. It is easy to suggest a large number of structures which satisfy the valence rules (hydrogen being univalent, and carbon tetravalent), but some of them seemed inconsistent with the properties of the substance, and how is one to decide between the different possibilities? One structure suggested was

It tuned out, however, to be impossible to explain the properties of benzene, and the structures of some of its derivatives, on the basis of such a structure. One difficulty is that double bonds between carbon atoms (four of which are in this structure) undergo rapid chemical reaction with certain substances such as bromine. Benzene, however, does not do so.

One of Kekulé's great contributions was to suggest, in 1865, that in benzene the carbon atoms are at the corners of a regular hexagon. Today his structure is usually represented as

This, however, was still not entirely satisfactory. For example it fails to explain why benzene does not react rapidly with bromine and other substances. In 1874 Kekulé suggested that all the difficulties were avoided if one assumed that a dynamic equilibrium existed between two equivalent forms of the structure in which the double and single bonds are interchanged:

The idea of dynamic equilibrium between equivalent molecules had previously been emphasized by Williamson. Kekulé's idea was that because of the equilibrium, with double bonds changing places with single bonds, none of the bonds would behave like double bonds. That idea, however, was still not satisfactory.

A better way of looking at the benzene ring, due to a great extent to the American chemist Linus Pauling (1901–1994), is to think of it as existing in a hybrid or resonant state between the two forms, instead of rapidly changing from one to the other. One way of expressing this is to say that each of the carbon atoms has an order of 1.5. A single bond is said to have an order of 1, and a double bond of 2; the carbon–carbon bonds in benzene are half way between single bonds and double bonds.

## THREE-DIMENSIONAL STRUCTURES

For some time little attention was paid to the actual shapes of molecules; they were simply represented as if they were flat, as has been done so far in this chapter. In 1874 an important contribution was made by the Dutch chemist Jacobus Henricus van't Hoff (1852–1911), whose pioneering work in thermodynamics was mentioned in Chapter 2 (Fig. 2.26). Although his main work was in physical chemistry, van't Hoff was at first an organic chemist. Some of his university studies were under Kekulé at the University of Bonn, and he later worked with the French chemist Charles Adolph Wurtz (1817–1884) at the Sorbonne in Paris. Before submitting his thesis to the University of Utrecht, van't Hoff published privately a pamphlet on what has been called the tetrahedral carbon atom. His idea was that the four valence bonds issuing from a carbon atom do not lie in a plane, as in some of the previous structures shown in this chapter; instead they point towards the corners of a regular tetrahedron (Fig. 7.1). This idea of a tetrahedral carbon atom was

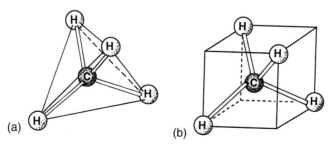

**Fig. 7.1** (a) A model of a simple carbon compound (methane, $CH_4$) showing the four bonds pointing to the corners of a regular tetrahedron. (b) Another representation of the tetrahedral arrangement. The carbon atom is placed at the centre of a cube, and the bonds point to four of its corners.

not entirely original with van't Hoff. Several chemists had considered the idea previously, but van't Hoff developed the idea in detail, and explained on its basis many results that previously could not be understood. (A similar explanation, but with not quite so much detail, was put forward independently and at about the same time by the French chemist Joseph Achille Le Bel (1847–1930).)

The main inspiration for van't Hoff's ideas—according to his own accounts—was the existence of what are called geometrical isomers. These are substances that have the same numbers of atoms in their molecules, but differ in the way they are arranged. Suppose, for example, that two of the hydrogen atoms in methane, $CH_4$, are replaced by chlorine atoms, to give $CH_2Cl_2$. If the four bonds were in the same plane, there would be two possible geometrical isomers of the molecule (Fig. 7.2(a) and 7.2(b)). If, on the other hand, they are arranged tetrahedrally there would be only one possible molecule (Fig. 7.2(c)). In fact, only one molecule of this formula has been discovered.

A particularly important contribution made by van't Hoff in his pamphlet was to relate the three-dimensional structures of organic compounds to what is called their optical activity. Reference has been made in Chapters 3 and 5 to the fact that light can be polarized. This means that the vibrations, instead of being in all directions at right angles to the path of the light (Fig. 3.8(a), p. 78), are in only one direction (Fig. 3.8(b)). Earlier investigators had found that if plane-polarized light is passed through certain chemical compounds, the plane of polarization is rotated (Fig. 7.3). Van't Hoff pointed out that, on the basis of his tetrahedral carbon atom, a molecule in which four different atoms or groups of atoms are connected to a carbon atom can exist in two different forms that are mirror images of one another; these are referred to as two different configurations of the molecule. Van't Hoff's own representation of this is shown in Fig. 7.4, in which the four groups are represented as $R_1$, $R_2$, $R_3$, and $R_4$. It can be seen that if any two of the groups are

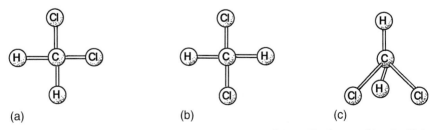

(a)                    (b)                    (c)

**Fig. 7.2**  (a) and (b) The two possible structures of $CH_2Cl_2$ that could exist if the bonds from the carbon atom were in a plane. If these molecules existed we would call them geometrical isomers. (c) The actual structure of the molecule, with the bonds pointing to the corners of a regular tetrahedron. There is now only one possible structure; if we put one of the chlorine atoms in another position it is still the same molecule.

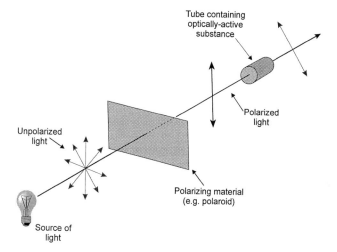

Tube containing
optically-active
substance

Polarized
light

Unpolarized
light

Polarizing material
(e.g. polaroid)

Source of
light

**Fig. 7.3** When plane-polarized light is passed through certain materials, said to be optically active, the plane of polarization is twisted, to the right or to the left.

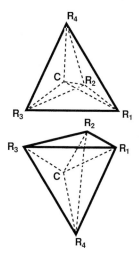

**Fig. 7.4** Van't Hoff's diagram showing the two ways in which four different groups $R_1$, $R_2$, $R_3$, and $R_4$ can be arranged around a carbon atom. These two ways are referred to as configurations. The lower structure cannot be superimposed on the upper one, but is its mirror image. From van't Hoff's *La chimie dans l'espace*, published in 1875.

identical with one another, it is no longer possible to have two different configurations. The important point made by van't Hoff was that whenever there are two different structures (called *configurations*), that are mirror images of one another, there will be optical activity. That is, when polarized light passes through them, the plane of polarization is rotated. Optical activity thus arises when a carbon atom is attached to four different groups. Such a carbon atom is conveniently referred to as an asymmetric carbon atom. One form will rotate the plane of polarization in one direction, while the mirror-image form will rotate it in the opposite direction.

Van't Hoff obtained strong confirmation of his ideas by applying them to some earlier work on tartaric acid done by the great French chemist Louis Pasteur (1822–1895). In tartaric acid there are two asymmetric carbon atoms, and van't Hoff showed that the different configurations that Pasteur had discovered were consistent with his theory.

Van't Hoff's ideas also explained the isomerism found when a molecule contains a double bond between carbon atoms. A simple example is shown by maleic acid and fumaric acid (Fig. 7.5). It had previously puzzled chemists that although the molecules apparently have the same arrangements of atoms, the two acids are different from one another, particularly in physical properties. The explanation is that when carbon is connected to three other atoms, by two single bonds and one double bond, the bonds lie in a plane, and that there is no easy rotation about the double bond. The two structures in Fig. 7.5 are therefore different from one another, and we call them geometrical isomers. In maleic acid the two COOH groups (these are acidic groups) lie on the same side of the double bond (the *cis* arrangement), while in fumaric acid they are on opposite sides (the *trans* arrangement).

When van't Hoff presented his PhD thesis to the University of Utrecht in 1874, he made no mention of the pamphlet he had published the year before on the tetrahedral carbon atom. His thesis instead described a rather mundane piece of research in organic chemistry. The omission of the structural work seems at first sight surprising, but by the time van't Hoff submitted his thesis he had realized that his pamphlet was being violently attacked in some quarters. Scientists today find this rather remarkable, since van't Hoff was so obviously right in his proposals. Several prominent chemists agreed with him at once, but a particularly vitriolic attack on him was made by the famous, but rather unpleasant, organic chemist Hermann Kolbe (1818–1884).

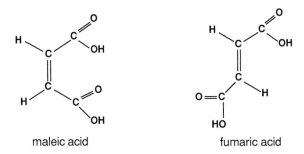

maleic acid                     fumaric acid

**Fig. 7.5**  The structure of maleic and fumaric acids. These are called geometrical isomers. When a double bond and two single bonds emerge from a carbon atom, the three bonds lie in a plane.

What is remarkable is that Kolbe did not attack van't Hoff on scientific grounds at all, but by invective, referring to his ideas as 'witchcraft'. Apparently Kolbe thought it somehow improper even to think in terms of three-dimensional structures. He went so far as to sneer at the fact that van't Hoff was, at the time of one of his attacks (which was published in 1877), teaching at a veterinary college rather than at a university; it is difficult to understand the relevance of such a personal criticism.

In view of the intensity of the criticisms of his tetrahedral carbon atom, van't Hoff was undoubtedly wise not to introduce the ideas into his thesis, as they might well have prejudiced the examiners against him. A decade later Svante August Arrhenius (1859–1927) was not so cautious and got himself into trouble. His 1884 thesis on the conductivities of electrolytes, now recognized to have been of great importance although at the time not presented clearly enough, was not appreciated by his examiners, who gave it such a poor rating that Arrhenius might never have been able to continue his research career; only outside support, from van't Hoff among others, made this possible for him. This was fortunate, as Arrhenius went on to have a most distinguished scientific career.

In spite of the hostility of Kolbe and a few other scientists, van't Hoff's ideas about the shapes of organic molecules were soon generally accepted. Chemists began to draw three-dimensional structures for many of the simpler molecules containing carbon. With larger molecules, however, there were still difficulties. For example, it began to be recognized that a class of molecules called proteins have an important role in living systems. Some of these were found to contain so many atoms that it was impossible, by ordinary chemical techniques, to discover even how the atoms were connected together; deducing their detailed three-dimensional structures from their chemical behaviour was quite impossible. What was needed for this purpose was some new physical technique. Several such techniques were developed over the years; the one that turned out in the end to be most fruitful involved the use of X-rays, as will be explained later in this chapter.

## CRYSTALS

Since structural determinations with X-rays can be done satisfactorily only with crystals, we will first consider some important aspects of crystals. In some solids, such as glasses and pitch, there is no orderly arrangement of atoms or molecules, and we call them amorphous (from the Greek ά-μορφή, a-morphe, without shape). Other solids have a completely regular molecular structure at the atomic level, and are called crystals (Greek κρύσταλλος, krystallos, clear ice). They then have a regular external form, and unlike glasses (which gradually become softer when they are heated) crystalline substances melt sharply at a fixed temperature.

Crystalline materials have been known for several thousand years, but little was understood about them before the seventeenth century. In 1669 the Danish investigator Nicholaus Stensen (1638–1686), whose name is often latinized to Steno, discovered that in different crystals of quartz, the angles between the faces are always the same. This is at first sight surprising, since some substances occur in a variety of crystalline forms which look quite different from one another.

In 1678 the Dutch physicist Christiaan Huygens (1629–1695) suggested that in crystals there was a regular packing of particles in layers. This idea was extended in 1784 by the French mineralogist René Just Haüy (1743–1822). After accidentally shattering a crystal and noting the similarity between the shapes of the fragments, he suggested that the crystal was formed from tiny cubes, the orderly internal arrangement of these building blocks producing the regular external faces of the crystals. Haüy is today recognized as the founder of the science of crystallography.

Much mathematical work has been done on the ways in which atoms can be arranged in crystals. The science that is concerned with symmetry, such as that found in crystals, is called group theory. One of the conclusions drawn from it is that exactly 230 arrangements are possible in crystals; these have all been worked out, and examples of them discovered. Important contributions to the mathematics of crystallography were made by Kathleen Lonsdale (1903–1971; Fig. 7.6), who was born Kathleen Yardley in Newbridge, Southern Ireland. She attended Bedford College for Women, one of the constituent colleges of the University of London. During her first year she concentrated on mathematics but switched to physics and obtained her BSc degree in 1922 at the age of 19. She headed the honours list in the degree examinations, and one of her examiners was Sir William Bragg who was then professor of physics at University College, London. He was so impressed with her work that he offered her a place in his research team, an offer she accepted enthusiastically. In the following year Bragg was appointed director at the Royal Institution, and she followed him there.

One of Kathleen Yardley's first tasks at the Royal Institution was to make a detailed systematic study of the theory of crystal symmetry. She worked in collaboration with the chemist W T Astbury, and they paid particular attention to deriving crystal structures from the results that were beginning to be obtained in the new science of X-ray diffraction. Her education had been in physics and mathematics, and on this problem she was highly successful. In 1924 Yardley and Astbury published an important paper with the title 'Tabulated data for the examination of the 230 space-groups by homogeneous X-rays'.

In 1929, X-ray crystallographers from a number of countries met in Zürich to exchange ideas and arrange for the compilation of tables for crystal structure determination. Kathleen Lonsdale (as Kathleen Yardley now was, having married Dr Thomas Lonsdale) was appointed a member of the editorial

**Fig. 7.6** Dame Kathleen Lonsdale (1903–1971). She was born Kathleen Yardley, and at the Royal Institution and later at University College, London, she was one of the great pioneers of X-ray crystallography. In 1945 she and the biologist Marjorie Stephenson (1895–1948) were the first women to be elected Fellows of the Royal Society. Dorothy Hodgkin was the third woman to be elected, two years later.

group for the tables that were to be prepared. Volume 1 of the *International tables for the determination of crystal structure* was published in 1935. After the war it was decided to begin a new series, and Kathleen Lonsdale was appointed the first chairman of the new Commission on Tables. The new series was entitled *International tables for X-ray crystallography*. Five volumes of this appeared from 1952 onwards, and have proved to be of great value to crystallographers. Kathleen Lonsdale was one of the principal editors of all of these volumes. For this, and other important scientific work she carried out (to be discussed later in this chapter), she was appointed Dame of the Order of the British Empire in 1956.

## RÖNTGEN AND THE DISCOVERY OF X-RAYS

The last decade of the nineteenth century, especially 1895–1898, was a remarkable one for physics. X-rays were discovered in 1895, radioactivity in 1896, and the electron (Chapter 6) in 1897. In 1898 Marie and Pierre Curie discovered the elements radium and polonium. The discovery of X-rays, on

8 November 1895, by Wilhelm Konrad Röntgen (1845–1923) was accidental. Röntgen had produced an electric discharge in a vacuum tube which was about 1 metre long, the pressure being about one thousandth of a torr. It happened that he had in his hand a small screen covered with barium platinocyanide, a fluorescent material which he was using as a detector for radiation. Although this detector was concealed by his hand some distance from the discharge tube, Röntgen was startled to see that it gave off fluorescence. During the next few days he feverishly made further investigations, and soon found that photographic plates which were well protected from light became fogged and spoilt if they were in the neighbourhood of a discharge tube, as if they had been exposed to light. He realized that he had discovered something unusual—radiation that could penetrate material that would not allow ordinary light to penetrate it. Knowing that his discovery would create something of a sensation he was extremely careful with his experiments, making sure that there was no mistake. Finally, on 28 December, he submitted his first paper on the rays, and published again on the same subject in March 1896 and in March 1897. Although he published many subsequent papers on other topics, he never again wrote on this new radiation. In Germany to this day the radiations are usually called Röntgenstrahlen; elsewhere they are called X-rays.

In his first paper on his rays Röntgen described the initial observations he had made with the fluorescent screen, and also said that he had obtained photographs of 'the shadows of the bones of the hand…of a set of weights enclosed in a small box….and so on'. On 13 January Röntgen gave a demonstration before Kaiser Wilhelm II in Berlin. This was the only demonstration he ever gave, as he shunned publicity. His work on X-rays was done at the University of Würzburg, where he had been professor of physics since 1888, later serving as its rector. In 1901 he was awarded the first Nobel Prize in physics for his discovery of X-rays, but his reticence and dislike of lecturing were such that he left before giving the lecture that is expected of the award winners. He declined invitations to return to Sweden and to give a lecture on a later occasion.

The announcement of the discovery of X-rays was received by the scientific community with intense interest. Complete translations of Röntgen's first paper appeared in *Nature* on 23 January 1896 and in *Science* on 13 February 1896. In 1896 *Nature* also published an interesting X-ray photograph of a hand wearing two rings, taken by Walther Nernst in his laboratories at Göttingen. The intensity of the X-rays must have been high, as the flesh is only just visible, while the rings and bones are very dark.

It is interesting to contrast the excitement over the discovery of X-rays with the initial indifference when radioactivity was discovered by Antoine Henri Becquerel (1852–1908) only four months later; the press took no notice at all, and the interest in it was not really aroused until Marie and Pierre Curie began to have success with their investigations.

It was soon found that the X-rays were produced at greater intensities when rays from a discharge tube strike certain substances such as metals. They were found to have shorter wavelengths than visible light (Fig. 5.16).

The important medical use for X-rays, for the examination of bones, was quickly recognized. Only three months after Röntgen's discovery, a surgeon in Dartmouth, Massachusetts, set a patient's broken arm with the use of X-rays. By a curious coincidence, later in the same year five-year-old Lawrence Bragg, who was to play such a key role in X-ray crystallography, had his shattered left elbow set with the help of X-rays. His father, W H Bragg, professor of physics at Adelaide, had set up X-ray apparatus in his laboratory. The big induction coil buzzed loudly, sparks crackled and the tube emitted a weird green glow. A photograph was taken of the boy's elbow, and it showed clearly the location and extent of the injury. This was the first use of the rays in Australia.

It is amusing to recall that at first the properties of X-rays produced some alarm amongst the public. In particular, women believed that men would be able to carry X-ray machines about with them and see too many details of feminine anatomy.

It was not at first realized that X-rays must be treated with care, as they can have injurious effects. An early suggestion of this appeared in a short report in *Nature* in 1896, in which a worker reported that the rays could be harmful to one's vision. For a time, however, in spite of warnings, many public demonstrations of X-rays were given without adequate protection. In 1896 a member of the academic staff at Columbia University gave demonstrations over a period of weeks at Bloomingdale's, the New York department store, but later suffered severe skin damage. Even when precautions were taken, they were usually inadequate. Many members of the medical profession paid scant attention to the need to take great care, and were to pay a heavy price later. A clear idea of the casual way in which early X-ray photographs were taken is provided by an excellent detective story, *The eye of Osiris* (American title: *The vanishing man*), published in 1911 by R Austin Freeman, himself a medical practitioner.

As so often happens with an important scientific discovery, others had observed the effects previously but had failed to recognize it. After Röntgen's discovery Sir William Crookes realized that years before he had returned to the manufacturer some photographic plates with the complaint that they were fogged; he later remembered that they had been close to a discharge tube, and had been affected by the X-rays emitted. Even more remarkable is the fact that the Revd Frederick John Smith (1848–1911) had been warning people for some years that photographic plates would be fogged unless they were kept well clear of discharge tubes, even if they were well protected from light; Smith, who later changed his name to Jervis-Smith, was lecturer in experimental mechanics at Oxford. Oliver Lodge, among others, was aware of this precaution, but no one apparently thought of investigating the curious phenomenon further—until Röntgen did so in 1895.

## THE BIRTH OF X-RAY CRYSTALLOGRAPHY

An important application of X-rays, and the particular concern of this chapter, is concerned with the investigation of the structure of molecules. It was mentioned in Chapter 3 that instead of using a prism to produce a spectrum, it is possible to diffract the light by means of diffraction grating, which may be a sheet of glass on which closely-spaced lines have been etched. For effective diffraction, the spacing between the lines must be similar to the wavelength of the light. In the case of visible light, the lines can be produced mechanically, but for X-rays the wavelengths are too small for this to be possible. However, the wavelengths of X-rays are similar to the distances between neighbouring atoms or ions in crystals. This important point was noticed in 1912 by the German physicist Max Theodor Felix von Laue (1879–1960), who realized that it should therefore be possible to use crystals as gratings for the diffraction of X-rays. With his colleagues W Friedrich and Paul Knipping (1883–1935), von Laue confirmed this prediction experimentally, and for this contribution he was awarded the 1914 Nobel Prize for physics.

In their diffraction experiments the German investigators had used 'white' X-rays, having a wide range of wavelengths. Their diffraction patterns were therefore complex, and could be interpreted only with difficulty even for the simplest crystals. Soon afterwards another X-ray diffraction technique was introduced by W H Bragg and his son W L Bragg, and in the end this proved to be more useful.

William Henry Bragg (1862–1942; Fig. 7.7) was born in Cumberland and educated at Cambridge. He first became professor of mathematics and physics at the University of Adelaide, serving there from 1886 to 1908. In later life Bragg enjoyed telling that when he took up the position in Australia he knew hardly any physics at all, since at Cambridge he had studied nothing but mathematics and had never worked in a physics laboratory; on the boat he learned as much physics as he could to give his lectures. Bragg is unusual for the fact that his research career lay largely dormant until he was in his forties. His life in Australia was rather a leisurely one, and he published only three unimportant papers during the first 18 years he was there.

Beginning in 1903, when he was 41, he became interested in radioactivity and X-rays, and decided to begin research in experimental physics. In 1909 he was appointed professor of physics at the University of Leeds, and there carried out some investigations on the ionizing radiation emitted by radioactive substances, and on X-rays. In 1912 (then aged 50) he learnt of von Laue's results, and so became interested in X-ray diffraction. At first he was strongly of the opinion that X-rays were not a form of electromagnetic radiation, but involved a stream of particles. He interpreted von Laue's result in terms of the collisions of the particles with the atoms in the crystal, but many physicists disagreed with this interpretation. It is ironic that at just about the same time Albert Einstein was pointing out that electromagnetic radiation

**Fig. 7.7** William Henry Bragg (1862–1942). With his son William Lawrence Bragg he pioneered the analysis of structure by X-ray crystallography. In particular, he was involved in the first studies of organic compounds.

does show particle properties in some experiments (Chapter 8). However, in the particular experiments that Bragg was concerned with, the diffraction of X-rays by crystals, it is the wave properties of the radiation that are significant; Bragg was mistaken in pressing his particle interpretation.

In 1912 Bragg's son William Lawrence Bragg (1890–1971; Fig. 7.8) was a research student in Cambridge, working under the direction of J J Thomson on a problem that had nothing to do with X-rays. However, as a result of discussions with his father he too became interested in von Laue's results. His conclusion, in contrast to his father's, was that X-rays are not particles but are electromagnetic, although he discreetly avoided disagreeing with his father in public. There was, in fact, a discrepancy in von Laue's results that led to some confusion. Von Laue had examined the structure of the mineral zinc blende (zinc sulphide, ZnS) and had concluded that it had a simple cubic lattice—similar to that later found for sodium chloride (Fig. 7.9). This result did not, however, follow directly from the X-ray patterns, and in fact was wrong; ZnS actually has the more complicated structure shown in Fig. 7.10. It was because of this anomaly that W H Bragg first thought that X-rays could not be electromagnetic; later, when he and his son had obtained the correct structure, he changed his mind.

In Cambridge Lawrence Bragg quickly developed the now well-known 'Bragg equation', which relates the angle at which the rays impinge on the crystal to give diffraction, to the wavelength of the X-rays and the distance

**Fig. 7.8** William Lawrence Bragg (1890–1971) derived the fundamental equation used in the analysis of crystal structures by the use of X-rays. His career spanned the earliest elucidation of simple structures to the study of large molecules like proteins and DNA.

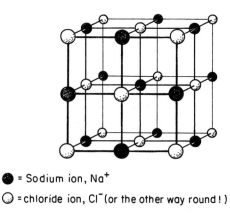

● = Sodium ion, Na$^+$

◎ = chloride ion, Cl$^-$ (or the other way round!)

**Fig. 7.9** A model of the sodium chloride (NaCl) lattice. For clarity the ions are shown held together by 'sticks' but in reality it is electrical attractions that hold them together. Each sodium ion is surrounded symmetrically by six chloride ions, and each chloride ion by six sodium ions. No individual sodium chloride molecules can be identified—to the indignation of some chemists!

between the particles in the crystal. The equation, a simple one, was communicated to the Cambridge Philosophical Society in November 1912, and the paper was published in February 1913 (when the author had still not reached his 23rd birthday). Lawrence Bragg realized at once that his equation allowed

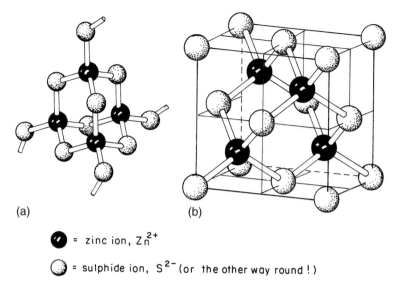

(a)                                  (b)

⬤ = zinc ion, $Zn^{2+}$

◯ = sulphide ion, $S^{2-}$ (or the other way round!)

**Fig. 7.10** (a) The structure of zinc blende, ZnS. Each sulphur ion is surrounded tetrahedrally by four zinc ions, and each zinc ion by four sulphur ions. Again the bonds, represented by sticks, are electrostatic. If all the atoms were replaced by carbon atoms, the result would be the diamond structure (Fig. 7.11(a)). (b) The same structure, showing how it fits into a cubic crystal lattice. This particular cubic lattice is called a face-centred lattice; there is an atom at the centre of each face of the cube.

him to measure the distance between the layers in a crystal. He chose rock salt (sodium chloride, NaCl) as a simple crystal with which to work. Instead of using X-rays that had a wide range of wavelengths—as had been done by von Laue—Bragg used X-rays having a narrow range. In this way he obtained simpler diffraction patterns which were easier to analyse.

Until then chemists had taken it for granted that a crystal like sodium chloride would contain sodium chloride (NaCl) molecules, in which sodium and chlorine atoms were held together by a chemical bond. Bragg's X-ray work, however, led to the structure shown in Fig. 7.9. There are no individual NaCl molecules present; instead there are positively charged sodium ions, $Na^+$, and negatively charged chloride ions, $Cl^-$. These are arranged in a particular kind of crystal lattice, in which each $Na^+$ ion is surrounded by six $Cl^-$ ions, all of which are at the same distance from it. Each chloride ion is surrounded equidistantly by six sodium ions.

This conclusion caused chemists much astonishment, and some could not believe it at all. Lawrence Bragg was asked to make sure that a sodium ion was not nearer to one of the six chloride ions surrounding it than to the other five; then one could talk about sodium chloride molecules in the crystal. But Bragg was quite firm; his results excluded that possibility.

One particularly vigorous opponent of the structure was Henry Edward Armstrong (1848–1937), who from 1884 to 1911 was professor of chemistry at the Central Technical College, one of the many colleges which eventually became absorbed into the Imperial College of Science and Technology which is part of the University of London. Armstrong, a large man who always expressed himself forcefully, was a distinguished organic chemist who also did a considerable amount of work in electrochemistry. When Svante Arrhenius had proposed in 1887 that substances like sodium chloride dissociated into $Na^+$ and $Cl^-$ ions when dissolved in water, Armstrong was harshly critical, and even lampooned the idea in two fairy tales he wrote, *A dream of fair Hydrone*, and *The thirst of salt water*. But Arrhenius was right, and Armstrong wrong.

It is therefore not surprising that when Bragg suggested that there is ionic dissociation even in solid sodium chloride, Armstrong was even more derisive. He sent a letter to *Nature*, entitled 'Poor common salt', and it read as follows:

'Some books are lies frae end to end' says Burns. Scientific speculation (save the mark) seems to be on its way to this state! .... Prof W L Bragg [he was not to be a professor until 1919] asserts that 'In sodium chloride there appear to be no molecules represented by NaCl. The equality in numbers of sodium and chlorine atoms is arrived at by a chess board pattern of these atoms; it is a result of geometry and not of a pairing-off of the atoms'.

This statement is more than 'repugnant to common sense'. It is absurd to the *n*...th degree, not chemical cricket. Chemistry is neither chess nor geometry, whatever X-ray physics may be. Such unjustified aspersion of the molecular character of our most necessary condiment must not be allowed any longer to pass unchallenged.... It were time that chemists took charge of chemistry once more and protected neophytes against the worship of false gods; at least taught them to ask for something more than chess-board evidence.'

Bragg's conclusion was quite correct, and Armstrong's criticism makes strange reading today. It is surprising that a man of his undoubted ability in certain fields should have published such an unscientific statement—and even more surprising that a distinguished journal like *Nature* would publish it. No rational arguments are put forward, and the reference to chemical cricket is hardly relevant or persuasive. Armstrong was aptly described by an obituarist, Sir F Keeble, as having 'all the spare parts of genius but not the long patience to put them together'. Within a few years most scientists had become convinced by Bragg's evidence, but inevitably there were a few who held out. Even as late as the 1930s Louis Kahlenberg (1870–1941) was still insisting in his chemistry lectures at the University of Wisconsin that ions do not exist, either in solids or in solution.

J J Thomson had assigned a different research problem to Bragg, but on learning of the successful X-ray work he encouraged him to continue it and even suggested experiments to him. However, Bragg ran into trouble with the laboratory steward at the Cavendish Laboratory. Laboratory stewards, who are in charge of the equipment in some British university laboratories, were at one time (and perhaps sometimes still are) laws unto themselves, and could not be tamed even by the head of the department. Finding himself unable to get some of the equipment he needed for his X-ray work, Bragg decided to leave Cambridge and work with his father at Leeds.

W H Bragg had been working on somewhat different X-ray problems. In particular, he had constructed the first X-ray spectrometer, an instrument that would improve the procedures for examining the structures of crystals. This in itself was a remarkable technical achievement. In those early days the production of X-rays having a narrow band of wavelengths was exceedingly difficult, and W H Bragg showed great skill in overcoming the problems. The spectrometer he built made it easy to carry out the diffraction experiments. Father and son collaborated on examining a number of inorganic crystals, including ZnS (Fig. 7.10), and establishing their structures. By the end of 1913, the Braggs had reduced to a standard procedure the examination of lattice spacings in inorganic crystals. W L Bragg later remarked that with his father's spectrometer the work was easy—like picking up nuggets in a gold field; one did not even have to dig for them.

One of the structures established by the Braggs, in 1914, was that of diamond. The structure, shown in Fig. 7.11(a), is a simple one, consisting of carbon atoms surrounded tetrahedrally by four other carbon atoms. A whole diamond is therefore really a single molecule; it cannot be said to be made

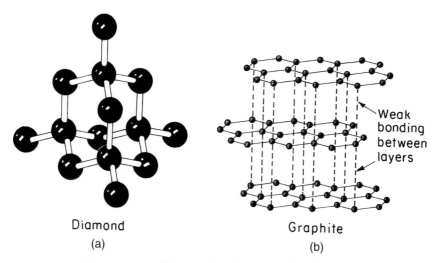

Diamond
(a)

Graphite
(b)

Weak bonding between layers

**Fig. 7.11**   (a) The structure of diamond, as determined by the Braggs in 1914. Each carbon atom is surrounded tetrahedrally by four other carbon atoms. (b) The structure of graphite, another form of carbon. The hexagonal rings of carbon atoms are similar to those in benzene rings. The layers can easily pass each other, so that graphite is a lubricant. This structure was established in 1924 by J D Bernal, by X-ray crystallography.

up of smaller molecules. In view of van't Hoff's work forty years earlier it seems obvious that this should be the structure of diamond, and yet chemists do not appear to have thought of such a structure before. Another form of pure carbon is graphite, which has the structure shown in Fig. 7.11(b). Because it exists in layers which can easily slide past each other, graphite is a good lubricant.

For the work they did on the structures of various crystals the Braggs shared the 1915 Nobel Prize for physics. This joint award was remarkable in two ways; it is the only time a father and son team have been awarded the Prize, and Lawrence, at 25, was the youngest person ever to receive a Nobel Prize. He also holds the record in being a Nobel Laureate for 56 years. When they received the award it was natural for others to assume that W H Bragg had led the research, and he later had difficulty in convincing people that the basic ideas were not his but his son's. W H Bragg was a modest man, who always took great pride in the achievements of his 'boy', as he called his son even after he was knighted in 1941. Besides being an excellent scientist, W H Bragg gave a great many popular scientific lectures, at the Royal Institution and elsewhere. One of them had the intriguing title 'Why people sing in bathrooms'.

## THE STRUCTURE OF MINERALS

When Lawrence Bragg heard in 1915 that he had been awarded a Nobel Prize he was in France setting up a sound ranging station, for gun location, near Ypres. The village curé with whom he was billeted produced a bottle of *Lachrimae Christi* with which they celebrated the award. Fifty years later Bragg visited Stockholm on the occasion of the Nobel Prize awards, and his own success was then celebrated more formally.

After the war, in 1919, Lawrence Bragg was appointed professor of physics at Manchester University. Having found the study of simple inorganic crystals quite straightforward, he wanted to work on something more demanding. He came to an agreement with his father that he would confine his efforts, at least at first, to inorganic compounds—those that do not contain carbon— leaving the organic compounds to W H Bragg, who in 1915 had moved from Leeds to University College, London. One of his early achievements in Leeds was to compile a table of sizes of common ions, based on their distance apart in a large number of salts. It was soon found that he had made all the negative ions too small and the positive ions too large, but this was soon corrected, particularly by the distinguished Swiss–Norwegian crystallographer Victor Moritz Goldschmidt (1888–1947).

At the suggestion of the mineralogist Sir Henry Miers, who was Vice-Chancellor of the University of Manchester, Lawrence Bragg decided to tackle the problem of the structures of minerals, the main solid constituents of planets including the earth. A great deal of effort had already gone into investigating the structures of minerals, but they proved to be very complicated. The atoms in minerals do not always combine together in simple proportions as they do in simple compounds, and odd elements often appear quite unsystematically.

Lawrence Bragg later admitted that the X-ray photographs of minerals often looked extremely formidable. A few, however, could be analysed easily. The mineral beryl, which contains the elements beryllium (Be), aluminium (Al), silicon (Si), and oxygen (O), having the formula $Be_3Al_2Si_6O_{18}$, proved easy, and its structure was established by W L Bragg and J West in a single afternoon. Within the next few years Bragg and his colleagues had established the structures of a considerable number of minerals.

What they found with many minerals was that a basic unit existed, consisting of a silicon atom attached tetrahedrally to four oxygen atoms (Fig. 7.12). Other atoms could fit into lattice spaces, and different atoms could replace each other.

Lawrence Bragg was also one of the pioneers in the application of Fourier analysis to the examination of crystal structures. This is a mathematical procedure, originally suggested by Jean Baptiste Joseph Fourier (1768–1830) for the analysis of the conduction of heat. A Fourier series is an infinite series of terms which are constants multiplied by trigonometric functions. It can be used to give an approximation to the kind of mathematical functions that are involved in X-ray analysis, and provides a neat way of analysing the rather complicated patterns that can be found in all but the simplest crystal structures. In 1929 Lawrence Bragg was successful in applying the method to the analysis of minerals.

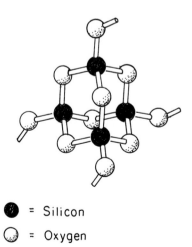

● = Silicon

○ = Oxygen

**Fig. 7.12**   The type of structure found in some silicates and many other minerals. Each silicon atom is surrounded tetrahedrally by four oxygen atoms, and each oxygen atom is connected to two silicon atoms.

In 1934 a useful procedure was suggested by A Lindo Patterson (b. 1902). Born in New Zealand, Patterson received his early education in Canada and England; after taking his degree at McGill University in Montreal he worked for two years with W H Bragg at the Royal Institution, later obtaining his PhD degree at McGill in 1928. For some years he taught at Bryn Mawr College in Pennsylvania, and it was there that he introduced an effective Fourier procedure for the analysis of X-ray diffraction patterns. His method allowed the construction, from the data, of a so-called Patterson diagram, which consists of contour lines connecting positions of equal electron density; we will see some examples of Patterson diagrams later in this chapter.

Until well after World War II, the calculations that had to be made to obtain structures from X-ray data were extremely laborious and tedious. In the 1930s Henry Solomon Lipson (1910–1991) and his former student Cecil Arnold Beevers, who were at the University of Manchester, introduced a helpful aid by producing what came to be called the 'Beevers–Lipson strips'. Certain routine calculations were taking them about 90 minutes each time, and they found that sometimes they later needed the same calculations for another problem. They began by recording the results in a notebook, but realized that it would be better to summarize the results on strips of paper that could be kept in a box in the right order (and which they had to be careful not to drop). The fame of these strips spread, and other crystallographers asked for copies. It cost £200, then a very large sum to a scientist, to print a number of copies of them, and over the years Beevers and Lipson were able to recover the cost—and even make a small profit—by selling the boxes of strips for £5 each. About 500 sets were sold up to about 1970, when they became obsolete with the advent of computers.

The heavy computational work was reduced if one initially had some idea of the structure, and the early workers showed considerable ingenuity in relating what they already knew to their X-ray results. Sometimes they were helped by inspired intuition, and even by guesswork.

Metals and alloys were also studied by W L Bragg at Manchester. In 1937 he accepted the post of Director of the National Physical Laboratory at Teddington, near London, but within a year was invited to succeed Lord Rutherford as Cavendish professor of physics in Cambridge. To a great extent his work there was necessarily administrative, but he was remarkably successful in attracting to the Cavendish laboratories a number of outstanding scientists, some of whose work will be mentioned later in this chapter. His final appointment, in 1954, was as director of the Royal Institution, a post which his father had held until his death in 1942. In that position he was required to reside in the official apartments in Burlington House. Having always been an enthusiastic gardener, and finding the lack of a garden frustrating, Bragg decided to take every Thursday afternoon off. He gained employment as part-time gardener to a lady who lived near the centre of London, asking her to call him Willy. All went well until one afternoon a lady visiting the house asked her hostess 'Why in the world is Sir Lawrence Bragg working in your garden?'

## ORGANIC MOLECULES

Just as Lawrence Bragg had been a pioneer in the X-ray analysis of inorganic crystals, his father W H Bragg played a great role in initiating the work on organic structures. After World War I he was still at University College, London, and he was knighted in 1920. In 1923 he was appointed director of the Royal Institution, which for some years had been somewhat moribund. There he soon assembled a highly effective team of research workers, many of whom were to make distinguished contributions to the examination of the structure of crystals. Two members of the team, Kathleen Yardley (Lonsdale) and A Lindo Patterson, have already been mentioned. Other members of the team were W T Astbury, J D Bernal, and J M Robertson, all of whom are mentioned again later in this chapter.

In 1927 Kathleen Yardley married Thomas Lonsdale and moved to Leeds, where Thomas worked for the Silk Research Association, then housed at the University of Leeds. With her husband's strong encouragement Kathleen Lonsdale continued her experimental work at the University, with the aid of a grant from the Royal Society. Christopher Kelk Ingold (1893–1970) was then professor of chemistry at the University of Leeds, and he gave her some beautiful crystals of hexamethyl benzene. With them she made one of her more important contributions by establishing its structure by X-ray analysis. She found that the benzene ring was hexagonal and flat (Fig. 7.13), and she gave its exact dimensions. This result had at the time been inferred by organic

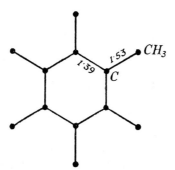

**Fig. 7.13**    The diagram given by Kathleen Lonsdale for the structure of hexamethyl benzene. Chemists had previously concluded that the benzene ring was flat, but this was the first direct proof that this was really the case. The numbers are Lonsdale's values for the lengths of the carbon–carbon bonds, expressed in Ångstrom units; $1 \text{ Å} = 10^{-10}$ m.

chemists, but there was no direct proof. Ingold's comment on her work was 'The calculations must have been dreadful but one paper like this brings more certainty into organic chemistry than generations of the activity by professionals'. Lonsdale herself insisted that she 'knew no organic chemistry and very little of any other kind'.

In 1929 Kathleen Lonsdale resumed her position at the Royal Institution, and in 1946 she was appointed Reader in crystallography at University College, London. Three years later she became professor of chemistry and head of the Department of Crystallography. She then, at the age of 43, began to take part in university teaching and in the development of her own research school. All of her work had a profound influence on the development of X-ray crystallography. Her later work covered a wide range. She had a special interest in the structure of diamonds, and by making important improvements in the techniques of X-ray crystallography she was able to measure the carbon–carbon distances in diamonds to seven figures.

Another who made important examinations of the structures of organic molecules was the Scotsman John Monteith Robertson (1900–1989; Fig. 7.14). He spent his boyhood in Auchterarder, Perthshire, and attended the University of Glasgow, obtaining his PhD degree in 1926. he was at the Royal Institution from 1926 to 1928, and W H Bragg suggested that he study the structures of naphthalene, anthracene, and various other organic compounds. Robertson's final appointment, from 1942 to 1970, was as professor of chemistry at the University of Glasgow.

**Fig. 7.14**  John Monteith Robertson (1900–1989), who for many years was professor of chemistry at the University of Glasgow. By X-ray methods he established the structures of naphthalene, anthracene, and many substances of biological interest.

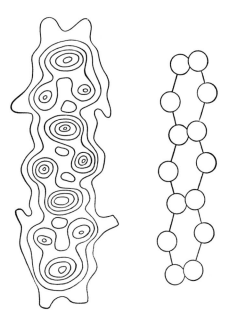

**Fig. 7.15**   The structure of anthracene, as shown by John Monteath Robinson. Also shown is the electron-density contour diagram obtained by Robertson from the X-ray patterns.

By the application of Fourier analysis to X-ray data he obtained from anthracene, Robertson obtained the electron-density (Patterson) diagram shown in Fig. 7.15, which also shows the structure derived from it. When Robertson established this structure in 1936, anthracene was the largest organic molecule that had been analysed in this way. In 1960 he examined the structure of limonin, which is the bitter principle of citrus fruits. Although the compound had been known since 1841, organic chemists had found it impossible to establish its structure. Later Robertson was successful in obtaining the structures of a number of other compounds, many of them found in biological systems.

## PENICILLIN AND VITAMIN B$_{12}$

Outstanding contributions to X-ray crystallography were made by Dorothy Hodgkin (1910–1994; Fig. 7.16), who won a Nobel Prize for her work. She was born Dorothy Crowfoot in Cairo of British parents, and after a somewhat irregular schooling went to Somerville College, Oxford, in 1928. Her choice of college was appropriate, as it had been named in honour of another distinguished scientist, the mathematician and theoretical physicist Mary Somerville (1780–1872), whom she resembled in many respects. Mary

**Fig. 7.16** Dorothy Crowfoot Hodgkin (1910–1994), whose elucidation of the structures of penicillin and vitamin $B_{12}$ led to her being awarded the 1964 Nobel Prize. In the following year Queen Elizabeth appointed her a member of the Order of Merit. This is a particularly exclusive British honour, restricted to 24 members, and the only previous female member had been Florence Nightingale. Hodgkin had modestly expressed a wish not to be given any honour that carried a title, and was pleased to be made a member of the order rather than become Dame Dorothy. She later established the structure of insulin.

Somerville, Kathleen Lonsdale, and Dorothy Hodgkin were all able to combine the responsibilities of family life with outstanding contributions to science, and all exerted a profound influence in many other ways. After taking her Oxford degree in 1932, Dorothy Crowfoot (Hodgkin) decided to move to Cambridge to work with J D Bernal, who was just starting to do X-ray work on the structure of protein molecules and other substances of biological importance. There, working for her PhD degree, she took some of the earliest X-ray photographs of proteins, but she and Bernal realized that the time was not yet ripe for their analysis.

In 1937 Dorothy Crowfoot returned to Somerville College as a tutor in chemistry. In the same year she married Thomas Hodgkin, a historian, who realized that his wife was more creative than he was; as a result he did everything in his power to help her, particularly with the raising of their family. (One is reminded of Dr Somerville, who was similarly helpful to Mary Somerville.) The Hodgkins had three children, and eventually nine grandchildren and some great-grandchildren.

When Dorothy Hodgkin arrived in Oxford the University had no X-ray crystallographic equipment, and her first task was to assemble some. Early in the Second World War she became interested in penicillin, which had been discovered in 1928 by the Scottish bacteriologist Sir Alexander Fleming (1881–1955). This substance was recognized to have enormous potential for the prevention of bacterial infection, and intensive work on it was carried out at Oxford with a view to finding out the structure of the molecule, so that it could be synthesized. The leaders of this research were the Australian-born pathologist Howard Walter Florey (later Baron Florey, 1898–1968) and the German-born biochemist Ernst Boris Chain (1906–1979). (Fleming, Florey, and Chain shared the 1945 Nobel Prize for medicine and physiology for this work.) It was soon realized that the structure of the penicillin molecule was quite unusual, and difficult to establish by the ordinary methods of organic chemistry.

Crystals of penicillin were provided to Dorothy Hodgkin, who found that there were serious difficulties with the X-ray analysis. One problem was that the organic chemists had been unable to discover what chemical groups were present. Hodgkin and her assistants concluded from the X-ray patterns that part of the molecule was a beta-lactam ring, which is a ring of three carbon atoms and one nitrogen atom. That conclusion created something of a controversy, as the organic chemists felt sure that such a structure was highly improbable. In the end she was proved to be right. The later analysis of the X-ray patterns was carried out with the aid of a primitive International Business Machine (IBM) digital computer, the data being punched on cards. By day the computer tracked ships' cargoes, and by night it analysed the structure of penicillin. As the work was highly classified until after the war, the penicillin data were disguised as ship-cargo data, which would have confused the enemy in case of a breach of security. Dorothy Hodgkin was unusually skilful in this type of investigation; she had an uncanny skill at being able to make guesses, which often turned out to be right, and which led—only after much effort which demanded great patience—to the final structure.

Before the end of the war, the complete structure of penicillin had been established, the first time that X-ray analysis had been successful before a structure could be discovered by conventional chemical methods. Fig. 7.17(a) shows the electron-density contour diagram obtained for the molecule, with the outline of the structure superimposed on it. Figure 7.17(b) shows a 'ball and stick' model of the molecule. As it turned out, the establishment of the structure did not immediately affect the supplies of penicillin, since improvements in fermentation techniques before the end of the war had led to its production in large quantities. During the final months of the war penicillin was available for the treatment of wounds, and later it became available in general medical practice at a reasonably low cost. Dorothy Hodgkin's establishment of the structure became of great importance later, when chemists extended their efforts to the synthesis of modified forms of the penicillin molecule.

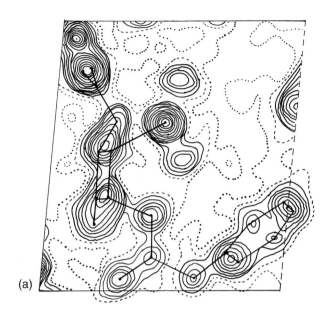

(a)

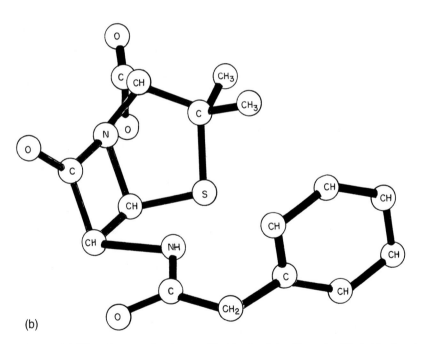

(b)

**Fig. 7.17** (a) The electron-density profile obtained by Dorothy Hodgkin for the penicillin molecule, with the structure superimposed on it. (b) A model of the penicillin molecule, based on the X-ray work. The β-lactam ring, which the organic chemists thought could not be present, is the four-sided ring seen in half profile at the left.

In about 1948 Dorothy Hodgkin began work on the structure of the vitamin $B_{12}$ molecule, which is about four times as large as the penicillin molecule and is one of the most complicated molecules of a non-protein nature. It contains one cobalt atom and 180 other atoms, and the organic chemists had been able to discover little of its structure. The X-ray analysis of the structure by the computational methods previously used would have taken an enormous amount of time. Fortunately Hodgkin was able to make use of a National Bureau of Standards Western Automatic Computer (SWAC), on the campus of the University of California at Los Angeles (UCLA). This made it possible for her and her students to complete the structural analysis by 1956.

The discovery of the structure of so large a molecule as vitamin $B_{12}$ was widely recognized as an enormous achievement, and Sir Lawrence Bragg summarized the general admiration with his comment that Dorothy Hodgkin had 'broken the sound barrier'. Since she had published the penicillin work under the name of Dorothy Crowfoot, and the vitamin $B_{12}$ work as Dorothy Hodgkin, it was widely believed at first that there were two brilliant X-ray crystallographers at Oxford.

It was mainly as a result of the vitamin $B_{12}$ work that Dorothy Hodgkin was awarded the 1964 Nobel Prize for chemistry. After receiving the Prize she attacked an even more challenging problem, the structure of insulin, which is one of the smallest of the protein molecules. We will defer discussion of her work on insulin until later in this chapter.

Kathleen Lonsdale and Dorothy Hodgkin had other things in common besides being great women crystallographers. They both involved themselves in the cause of world peace. In the First World War Lonsdale even spent a month in jail for refusing to register for military service, a formality in her case since she had three children and would not have been required to serve. Both Lonsdale and Hodgkin were active members of the Pugwash movement which is concerned with international harmony, both expressed their opposition to the Vietnam war, and both travelled widely in the cause of peace, visiting both China and Russia. On one occasion Dorothy Hodgkin wrote to British prime minister Margaret Thatcher, who had been her student at Oxford, gently chiding her for criticizing Russia before she had visited the country. As a result Thatcher did visit Russia, and found to her pleasant surprise that she could establish close accord with Mikhail Gorbachev and other Russian leaders.

It is a curious irony that both Kathleen Lonsdale and Dorothy Hodgkin found it much easier to enter Russia and China than that purportedly democratic country the United States. Both, in fact, were for many years excluded from entering the US except by special waiver from the Attorney General. On one occasion Kathleen Lonsdale was told by a US official that the three objections to her entering were, in increasing order of importance, her jail term, her visits to China, and her visits to Russia. Dorothy Hodgkin's exclusion is all the more remarkable in view of the fact that her husband, unlike her, was an avowed communist, and yet he had no difficulty in entering the United States. Apparently Dorothy's gently expressed opinions about world peace was considered by some officials to make her more a threat to the stability of the US government than a member of the Communist party.

## THE STRUCTURE OF FIBRES

Information of great interest and practical importance has come from the study of various kinds of fibres. Some of these, such as hair and wool (which are proteins) and natural rubber, are found in nature. Many of the fibrous materials we use today are synthetic, i.e., are manufactured; examples are the rubber that is used for tyres, and nylon.

One of the pioneers in the study of fibres was William Thomas Astbury (1889–1961), who besides being a capable scientist was what is known as a 'character'. He was born near Stoke-on-Trent, the son of a potter's turner, and was always proud of his humble birth. He won a scholarship to Cambridge where he studied chemistry, physics, and mathematics. He made particular study of crystallography, and in fact knew more about that subject than either Sir William or Sir Lawrence Bragg. As a result of his extrovert and blunt manner he was referred to as a 'John Bull of an Englishman', an *enfant terrible* (by Dorothy Hodgkin), and a 'card' (by Desmond Bernal and others). Many people, although admiring his ability, found him overbearing and overconfident in his own ideas. I met him as a graduate student at Princeton, with my research director Henry Eyring. We found it quite difficult to converse with him, as he talked all the time and seemed to have no interest in what anyone else had to say—and Henry Eyring really did know a lot about fibres.

Astbury's work on fibres began when he was with Sir William Bragg at the Royal Institution in the 1920s, and it was somewhat by accident. Bragg did not give his colleagues much direction, leaving them to choose their own problems, but he gave many popular lectures, and often asked his assistants to prepare demonstrations for him. Often they did not enjoy doing this, especially as sometimes they had to devote several weeks to the work. For a lecture called 'The imperfect crystallization of common things', Bragg asked Astbury in 1926 to take X-ray photographs of fibres such as wool and silk.

This Astbury did very effectively, and soon became fascinated by the biological implications of the work. he was one of the first people to work in the field of 'molecular biology', a term which he perhaps coined himself in the 1920s; it was also used from that time by Bernal. In 1928 the University of Leeds had an opening for a lecturer in textile physics, and Bragg persuaded Astbury to accept it, which he did a little reluctantly. He was successful in his research on fibres, and in 1945 was appointed the first professor of biomolecular structure at the University. Astbury himself wanted the title to be professor of molecular biology, but the authorities thought that he was not really qualified to be called a biologist.

During the first few years in Leeds he did important work—perhaps his most important work—on the structure of fibres related to wool. The city of Leeds was heavily involved in the wool industry (as opposed to Liverpool and Manchester which were more concerned with cotton), and Astbury always felt it important to work on problems of local interest. An important constituent of wool is a fibrous protein called keratin (Green κέρας, keras, horn), which also occurs in skin, hair, nails, and feathers.

By this time chemists had recognized that proteins have a role of unique importance in living systems. They are essential components of all biological structures, such as blood and living cells. Aside from their structural role, some proteins, the enzymes, act as biological catalysts, which means that they help to bring about many biological processes, such as those involved in respiration and digestion. It was found that they could be broken down, by a procedure called hydrolysis, into relatively small molecules known as amino acids. The simplest amino acid is glycine which has the structure

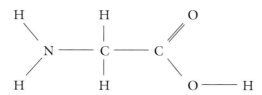

Nearly all proteins are made up of various numbers of 20 different amino acids; the other 19 are similar in general structure to glycine but have one of the central hydrogen atoms replaced by a specific group of atoms, which chemists refer to as side chains.

By the 1920s many, but not all, chemists had accepted the ideas about proteins that had been put forward in 1901 by the great German organic chemist Emil Hermann Fischer (1852–1919), who was then professor in Berlin. Fischer suggested that protein molecules are composed of a chain of amino acids connected together by what is called the peptide linkage:

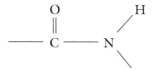

Fischer actually linked together a number of amino acids, to form what are called polypeptides. He received the 1902 Nobel Prize for chemistry, particularly for his important work on sugars.

By chemical means it is possible to find out which amino acid molecules are present in a given protein molecule, and in what numbers. With more difficulty, chemists were later able to discover the order in which the amino acids occur in a protein molecule.

Soon after Astbury began his work on the protein keratin he obtained some excellent X-ray diffraction photographs of it in its normal, unstretched, form, which he called the $\alpha$ (alpha) form. Keratin can be stretched when it is damp, and Astbury found that in steam he could stretch the strands to almost twice their normal length, and could again obtain good X-ray photographs. He called the stretched form the $\beta$ (beta) form.

He realized that even in the alpha form the polypeptide strands are not completely extended, but are folded in some way. The specific folding he suggested for the alpha and beta forms are shown in Fig. 7.18. Here he went wrong, as he insisted on folds in two dimensions. We will see in the next section that helical structures, in three dimensions, occur in proteins; characteristically, Astbury was slow to accept these three-dimensional forms when they were proposed in the 1950s.

Nevertheless, Astbury's ideas were important, if only because they led people to try to prove him wrong, and in that way to arrive at the right structure. This happens quite often in science, and scientists generally agree that it is better to form a hypothesis, even if it proves wrong, than to suggest no explanation for one's results. No one need be embarrassed at suggesting something sensible that is later disproved.

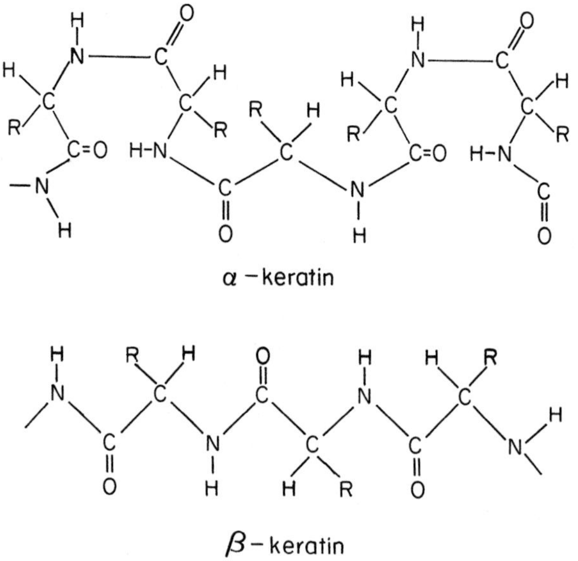

$\alpha$ – keratin

$\beta$ – keratin

Fig. 7.18   William Astbury's idea, which proved to be wrong, of the structures of certain fibres in their contracted (alpha) and extended (beta) forms. He went wrong in assuming the structures to lie in two dimensions instead of three, as in the helical forms suggested later.

Astbury's ideas about the two forms of keratin are nicely summarized in a limerick written by A Lindo Patterson:

Amino acids in chains
Are the cause, so the X ray explains,
Of the stretching of wool
And its strength when you pull,
And show why it shrinks when it rains.

Astbury always kept the wool industry informed of what he was doing, and he was particularly concerned with the practical consequences of his results. He showed how a permanent wave in hair is related to the alpha–beta transformation. If this occurs only on one side of the hair it produces a curl; if it is done alternately on both sides it produces a wave. Astbury even put a permanent wave into a cow's horn. This was not original with him; Pliny the Elder (AD 23–79) reported in his *Historia naturalis* that cattle thieves often did this to change the shapes of cows' horns so that the stolen cows could not be recognized.

Astbury did much work on other proteins besides keratin. He studied myosin, an important component of muscle which is concerned with muscular contraction. He also studied skin keratin (which he called epidermin), and fibrinogen, which is concerned with the clotting of the blood. He concluded that these and some other proteins behave in much the same way as far as the alpha-beta transformation is concerned.

He found, however, that some proteins behave differently. Collagen, for example, is inextensible. This is one of the connective tissue proteins, and it is also important in the leather industry. For many years Astbury made detailed investigations of it, and related his findings to the technical work that is involved in the tanning industry. He also made important early contributions to the investigations of the nucleic acids and other biological substances.

## THE THREE-DIMENSIONAL STRUCTURES OF PROTEINS

Besides Astbury, several others began in the 1930s to consider the detailed three-dimensional structures of protein molecules. One of these was the Irish-born John Desmond Bernal (1901–1971; Fig. 7.19), one of the most colourful figures in the history of science. His friends called him 'Sage', as he seemed to know all about everything—not only science, but art, politics, and women. His appearance was striking; his pallid complexion contrasted with his flaming red hair which matched his communist politics. He lived life to the full, marrying one woman and remaining married to her until his death, but maintaining two other households with other women. He said himself that his biography should be printed on paper of various colours, and in his obituary Dorothy Hodgkin suggested that the appropriate colour for his personal life should be yellow or purple.

**Fig. 7.19**   John Desmond Bernal (1901–1971). His many contributions include the structure of graphite (1924), an important theory of the structure of water and of ionic solutions, and early studies of proteins and other substances of biological interest.

After taking his degree from Cambridge, Bernal was on the staff of the Royal Institution from 1923 to 1927, and worked with Sir William Bragg. One of his achievements there was to discover the structure of graphite (Fig. 7.12(b)). In 1927 he was put in charge of the crystallography unit at Cambridge, and Dorothy Crowfoot (Hodgkin) worked with him on various problems from 1932 to 1935. In particular they took X-ray photographs of pepsin and other proteins (enzymes) that act as catalysts in that they speed up the digestive processes. These proteins are classified as globular proteins, in contrast to fibrous proteins. In the globular proteins the molecules do not exist in an extended form but are wound up into a ball, something like a ball of wool. An important point to realize, however, is that the winding up is not random; the molecules of a given globular protein molecule are all arranged in exactly the same way; they are said to have the same *conformation*.

In the 1930s the chances of making a complete analysis of the X-ray photograph of a protein molecule appeared to be slim, in view of its complexity. In any protein molecule there are many more interatomic distances to be determined than in an inorganic crystal or even a mineral. A further complication is that proteins proved difficult to crystallize; indeed for a time many chemists had believed that it would be impossible to crystallize them.

The problem had begun to look a little simpler by 1928, when the American biochemist James Batcheller Sumner (1887–1955), at Cornell University, crystallized the enzyme urease. At first there was scepticism about his achievement, especially about his announcement that urease was a protein.

However, within a few years other enzymes had been crystallized, including pepsin, trypsin and chymotrypsin, all of which are concerned with digestive processes. They also were found to be proteins. The protein haemoglobin, the major component of the red blood cells and concerned in the transport of oxygen, was also crystallized. Early X-ray crystallographers had tried to make diffraction pictures of some of these crystals, after they had been dried, but the records were much too confused to be interpreted.

In 1934 an important breakthrough was made by J D Bernal and Dorothy Crowfoot (Hodgkin). They discovered that if instead of drying pepsin crystals they left them in contact with the solution from which they had crystallized, the resulting X-ray patterns were sharp enough for interpretation to be a possibility—although only in the distant future, after other problems had been overcome.

In 1935 Isidor Fankuchen (1905–1964), from Brookyln, New York, joined Bernal's laboratory, and they were given some crystals of tobacco mosaic virus. This was of particular importance, because a virus is close to being a living system; alone it cannot reproduce itself, but it can do so if it is inside a living cell. Tobacco mosaic virus was the first virus to be crystallized, by Wendell Meredith Stanley (1904–1971) at the Rockefeller Institute in New York (now Rockefeller University) and independently by Frederick Charles Bawden (1908–1972) and N W Pirie at Cambridge. Bawden and Pirie also established that the virus contains ribonucleic acid (RNA), which had previously been found in all living cells. The fact that it was found in a virus was of particular significance. Bernal and Fankuchen took X-ray diffraction photographs of the virus, and were able to establish that it had a regularly repeating pattern along its length. The RNA in the virus was combined with protein material to form the whole organism.

In 1936 Bernal was joined by Max Perutz (b. 1914), whose subsequent career was entirely at Cambridge. Perutz devoted himself particularly to investigating the structure of haemoglobin, the molecule of which contains four iron atoms. He put immense effort into analysing the X-ray patterns, and at first he found difficulties that appeared to be insuperable. Too little had been discovered by other means about the structure.

When Lawrence Bragg arrived in Cambridge as Cavendish professor in 1938 he was at first much preoccupied with administrative problems. For many years the emphasis in the Cavendish Laboratory had been on nuclear physics. There was, in fact, some disappointment about the appointment of Bragg, as it was thought that the importance of nuclear physics would wane. Bragg, however, was careful not to allow this; nuclear physics remained predominant, but X-ray crystallography was added. There was also important work in the Cavendish Laboratory on mathematical physics, metal physics, low temperature physics, and radio physics, which included the important new field of radio astronomy. As head of such a laboratory Bragg obviously had to spread his interests widely.

Bragg soon learnt from Perutz about his work on haemoglobin, and became enthusiastic about it. Also at Cambridge was John Cowdery Kendrew (b. 1917), who in 1948 began to work on myoglobin, a protein which occurs in muscle. Bragg did much work in collaboration with Perutz and Kendrew, and became interested in the general problem of how polypeptide chains are folded in the various protein structures. The work at Leeds University of Astbury on the proteins that occur in fibres and in hair had suggested that the chains are folded into helices. (A well-known example of a helix is a corkscrew. Helices are often called spirals, but this is not strictly correct; a spiral is really a two-dimensional representation of a helix.) Bragg, Kendrew, and Perutz pursued this idea in a paper they published in 1950, in which they suggested various helical arrangements that the protein chain could assume.

Unfortunately, their suggestions were incorrect, as they had gone wrong on two important matters. In the first place, they had failed to realize that in the peptide linkage, the structure of which was shown earlier, all of the atoms lie in the same plane. This is something that chemists were aware of, but which physicists were apt not to know about. Bragg later said that he could never forgive himself for having been ignorant of this feature of the peptide linkage.

The other matter on which Bragg and his colleagues went wrong was that they assumed that there must be an integral number of amino acids in one complete turn of the helix. In other words, they considered the possibility that, for example, a chain of four amino acids would fit into one of the helix; they also considered five and six. Their reason for doing this was that the X-ray evidence indicated a helix without any bulges, which seemed to require an integral number of units in one turn; actually, this evidence was misleading.

The distinguished American chemist Linus Pauling, at the California Institute of Technology, was well aware of the planarity of the peptide linkage, and did not see that there was any need to assume an integral number of amino acids in each turn. In the spring of 1948 he was visiting professor at Oxford, and one day he was in bed with a cold. To relieve his boredom he amused himself by cutting out strips of paper in the shape of the peptide linkage, and fitting them together. He found in this way that when he placed emphasis on the attractive forces that exist between various parts of the chain, the chain naturally assumed the form of a helix that had 3.7 amino acids per turn of the helix; it is illustrated in Fig. 7.20. When Pauling returned to California he elaborated his ideas with one of his colleagues, Robert B Corey, and they published their helical structure in 1950, giving details in the following year. This suggestion turned out to be the correct one. Pauling's 1954 Nobel Prize for chemistry was partly in recognition of his work on protein structure, but he had made many other contributions of great importance.

This type of helical structure is particularly important in the fibrous proteins which had been so carefully studied by Astbury, although the three-

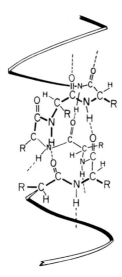

**Fig. 7.20** The type of helical structure that often occurs in proteins. This is one of the possible structures proposed in 1951 by Linus Pauling and Robert B Corey.

dimensional structure had eluded him. The helical structure nearly always occurs, but to a smaller extent, in the globular proteins. Figure 7.21 shows in a purely schematic way the kind of structure that occurs in the enzyme chymotrypsin; very little of the structure is helical.

The helix is usually a right-handed one, such as occurs in an ordinary corkscrew. Very occasionally, however, a left-handed helix is found; as we will see, there is a region of left-handed structure in the insulin molecule, the structure of which we consider later (Fig. 7.23).

In 1953 Perutz made a contribution of vital importance that was of great help in all later work. To explain it, a little more should be said about the difficulties of discovering, from X-ray data, molecular structures of any complexity. If the structure of a molecule is known exactly it is quite straightforward, using the Bragg equation and the Fourier functions, to calculate the electron density map for the molecule. If, therefore, one were to guess a structure, one could confirm it by comparing the electron density diagram with the one obtained experimentally. If one had not got the exact structure there would be discrepancies, and one could make suitable adjustments. This explains why it is easier to ascertain a structure if one starts with a good idea of what the structure will be.

In obtaining the electron-density diagrams from the structure one has to take the squares of certain terms which appear in the Fourier function. Suppose, on the other hand, that one wants to find the structure from the experimental results. One now has to take the square roots of functions, and this is where the difficulty arises. When one takes a square root there are

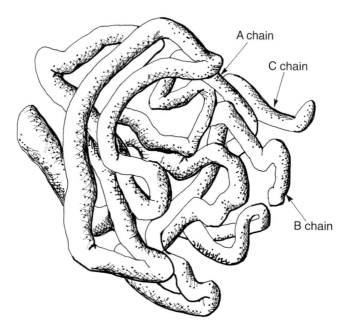

A chain

C chain

B chain

**Fig. 7.21**  The type of structure that occurs in a molecule of chymotrypsin, which is one of the digestive enzymes (a protein). Three separate polypeptide chains occur in the molecule. The molecular weight of the molecule is 24 800, and there are 245 amino acids in it.

always two possibilities—one can take the positive sign or the negative sign. The square root of 4, for example, can be 2 or –2. This presented a great problem to X-ray crystallographers, and in technical language it is referred to as the *phase problem*.

The important contribution that Perutz made in 1953 was to overcome the phase problem by what has been referred to as the method of *isomorphous replacement*. The method involves the use of a heavy atom, which means an atom that has a lot of electrons round it so that it produces a stronger effect on the X-ray photograph. If such a heavy atom, such as an atom of mercury, can be introduced into a protein without changing the crystal structure, and X-ray patterns are obtained, the interpretation becomes easier. One can obtain two X-ray patterns, one with the mercury atom present, and one without it—or with another atom there. One can then look at the difference between the patterns and from it deduce where the mercury is in the molecule. From that information one can make further deductions, and may be led to the final structure. If more than one replacement can be made, so much the better. Good use of the method was later made by Dorothy Hodgkin in her work on vitamin $B_{12}$ which we mentioned earlier.

The method of isomorphous replacement was not original with Perutz, but it was he who first demonstrated its value by applying it to the haemoglobin problem. Kendrew also applied the method to the myoglobin molecule, and in 1960 he found its structure. The haemoglobin molecule has many of the same features as the myoglobin molecule, but is four times as large. With the knowledge of the structure of the myoglobin molecule, Perutz was able in the same year to reveal the complete structure of the haemoglobin molecule, which is shown in Fig. 7.22. For this work Kendrew and Perutz shared the 1962 Nobel Prize for chemistry.

Seven years later Dorothy Hodgkin announced the structure of insulin. This substance, important in the treatment of diabetes, was discovered at the University of Toronto by Frederick Grant Banting (1891–1941) in association with J J R Macleod, Charles H Best, and J B Collip; Banting and Macleod were awarded the 1923 Nobel Prize for medicine and physiology. In the 1930s, in Bernal's laboratory at Cambridge, Dorothy Hodgkin had taken X-ray photographs of insulin, but realized that analysis was then impossible. The situation was changed after 1953 when Frederick Sanger (b. 1918) discovered the sequence of amino acids in insulin, for which he was awarded the 1958 Nobel Prize for chemistry (he later shared the 1980 Prize for his work on the DNA molecule). The insulin molecule has 777 atoms, so that it is smaller than the myoglobin molecule. Its structure, however, was more

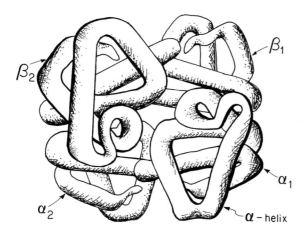

**Fig. 7.22**   A schematic diagram of the haemoglobin molecule, as established in 1960 by Max Perutz. This diagram is intended merely to give some idea of the general shape of the molecule. The molecule, which has a molecular weight of 65 000, consists of four polypeptide chains, $\alpha_1$, $\alpha_2$, $\beta_1$, and $\beta_2$, The chains $\alpha_1$ and $\alpha_2$ are identical, and are made up of 141 amino acid molecules. The chains $\beta_1$ and $\beta_2$ are identical chains of 146 amino acids. Each of the four chains is attached to a haem group, which has an iron atom at its centre. The polypeptide chains are in some places wound into alpha-helices (Fig. 7.20); one example of a region where there is alpha-helix structure is indicated.

difficult to determine, partly because isomorphous replacements by heavy metals could not be made. Nevertheless, in 1969 Dorothy Hodgkin and her co-workers were able to announce the complete three-dimensional structure of insulin. This again was recognized as a remarkable achievement, in view of the size and complexity of the molecule. Fig. 7.23 shows in a schematic way the general structure of the insulin molecule.

In more recent years, new techniques such as nuclear magnetic resonance spectroscopy (NMR) have been introduced for the study of proteins, and computers are faster. As a result we now know the structures of a great many proteins, including many enzymes. An average of three new structures a day are now being added to the data bank. In the case of enzymes we can often identify the chemical groups that perform the catalysis.

## DNA

At about the same time that Bragg, Perutz, and Kendrew were beginning their protein work there was also much investigation of the structure of DNA (deoxyribonucleic acid). Some of this work was also done in the Cavendish

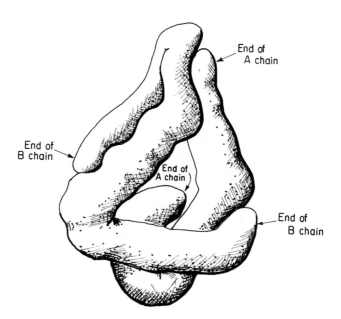

**Fig. 7.23**  A schematic diagram of the insulin molecule, as established in 1969 by Dorothy Hodgkin. This diagram merely gives some idea of the general shape of the molecule. There are two chains, connected together, consisting of helices. A small portion of the A chain is a left-handed helix. The B chain is largely a right-handed helix, which is the type usually found in proteins. Insulin is a protein of molecular weight 6000, and contains 777 atoms. It is composed of 51 amino acids.

Laboratory at Cambridge. DNA was then believed by many scientists to be the fundamental genetic material, although the matter was not quite certain until the structure had been discovered. More will be said about genes in the next section; here it may simply be noted that genes are portions of single DNA molecules. A single gene carries an enormous amount of genetic information, and it does so simply as the result of the order in which four chemical groups, known as bases, are arranged along the gene's molecular chain.

The story of how the structure of DNA was arrived at is a remarkable one, quite different from the stories of how the structures of other macromolecules were established. In all other cases, much painstaking experimental X-ray work was carried out, followed by detailed calculations (now made much easier by computers). In most cases the structure was finally obtained, with almost complete certainty, at the end of a careful experimental investigation. With DNA, on the other hand, the conclusion was reached making much less use of the X-ray data. Instead, a plausible molecular model was suggested on the basis of ingenious speculations based partly on the experimental results, but with much consideration given to how the various atomic groups fit together. In particular, the building of molecular models played a vital part.

Four people were primarily involved in the work. The person who perhaps did the most experimental work on DNA—and it was X-ray work of the highest quality—was Rosalind Elsie Franklin (1920–1958; Fig. 7.24); she did her work at King's College, London. Also doing excellent X-ray work on the problem at King's College was Maurice Hugh Frederick Wilkins, who was born in New Zealand in 1916; during the war he had done isotope separation work at the University of California. At the Cavendish Laboratory at Cambridge was Francis Harry Compton Crick (b. 1916), who after war service with the British Admiralty on the construction of mines was working for his PhD degree under the direction of Perutz. All of these three had received most of their education and experience in physics. The fourth person, the most unconventional of them, was the American James Dewey Watson (b. 1928), who as primarily a biologist.

Watson arrived at Cambridge in 1951, having taken his PhD degree in zoology at the University of Indiana in 1950. He was in Europe on a US fellowship, and was supposed to be working at the University of Copenhagen. However, being independent and not a little irresponsible, he moved to Cambridge, and for a time exasperated the administrators of his fellowship by keeping them in the dark as to where he was and what he was doing; his research grant was finally revoked, but he obtained some financial help at Cambridge. He was assigned by Bragg to carry out research on myoglobin under Kendrew's direction. However, to a great extent he attached himself to Crick, who had not yet obtained his PhD degree, and together they spent much time considering the DNA structure. Perutz and Kendrew seem to have been quite tolerant of the fact that Crick and Watson were neglecting the experimental work they were expected to do. Bragg was somewhat more

**Fig. 7.24**    Rosalind Elsie Franklin (1920–1958), a brilliant experimental X-ray crystallographer. She was educated at Newnham College, Cambridge, and after spending some time in France she obtained an appointment at King's College, London, where she worked in association with M H F Wilkins. There she obtained X-ray evidence that was crucial to Watson and Crick's discovery of the double-helical structure of DNA. She later worked at Birkbeck College, London, in association with J D Bernal, and carried out important work on tobacco mosaic virus. Had she lived longer she would almost certainly have won a Nobel Prize.

concerned; he felt that Crick was making little progress towards his PhD degree, and at first considered that both he and Watson were approaching the DNA problem in an unscientific way. Bragg was uneasy about the fact that Crick and Watson were not obtaining any DNA data themselves, but were relying on the King's College data generously supplied to them by Wilkins, and less willingly (and sometimes even unknown to her) obtained from Rosalind Franklin. At one stage Bragg told the two men that they should do no further work on DNA. But Watson had done no experiments on DNA and had done no more than think about its structure in the light of the experiments of others; he did not consider that Bragg's order prevented him from continuing to think about DNA. Later, when Crick and Watson were obviously making some progress towards arriving at the DNA structure, and especially when it was learnt that Pauling was also attacking the problem, Bragg encouraged them to continue.

Crick and Watson had been impressed by Pauling's conclusion that proteins can have a helical structure (Fig. 7.20), and they guessed that DNA would have the same kind of structure. Some of Rosalind Franklin's X-ray photographs were consistent with a helix. Crick and Watson were also aware of some of the DNA work done by biochemists. In particular Erwin Chargaff (b. 1905), at Columbia University, had shown that there was a definite relationship between the amounts of the four bases that occur in DNA. The bases are adenine (A), guanine (G), thymine (T), and cytosine (C), and it seemed that the number of units of A was always equal to the number of units of T, while the number of units of G was the same as the number of units of C. This was referred to as Chargaff's rule. The ratio of the numbers of A–T and G–C pairs, however, could vary widely.

Crick and Watson spent much time building molecular models, just as Pauling had done to arrive at his helical structure for proteins. They took care with the sizes of the various groups, to make sure that everything fitted together satisfactorily. In 1953 they proposed the DNA structure shown schematically in Fig. 7.25. To explain the replication they proposed a double helix—in other words, two helices intertwined. They saw from models they had built that the four bases are of different sizes, and that for the double helix to be of uniform width, an A unit can only approach a T, and a G must pair with a C. This pairing might have been inferred by them from Chargaff's rule, but Crick later said that it was not. They arrived at the pairing because only in that way would the helix be uniform in width; any other arrangement would give bulges along the chain, which was excluded by the X-ray work. It was only afterwards, Crick said, that they were reminded of Chargaff's rule and realized that the A–T and G–C pairing accounts for the fact that the numbers of A and T are equal, and the numbers of G and C are equal. Another important feature of the double helix, introduced mainly to explain how replication can occur, is that the chains run in opposite directions. For chemical reasons the ends of the chains are designed 5′, conventionally taken as the beginning of the chain, and 3′, the end of the chain.

Watson and Crick suggested that replication involves the unwinding of the helix (Fig. 7.26). Each single helix serves as a template for the formation of another helix. Wherever there was an A, a T would be selected as its partner, and vice versa; a G would always select a C, and a C would select a G. Helix-1 from the original double helix would thus bring about the formation of a new helix-2, while the original helix-2 would form a new helix-1. The two would twin, and replication would be complete.

The announcement of the double helix structure aroused great interest and even excitement, and the work has been called the most original of the twentieth century. Although the experimental evidence was slender, the model seemed to many people so plausible that it could hardly be wrong. There were some sceptics, and alternative structures were proposed over the next few years. However, as the experimental evidence accumulated, it became more and more likely that the Watson–Crick model was correct. Pauling had been working on the structure of DNA, but he got it completely wrong; he later admitted that he had not given the matter enough attention.

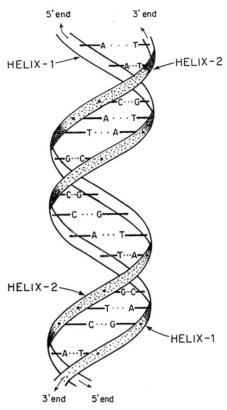

**Fig. 7.25** A schematic illustration of the helical structure proposed by Watson and Crick for DNA. The bases are T, A, C, and G, and they can only pair as T–A and C–G. The ends of the chains are usually designated 3′ and 5′, for chemical reasons. The two chains run in opposite directions, with the 3′ end of one chain near to the 5′ end of the other. The structure resembles a spiral staircase with the base pairs forming the steps.

In 1962 (the same year in which Perutz and Kendrew shared the Nobel Prize for chemistry), Crick, Watson, and Wilkins shared the Nobel Prize for medicine or physiology. Although he was not a co-author of the 1953 paper announcing the structure, it was recognized that Wilkins had made an important contribution in providing vital data without which the model could hardly have been proposed. Rosalind Franklin had, of course, provided even more of the data, but unfortunately she had died of cancer at the age of 37, and Nobel Prizes cannot be awarded posthumously. Not more than three people can share a Nobel Prize, but she might well have won one in another year.

Much has been said and written about Franklin's role and her recognition or lack of it. There is no doubt that she produced much valuable X-ray evi-

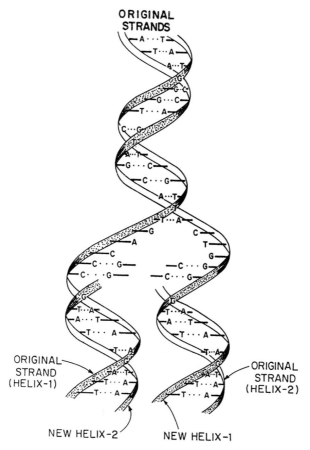

**Fig. 7.26** A diagram showing how replication can occur. If the double helix of the old DNA molecule unwinds somewhat into two helices, new single helices can be synthesized, each complementary to the helix on which it is growing. Helix-1 forms a new helix-2, and helix-2 forms a new helix-1. When the two new strands twin together, the resulting DNA molecule is identical to the original one.

dence, without which Watson and Crick could not have proposed their model. Her philosophy of how science should be done was, however, different from that of Crick and Watson. They had encouraged her to build models, but she did not like the idea and never built a model. (Wilkins was about to do so when the others arrived at their structure.) It was her style to defer drawing conclusions until enough experimental results had been obtained, and at the time of the Watson–Crick announcement she did not think that the evidence was adequate. It has been plausibly argued that after she had done a few more experiments she would have arrived at the double helix herself. Although she had been informed of the double-helix idea, she seems to have been reluctant to accept it. Not having built a model, she did not appreciate the point about how the bases must pair. It is certainly possible that if Watson and Crick had not published their model in 1953, she might have reached the same conclusion a year or so later.

The controversy about her role in the work and the recognition of her contributions was exacerbated by the publication in 1968 of Watson's book *The double helix*, which was supposed to explain how the DNA structure was deduced. In the book Watson was quite unfair to Rosalind Franklin, no longer able to defend herself. Although denigrating her work, Watson quite inconsistently provided the reader with much evidence of its excellence. In an epilogue to his book he responded to critics of his treatment of her by describing her in more reasonable terms, but did not withdraw anything he had previously said of her.

Much later, in his book *What mad pursuit* (1988), Crick wrote an account of the work which is much more reliable. He is much fairer to Rosalind Franklin, with whom he and his wife were on friendly terms. Crick wrote amusingly about confusions as to the authorship of the double-helix structure. He tells us that after the publication of Watson's book, he (Crick) was often congratulated on it. He also tells us that when he mentioned Watson to Nevill Mott, the new Cavendish professor, Mott replied 'Watson? But I thought your name was Watson-Crick'. Sir Nevill seemed to have trouble with people's names. Shortly after the Second World War, when I mentioned to him that I had worked on ballistics in Valcartier, Quebec, he started to address me as 'Dr Valcartier'. I let this go at first, but when he started to introduce me as Dr Valcartier to people who already knew me, I had no choice but to correct him—much to his amusement.

## THE ORIGIN OF SPECIES

To appreciate the great significance of what Watson and Crick achieved in discovering the structure of DNA, we should briefly step back in history for a century and a half or so. In Chapter 1 we discussed the distinction between hard and soft science. Before Watson and Crick discovered the helical structure of DNA, the theory of biological evolution was soft science, as it was based on plausible inferences which were not completely beyond reasonable doubt. When the structure of DNA became known, the subject rapidly hardened. Physics had made a powerful invasion into biology, and stringent physical tests could now be applied. Reliable predictions could be made, and genetic experiments could be carried out with a high degree of reliability.

The main credit for founding the science of evolution should go to Jean Baptiste Lamarck (1744–1829), whose name is now commonly heard with reference to something that is known to be wrong. This is unfortunate: he was a great pioneer, and his work exerted an important influence on others, including Darwin. Most of his work was done at the Natural History Museum in Paris. Up to his time it had been generally accepted that the various species of animals and plants were immutable, having been placed on earth in their present form. Lamarck suggested in about 1809 that instead there was a gradual process of development from the simple to the complex. He went wrong, however, in suggesting how the changes come about. He believed that changes brought about in the lifetime of an individual organism would be inherited by the next generation. In other words, he proposed the inheritance of acquired characteristics. His theory is simply explained by the example of why giraffes have long necks. In order to survive, they eat the leaves of tall trees, and therefore stretch their necks. The next generation would, according to Lamarckism, inherit the longer necks, and from generation to generation the necks would get longer.

The correct theory was given later in the century by Charles Robert Darwin (1809–1882), whose conclusions were based on what he observed during a five-year voyage as naturalist on a British naval vessel, the *Beagle*. On this voyage Darwin gained a detailed practical knowledge of the geology of many lands, and of the plants and animals found on them. He made a vast collection of geological and biological specimens which were shipped back to England. On his return home in 1836 he thought deeply about his results for many years, and published many of his observations. However, he was reluctant to publish his ideas about evolution, as he disliked controversy and knew that they would produce something of a sensation. At first he had thought that Lamarck's idea of the inheritance of acquired characteristics was correct, but later decided that an important change was required. Darwin's alternative explanation for the long necks of giraffes was that long ago, as a matter of chance, some giraffes had longer necks than others. The ones with the longer necks were more likely to survive, as they could reach more food, and were more likely to have offspring. Those with shorter necks could reach less food, and were less likely to survive long enough to produce offspring.

Darwin's decision to publish his theory of evolution was occasioned by his receipt in 1858 of a letter from Alfred Russel Wallace (1823–1913), a younger man who had also made a lengthy sea voyage. The letter revealed to Darwin that Wallace had come to much the same conclusions as he had about the way in which evolution occurs. Wallace was obviously about to publish his theory, which would have won him priority. In the end Darwin and Wallace came to an agreement which was a great credit to both of them. Papers written by Darwin and by Wallace were read at a meeting of the Linnean Society in London on 1 July 1858. Oddly, neither of the two men was present on this historic occasion, and even more oddly, the announcement appears to have been largely ignored by other scientists.

In the following year Darwin published his famous book *The origin of species by means of natural selection*, and in 1871 appeared his *The descent of man and selection in relation to sex*. Today it is Darwin rather than Wallace who is chiefly remembered for the theory, and Darwin's exposition of his theory was indeed more complete and was supported by more evidence than that of Wallace. Wallace always behaved with a saintly generosity; he held Darwin in high esteem, and when writing his reminiscences in his old age he went so far as to say that his chief contribution to the theory had been to cause Darwin to write about it. If Wallace had been less generous he would have published his theory right away and seized the credit. Unfortunately, Wallace rather discredited himself in his later years by giving his support to obviously fraudulent spiritualists. As a result he was unable to find employment, and had to appeal to his friends. Darwin was particularly helpful, and arranged for Wallace to receive a modest government pension.

Darwin's theory has often been referred to as the *survival of the fittest*, this expression having been invented by the philosopher Herbert Spencer (1820–1903). Spencer was a man of remarkable versatility who had been a railway engineer, teacher, and journalist before devoting himself to writing on science, psychology, and philosophy. At first Darwin did not like the expression survival of the fittest, preferring the term natural selection that was used in the title of his first book. His change of opinion was prompted by the realization that, as Spencer had clearly pointed out, natural selection was being misinterpreted as meaning that God controlled the process, the exact opposite of what Darwin meant. Darwin did not believe in any such control, although he was reticent to express his views as he knew that they would produce controversy.

When Charles Darwin put forward his theory, there was some hostility. He had been sensitive to the fact that some religious people, especially those who were inclined towards a fundamentalist position, would think that his theory was in conflict with their religious beliefs. He was therefore greatly encouraged by a letter he received from the Revd Charles Kingsley (1819–1875), who in 1860 was to become professor of modern history at Cambridge and is now remembered as an author, particularly of *Westward Ho!* (1855) and *The water babies* (1863). Kingsley admitted that the theory had required him to modify his religious ideas, but added that he found it

> just as noble a conception of Deity, to believe that He created primal forms capable of self development...as to believe that He required a fresh act of intervention to supply the *lacunae* which He Himself had made.

Darwin was delighted at this interpretation, and included it in the second edition of his book, which had to be issued at once, the first edition having sold out quickly.

Six months after *The origin of species* appeared, there was a famous confrontation between the distinguished biologist Thomas Henry Huxley (1825–1895) and the Bishop of Oxford, the Rt Revd Samuel ('Soapy Sam') Wilberforce. It occurred at the 1860 meeting of the British Association for the Advancement of Science in Oxford, in the new University Museum. The confrontation almost never happened, as Huxley had decided to leave the meeting earlier but changed his mind (Darwin was not present, as he disliked meetings).

On the evening of Saturday, 30 June 1860, the distinguished English-born American scientist John William Draper was scheduled to give a paper entitled 'On the intellectual development of Europe, considered with reference to the views of Mr Darwin and others, that the progression of organisms is determined by law'. We have already met Draper in Chapter 3, in connection with his pioneering work on photography; at the time of this meeting he was director of the medical school at New York University. By all accounts his one-hour lecture was ponderous and dull (as its title might suggest), but the audience listened to it respectfully. Bishop Wilberforce, who was one of the vice-presidents of the BAAS and was sitting up on the platform at the meeting, had asked permission to present his views at the end of Draper's talk, and this was granted. Bishop Wilberforce was a rather remarkable man, who had taken honours in mathematics at Oxford, and had some amateur knowledge of science. It is important to appreciate that Wilberforce, although a bishop, did not speak against the theory of evolution on religious grounds but on scientific ones. He had been primed in scientific objections to the theory by Richard Owen (1804–1892), a zoologist and palaeontologist who was jealous of Darwin, but Wilberforce did not know enough biology to understand the arguments properly. He was an eloquent and forceful speaker, and during the course of his comments he turned to Huxley and said

> I should like to ask Professor Huxley...as to his belief in being descended from an ape. Is it on his grandfather's or his grandmother's side that the ape ancestry comes in?

According to various accounts, the Bishop's remarks, which took half an hour or so, were greeted with much applause and cheering.

Unfortunately for him, he had made a serious error of judgement in choosing such a formidable antagonist. Huxley, aged thirty-five at the time of the meeting, already had a great reputation; for six years had been professor of natural history at the Royal School of Mines in London. He was far from amused by Wilberforce's question; he was angered by it, as he was an outspoken atheist, and had long despised the religious establishment that was still so strong in England and particularly at Oxford and Cambridge. Later he coined the word *agnostic* to describe his views. He resented the intellectual arrogance of classically-trained academics who despised science, who opposed the teaching of it, and who considered that a complete education was provided by the classics, ancient philosophy, and (particularly at Cambridge) perhaps a little pure mathematics. Huxley was seated in the audience beside Benjamin Collins Brodie (1817–1880), the Aldrichian professor of chemistry at Oxford and another unbeliever; before rising to reply Huxley murmured to him 'The Lord hath delivered him into mine hands', ironically choosing a scriptural metaphor. He forcefully demolished Wilberforce's objections, exposing his inadequate understanding of the subject. His irreverent reply to the Bishop included the passage

> I should feel it no shame to have risen from such an origin. But I should feel it a shame to have sprung from one who prostituted the gifts of culture and eloquence to the service of prejudice and of falsehood.

In those days bishops were not often spoken to like that, and there was quite a sensation. A number of ladies are reported to have fainted, as was fashionable in Victorian times. (But were they feints rather than faints? Anthony Trollope, who knew about such things, tells us that Victorian ladies, even when they collapsed to the floor and had to be revived with smelling salts, were shamming and not really fainting.)

Present in the audience were a large number of scientists who were Anglican clergymen, most of them members of the Broad Church and unfriendly towards Wilberforce who was a High Churchman. The atmosphere at the meeting, regrettably, was not that of a normally staid scientific meeting; it was more like a sporting event, with much clapping and cheering, particularly after Huxley's reply. The room was packed with people, over 700 of them; included among them were many of the leading scientists of the day and also many clerics who had teaching positions at Oxford. Afterwards many of them met together in small groups to discuss the controversy. One who was present was the mathematics don Charles Lutwidge Dodgson (1832–1898), in five more years to be famous as Lewis Carroll, the author of *Alice's adventures in wonderland*; he and his friend the chemist Augustus George Vernon Harcourt (1834–1919) met afterwards, and it appears that they felt that Huxley had won the day. Many of the other clerics had the same opinion. Wilberforce himself, it appears, was always sure that he had won the argument.

Huxley, who became known as 'Darwin's bulldog', was less than charitable about Wilberforce. On hearing in 1873 that the Bishop had died as a result of falling on his head from his horse, Huxley wrote to the Irish physicist John Tyndall 'For once, reality and his brain came into contact, and the result was fatal'.

At about the same time that the Huxley-Wilberforce debate occurred, an obscure monk, Gregor Mendel (1822–1884), was carrying out some remarkable experiments on plants in the garden of a monastery in Brünn, then the capital of Moravia and now (called Brno) in the Czech Republic. Between 1856 and 1863 he grew about 30 000 pea plants, artificially fertilizing certain plants that had special characteristics. For example, he crossed tall plants with short plants, and showed that all the plants of the first generation were tall, but that of the plants in the next generation some were tall and some short, in the ratio 3:1. His important conclusion was that each plant receives one factor from each of its parents, tallness being dominant and shortness recessive. These factors were later called genes.

At about the same time there was being discovered further evidence that helped to build up an understanding of how biological inheritance takes place. In 1858, two years before Darwin's *The origin of species* was published, the German pathologist Rudolf Virchow (1821–1902), working at the University of Berlin, showed that each living cell has a central concentration of material known as a nucleus. Soon afterwards it was found by various investigators that every cell nucleus contains structures, called chromosomes, which always come in pairs. The pairs double before cell division and are then shared between daughter cells. The obvious conclusion, which later turned out to be correct, was that the chromosomes are the carriers of hereditary factors.

Of particular importance were experiments performed from about 1908 onwards by the American biologist Thomas Hunt Morgan (1866–1945). He carried out, first at Columbia University and then at the California Institute of Technology (Caltech), an extensive series of investigations on *Drosophila*, the fruit fly, which has the advantage of breeding rapidly; a new generation is produced about every two weeks. Morgan confirmed and extended many of Mendel's findings, and observed that certain pairs of genes tend to be inherited together more frequently than other pairs. He concluded that the genes are individual units that are arranged in a particular order along the chromosome, like beads on a string. He was thus able to achieve a certain amount of mapping of genes along the chromosome.

It was realized in the nineteenth century that sometimes genes undergo a change, known as a *mutation*. Herman Joseph Müller (1890–1967), who had been a student of Thomas Morgan, worked among other things on the action of X-rays on cells, showing that they induce genetic mutations. For this work he was awarded the 1946 Nobel Prize for medicine. It was he in particular who called the public's attention to the dangers of various kinds of high-energy radiation. Previously, X-rays had been used much too freely in hospitals, without proper precautions; X-ray equipment was standard equipment in shoe stores, for testing whether a shoe fitted a customer. It may be remarked, incidentally, that since a gene in a reproductive cell that has undergone mutation is passed on to the next generation, a dose of X-rays is an environmental effect that may be inherited. Lamarck was therefore not entirely wrong in his suggestion that acquired characteristics can be inherited; he was just wrong in suggesting that this is the primary way in which evolution takes place.

An important development came in 1944 from the Rockefeller Institute in New York. There the Canadian-born Oswald Theodore Avery (1877–1955) and some of his colleagues were working on *pneumococci*, in particular on the transformation of a non-virulent form into a virulent form. They found in 1944 that the transformation is performed by DNA. This led to the suspicion that the genes are portions of DNA molecules, and this was subsequently found to be true of the genes of every living organism, plant and animal. Watson and Crick's later helical structure for DNA was therefore of central importance in molecular biology. This contribution won for Avery the Copley Medal of the Royal Society in 1945.

At about the same time, important work was being done on the *bacteriophages*, the large viruses that infect bacterial cells. One who worked in this field was the German-born biophysicist Max Delbrück (1906–1981), who had been educated as a nuclear physicist and had studied in Copenhagen with Niels Bohr; much of his work was at Caltech, where he was a colleague of Linus Pauling. Another who worked on bacteriophages was Alfred Day Hershey (1908–1997) of the Carnegie Institution in Washington. At the University of Indiania, the Italian-born Salvator Edward Luria (1912–1991), with whom Watson had worked before collaborating with Crick at Cambridge, obtained good electron micrographs of bacteriophages. Delbrück, Hershey, and Luria all obtained results that supported the conclusion that genes are portions of DNA molecules. Together they formed the Phage Group, committed to using bacteriophages to investigate genetics, and made many advances. They shared the 1969 Nobel Prize for physiology or medicine.

Until Crick and Watson's double-helix structure for DNA was proposed in 1953, the emphasis in biology had been on the overall behaviour of living systems, rather than on molecular structures. Geneticists, for example, had been mainly concerned with problems of inheritance of characteristics, and had made many important discoveries which were of great help in developing the molecular aspects of genetics.

It is interesting that those physicists who decided to investigate biological, and especially genetic, problems tended to follow one of two quite different paths. Some of them, often called the *conformationalists*, applied physical methods, notably X-ray analysis, to the study of molecular structure; Bragg, Perutz, and Crick took this approach. Others trained as physicists, typified by Delbrück, felt at first that the problems were too complicated for such an approach to lead to useful results. They preferred to make detailed studies of genetic behaviour, and were sometimes referred to as *informationalists*.

## MOLECULAR GENETICS

The helical structure proposed by Watson and Crick for DNA soon led to rapid advances in the field of molecular genetics, which is the application of physical methods to genetics. Molecular genetics is an important branch of molecular biology—so important, in fact, that today many scientists regard them as one and the same. Only a bare outline of molecular genetics can be given here. We have seen that the chromosomes, which occur in animal and plant cells and transmit hereditary information, consist of strings of genes, which are segments of the long DNA molecules. The DNA molecules in the nucleus of a single cell carry billions of bits of information; there is enough information capacity in a single human cell to store, several times over, all 30 volumes of the *Encyclopaedia Britannica*. It takes a difference of only one bit of information per thousand to distinguish one individual from another. A change of one bit per hundred takes us from one animal species to another, and a change of one bit out of ten changes an animal into a plant.

We now know that the 'working' molecules of a living system are not the chromosomes, but the protein molecules. Each protein has a unique sequence of amino acids. Many biological structures, such as muscle and skin, are made up primarily of proteins. It is mainly the proteins that are concerned with physiological activity. The protein haemoglobin transports oxygen from the lungs to the places where it is needed. The enzymes, which are proteins, are the biological catalysts. Most of the chemical processes which occur in living systems would occur very slowly if the appropriate enzyme were not present.

The problem that molecular geneticists had to attack was how the information contained in the genes is able to control the manufacture of proteins. The mystery of how this is done began to be solved in the 1960s. The order in which the bases occur in DNA is the critical feature. Suppose, for example, that, beginning with the 5′ end of a particular DNA molecule (Fig. 7.25), the order of the four bases T, C, A, and G is

TTCGGTCGC...

One possibility might be to consider the base in pairs; to indicate this we can put commas between the pairs:

TT,CG,GT,CG...

This, however, does not work; there are twenty different amino acids, and if we take four bases in pairs we can get only $4 \times 4 = 16$ possibilities:

TT, TC, TA, TG
CT, CC, CA, CG
AT, AC, AA, AG
GT, GC, GA, GG

Suppose that instead we consider triplets, so that with commas the sequence becomes

TTC, GGT, CGC...

With triplets we can get $4 \times 4 \times 4 = 64$ possibilities, so that we now have more than enough. This, however, turned out to be the right answer; often different triplets lead to the same amino acid.

Important contributions were made by the South-African born Sydney Brenner (b. 1927), who joined the staff of the Medical Research Council in Cambridge in 1957, and for a time worked closely with Crick. Brenner introduced the word *codon* to refer to a triplet of bases in DNA. The essential feature of the coding is that there is an exact correspondence between the sequence of the *codons* in the DNA of the gene and the sequence of amino acids in the protein that is produced. For example, the codon TTC specifies the amino acid called phenylalanine, GGT specifies glycine, and CGC specifies arginine. The sequence shown above would therefore lead to the synthesis of a protein which would begin with phenylalanine-glycine-arginine...

All 20 amino acids could be coded without using all of the combinations, but in fact several different codons lead to the same amino acid. Glycine, the simplest of the amino acids, is for example given by any of the following four codons:

GGT, GGC, GGA, and GGG.

The function of certain codons is to order the synthesis to stop.

A considerable number of people were involved in breaking the genetic code—that is, in relating the codons to particular amino acids. Crick and Brenner certainly made important contributions. Severo Ochoa (1905–1993) of New York University, and Arthur Kornberg (b. 1918) of Stanford University discovered the mechanisms of synthesis of DNA and RNA and for this they were awarded the 1959 Nobel Prize for physiology or medicine. Marshall Warren Nirenberg (b. 1927), of the National Institutes of Health in Bethesda, Maryland, Har Gobind Khorana (b. 1922), of the University of Wisconsin, and Robert William Holley (1922–1994) of Cornell University took some of the final steps in unravelling the genetic code, and shared the 1968 Nobel Prize for physiology or medicine.

Mention should be made of an interesting project begun in 1963 and successfully continued for many years. It was suggested by Sydney Brenner that it would be profitable to focus attention on a small nematode (roundworm) called *Caenorhabditis elegans*, which is only a millimetre long and which is composed of only 959 cells. This work became known as the Worm Project, and by intensive investigation it was possible by 1986 to work out the main genetic and physiological details of the organism. The Human Genome Project, centred in the US but supported internationally, aims to explore human genetics in a similar way.

The outline we have given has necessarily left out a great many details. The DNA never forms the protein directly, but does so by transmitting its information through an intermediate, messenger RNA (mRNA). Ribonucleic acid (RNA) molecules, of which mRNA is a special form, are also important constituents of cells. They are formed from DNA in such a way that the bases are arranged in an order that is equivalent to the order in which they appear in the DNA molecule, but some details of their structure are different.

One additional and important detail of protein synthesis should be emphasized. X-ray studies of protein molecules have always shown that a given protein normally exists in a given three-dimensional form; the one-dimensional strands are always folded in the same way. In the globular proteins (such as the ones shown in Fig. 7.21–7.23) the long molecules are folded up like a ball of wool, and this raises the question (a troubling one for the early workers) as to how a molecular template could possibly produce such a complicated three-dimensional structure, in a precisely specified pattern. The answer is that it does not do so directly. The protein molecule is synthesized as a long strand, and when it is surrounded by water molecules the forces are such that the molecular strand folds itself spontaneously into a particular shape. The sequence of amino acids specifies the folding that must take place in the protein. Thus, the biological synthesis of a protein initially creates the molecule as a long strand of amino acids, and it spontaneously folds into the unique three-dimensional form that is a necessary consequence of the particular sequence of amino acids.

## Science and religious belief

Scientists find it puzzling and regrettable that there still remains so much opposition from some members of the general public to the theory of evolution. Other great principles of science, such as the conservation of energy, the quantum theory, and the theory of relativity, are generally accepted, even if they are not widely understood. Yet the scientific evidence for natural selection, especially since molecular genetics entered the picture, is just as strong as for the other theories. The modern work has, it is true, slightly modified some early ideas relating to natural selection, but has strongly supported and strengthened the basic concept.

The objections to the theory of evolution are mainly based on religious fundamentalism, according to which certain writings are considered sacred and to be taken literally. However, a vast amount of scientific evidence, in the fields of geology, palaeontology, and biology, has now accumulated. All of it leads to the inescapable conclusion that ancient religious writings are not to be accepted as they stand. They were never intended to be scientific accounts of the creation of the universe or the creation of life in the universe; they are merely the views of the people who wrote them. Religious fundamentalism as applied to evolution no longer has any claim to serious consideration.

It may be helpful to trace briefly the development of tolerant ideas among thoughtful religious people. Even as early as the fourth century AD, scholars had begun to realize that the Old Testament's account of the creation of the universe and the beginning of life on earth should not be taken literally. Augustine of Hippo (354–430) was particularly concerned about this problem. He himself was a convert to Christianity, and was anxious to convert others. He found that his efforts were being frustrated by the fact that some of his colleagues were insisting on a literal interpretation of the book of Genesis. Some time after 404 AD he wrote a book *De Genesii ad litteram* (On Genesis and its literal interpretation), in which he scolded his colleagues for the methods they were using. He pointed out that some of the people they were trying to convert 'know something about the earth, the heavens, and the other elements of this world, about the motion and orbits of the stars...', and they would not be converted to Christianity by people who told them things they knew to be wrong. Augustine did not mince his words, saying that it is a 'disgraceful and dangerous thing for an infidel to hear a Christian...talking nonsense on these topics'. Later he said that 'reckless and incompetent expounders of Holy Scripture bring untold trouble and sorrow' by their statements.

Although such enlightened views were held so long ago, many people later adopted more primitive ideas. In the seventeenth century Archbishop James Ussher (1581–1656) even went so far as to conclude from a careful study of the Old Testament that the universe was created in 4004 BC. Others even thought that they could specify more accurately the exact day and even the time of day. Sixteen centuries later it is rather depressing to find that there are still some religious fundamentalists who are holding the opinions that St Augustine so strongly criticized. In spite of improved educational standards, many people seem to have escaped the advance of civilization.

Serious thinkers, of course, have left this kind of foolishness behind. As scientific evidence accumulated it became impossible to avoid the conclusion that the universe must be vastly older, and life began much earlier, than had been supposed. By the seventeenth century most thoughtful people, although for the most part adhering strictly to their religious beliefs, accepted the point of view that scientific conclusions are not to be influenced by religious beliefs. For example, the men who founded the Royal Society in England in the seventeenth century were nearly all religious men, and many of them were clergymen. It was specified in the Charter of the Society that the work they did was to 'advance the Glory of God, the honour of the King, .... and the general good of mankind'. At the same time it was clearly specified that their work was to be done 'by experiments (not meddling with Divinity, Metaphysics, Moralls, Politicks, Rhetoric or Logick)'. Clearly, science and religion were to be kept in their separate compartments. Inevitably, some of the Fellows of the Royal Society tried to intrude religious ideas into their discussions, but on the whole they were put in their place.

It was also during the seventeenth century that a group of clergymen formed themselves into a group called the Latitudinarians, making it clear that they rejected fundamentalist ideas. Not only did they accept scientific conclusions; they welcomed them as enhancing their sense of wonder at the ingenuity with which the universe had been constructed by the Creator.

Some of them, in fact, went so far as to base their religious beliefs on the scientific conclusions. This point of view, referred to as natural theology, was strongly maintained in the nineteenth century by the *Bridgewater treatises*, a series of volumes by scientists which were published from 1833 to 1836. They covered much of contemporary science and related it to the 'power, wisdom and goodness of God, as revealed in the creation'.

This position, however, was not universally accepted. Others, even as late as the nineteenth century, thought that science should be modified in such a way as to conform with the early religious writings. The so-called Scriptural Geologists, for example, tried to interpret their findings in such a way that they were consistent with what was said in the Old Testament about the Great Flood. Some biologists remained convinced that all the different animal and plant species were created independently, and that Noah had performed the stupendous feat of preserving all of them in his Ark.

Later in the nineteenth century, however, the most enlightened religious people had concluded that such interpretations were impossible. They accepted the position that religion and science are concerned with two entirely different aspects of life, and are not related to each other. Their point of view is well summarized in a sermon by the Revd Frederick Temple (1821–1902) in St Mary's Church, Oxford, on 1 July 1860, the day following the famous Wilberforce–Huxley confrontation. This sermon was preached at the official service held for the delegates at the BAAS meeting, and included the passage:

> The student of science now feels himself bound by the interests of truth, and can admit no other obligation. And if he be a religious man, he believes that both books, the book of nature and the Book of Revelation, alike come from God, and that he has no more right to refuse what he finds in one than what he finds in the other.

Temple had no doubt been present at the confrontation, and had it in mind when he said those words. He had been a mathematics tutor, and had some knowledge of science; since he later became Archbishop of Canterbury it is evident that the Anglican Church, at least, had no quarrel with the theory of evolution or with tolerant views such as these.

This point of view can perhaps be understood in terms of an analogy. An expert on medieval painting, for example, might also happen to have a deep understanding of the scientific theories of colour put forward by Isaac Newton, Thomas Young, James Clerk Maxwell, and Edwin Land. There would surely be no conflict between the two points of view; instead they would supplement one another. In the same way, enlightened religious people find that their understanding of science strengthens their belief.

The discussion so far has been concerned with the points of view held by thoughtful people, with no reference to official policies of individual churches. Unfortunately these policies have sometimes been sadly out of date. This, it must be said, is particularly true of the Roman Catholic Church. It is well known that the works of Copernicus were placed on the *Index librorum prohibitorum* (List of prohibited books), and that Galileo was ordered not to follow the Copernican theory that the sun is at the centre of the solar system. In 1635 Galileo was summoned to abjure the 'heresies' that his ideas about the solar system were supposed to be, and his ideas were officially condemned by the Roman Catholic Church until 1922.

After the publication of Darwin's *The origin of the species* in 1859, the position of the Vatican became even more repressive. In 1864 Pope Pius IX, a good but rather simple-minded man, issued an encyclical appended to which was a *Syllabus of errors*. This syllabus was a blanket condemnation of science in general; evolution was not mentioned explicitly, but there could be no doubt that good Roman Catholics were not to accept it. One biologist who had been converted to Roman Catholicism was bold enough to produce a modified version of the theory that he thought would be acceptable to the evolutionists and to the Church. He was St George Jackson Mivart (1827–1900), a man of some ability who had studied under Huxley. The sad result of all his efforts was that he incurred the enmity of Huxley and most other evolutionists, and that he was excommunicated by the Roman Catholic Church. In addition, his writings were placed on the *Index*.

Fortunately, there has recently been a relaxation of these condemnations. In 1982 the pontifical Academy of Sciences, which advises the Vatican on scientific matters, issued a statement that 'masses of evidence render the application of the concept of evolution to man and the other primates beyond serious dispute'. Finally, in 1996, Pope John Paul issued a statement accepting this point of view.

Why has it taken so long for this obvious conclusion to be reached by the Church authorities? I am afraid that in many ways the theory of evolution got off on the wrong foot. The initial 1860 BAAS debate had some unfortunate features. The two main protagonists happened to be a militant atheist (as Huxley then proclaimed himself to be) and a bishop, and this led not surprisingly to a complete misunderstanding. Many people jumped to the conclusion that the quarrel was between science and religion. But, as we emphasized earlier, Bishop Wilberforce opposed the theory of evolution not on religious grounds but on scientific ones. I think, in fact, that a religious objection to the theory, presented at that meeting, would have been ridiculed. No doubt many of those present at the debate, particularly the many clergymen present, gave some thought to reconciling their religion to the theory.

Only later did the religious attacks on the theory come out into the open. In particular, Pope Pius IX's encyclical of 1864 at once turned all conscientious Roman Catholics, and some other religious people, against the theory. It is regrettable that it took from 1864 to 1996 for the official ruling of the Roman Catholic Church to be reversed. Of course, in the meantime many thoughtful Roman Catholics had taken it for granted that there could no longer be any religious objection to the theory.

Some of the problems with the general acceptance of evolution are undoubtedly due to the way in which some biologists have presented their case. In my opinion, Huxley was unnecessarily aggressive in his attack on religion. It is true that he had to defend good science against an unreasonable attack by religious groups. But he would, I think, have done better if he had adopted a more tolerant attitude towards religious belief.

Today, one of the more prolific exponents of evolutionary theory is Richard Dawkins. I greatly admire the clarity of his expositions, but I criticize him on two points. I believe that he says too little about the evidence from molecular biology, which in my opinion is much more convincing than the evidence from soft science. Soft science, as I have discussed in Chapter 1, is of its nature more controversial than hard science. There are still arguments about the theory of evolution, between highly competent biologists, but they relate only to minor details. Dawkins brings these out very clearly and fairly in his books, but by so doing he provides ammunition for some readers, who misrepresent the minor disagreements as major confrontations about the validity of the whole theory. With the hard science—the molecular genetics—there are hardly any disagreements.

Also, I think that Dawkins employs bad tactics in attacking religious belief. A religious person who reads an attack on religion in one of Dawkins' books is apt to read no further; one does not discard religious belief so easily. The plain fact is that there are many intelligent people who hold religious beliefs and accept evolutionary theory.

The modern point of view is well brought out in a story told about Albert Einstein. After his theory of relativity appeared, the Archbishop of Canterbury, Randall Davidson, was told by someone that he should pay attention to the theory, as it would affect his religious belief. The Archbishop took this advice seriously, but found that he could not understand the theory at all. Later he happened to be sitting at dinner next to Einstein, and discussed the matter with him. He was relieved when Einstein said 'Do not worry; my theory is a scientific theory, and as such has nothing whatever to do with religion'.

Einstein wrote on the subject of science and religion in 1940. His article, published in *Nature*, includes the passage

> To be sure, the doctrine of a personal God interfering with natural events could never be *refuted*, in the real sense, by science, for this doctrine can always take refuge in those domains in which scientific knowledge has not yet been able to set foot.

Einstein then went on to say that he did not accept that way out, and it appears that he himself held no religious belief. The importance of the passage, however, is that it shows his tolerance towards those who hold a different view.

Tolerance, after all, should be the guiding principle in dealing with the relationship between science and religion. The great religions of the world, such as Christianity and Buddhism, have from their beginnings laid stress on the importance of being charitable towards different points of view, and it is regrettable that this lesson has so often been forgotten. Good scientists, too, are always conscious of the possibility that they may be wrong; science, in fact, would not have flourished if this had not been the case.

# Planck, Einstein, the quantum theory, and relativity

By the beginning of the twentieth century most scientists felt satisfied that all their important problems were more or less solved. The principles of mechanics had been well worked out, heat was understood to be a mode of motion, and light to have wave properties. The exciting discoveries that had been made towards the end of the century—radio transmission, the electron, X-rays, and radioactivity—seemed to be all that could ever be discovered. Obviously there were a few problems, but it seemed that all that was needed was to fill in a few details.

How wrong they were. Little did anyone guess that within only five years from the beginning of the new century there would be two new theories that would bring about radical changes in the way we think about science, and which were to have great practical consequences. These two theories were the quantum theory, introduced by Max Planck in 1900 and developed by Einstein in 1905, and the theory of relativity, due entirely to Einstein in 1905 and later.

Until the quantum theory was introduced it had been taken for granted that energy, including radiation, is continuous; in other words, that one could think of an atom or molecule as having any amount of energy. The essence of the quantum theory is that this is not the case; energy comes in small packets, or quanta (from the Latin *quantum*, how much?). The quanta are small even on the atomic scale, and on the ordinary scale of everyday living they are negligible; in driving one's car one does not have to worry about quantum transitions. In view of this it is amusing to see that journalists, politicians and others are now fond of talking about 'quantum leaps', which they seem to think are very large.

The idea behind the quantum theory is really very simple, and is nicely explained in terms of a pendulum. A familiar example of a pendulum is to be found in certain old clocks, and in a swing in a children's playground. Here we will consider something simpler, a ball on the end of a string (Fig. 8.1). We can start it swinging to a small extent, and we say that the amplitude of the swing is small; the amplitude is just the distance between the two extremities. If we hit the ball and give it more energy, we cause it to swing with a greater amplitude. The interesting thing about a pendulum is that whatever the amplitude, the frequency of the swing is always the same. If the amplitude is greater, the ball moves faster to make the longer trip in the same time.

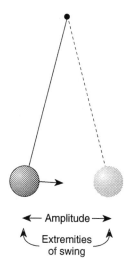

**Fig. 8.1**  A simple pendulum, consisting of a weight at the end of a string. The *amplitude* of the swing is the distance between the centre of the weight at each extremity. The *period* of the vibration is the time taken for the double swing, i.e. the time taken for the weight to move from one extremity to the other and back again. The *frequency* of the vibration is the number of times the weight performs the double trip in unit time (usually taken to be one second). The frequency is therefore the reciprocal of the period. The shorter the string the greater is the frequency, and vice versa.

We can speak of the *period* of the swing, the time that the ball takes to make the double trip, from one extremity to the other and back again. We can also speak of the *frequency* of the oscillation, which is the number of times the ball makes the double trip in unit time, which is usually taken to be one second. If the pendulum makes two complete swings (back and forth) in one second we would say that its frequency is 2 reciprocal seconds, which we write as $2\ s^{-1}$. We saw in Chapter 5 that the hertz (abbreviated Hz) means the same as a reciprocal second, so that the frequency can also be written as 2 Hz.

The *period* (0.5 s in our example) is the same whatever the amplitude, which means that it is the same however much energy we have given to the ball to make it swing. Obviously the frequency (2 Hz in our example) is also independent of the amplitude.

Another important thing we should know about a pendulum is that the frequency of its oscillation gets greater if we shorten the string. Anyone who has taken care of a pendulum clock will be aware of this. If the clock is losing time we correct it by adjusting the bob in such a way as to raise it, so as to make the distance between it and the suspension device a little shorter. The frequency of oscillation will then be greater, and the clock will go faster.

With a pendulum of visible size we take it for granted that, by adjusting the length of the string, we can make the frequency of oscillation anything we like. What the quantum theory says that this is not really the case; only certain energies are possible. However, these quantum restrictions are too small to matter in a clock; for them to be noticeable we must deal with matter at the atomic level.

Suppose, for example, that we imagine a pendulum of atomic dimensions. Now we find that only certain amplitudes of vibration, and therefore only certain energies, are possible. This is illustrated in Fig. 8.2, which shows a pendulum swinging with two possible amplitudes. According to the quantum theory, oscillations with amplitudes in between are just not allowed.

In everyday life we are quite unaware of these quantum restrictions. If we are pushing a child in a swing, it does not occur to us that there are any restrictions as to the motion we can impart. The reason is that when we are concerned with the motions that we can see with our eyes, the permitted quantum levels are so close together that they could not be detected by any instruments. It is only when we get to the atomic scale that the quantum effects are measurable. This explains, of course, why all the great scientists until the twentieth century were unable to find any evidence for the quantum theory. All their work had been done on much too large a scale.

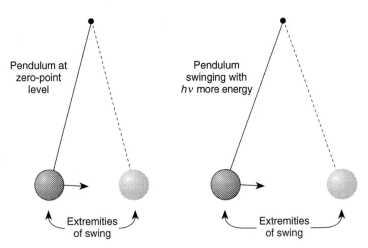

**Fig. 8.2** A hypothetical pendulum of atomic size. The vibrations are now quantized, and the two lowest possible energy levels are shown. Einstein and Otto Stern (1888–1969) proved in 1913 that it is not possible for an oscillator such as a pendulum to have zero energy, i.e, not to move; there must always be a certain minimum energy, called the *zero-point energy*. The difference between the lowest possible energy and the next one is $h\nu$, where $h$ is the Planck constant and $\nu$ is the frequency of oscillation of the pendulum. The energy at the zero-point level, relative to the hypothetical level corresponding to no oscillation, is $\tfrac{1}{2} h\nu$. An atom shows the same kind of vibrational motion.

## THE BIRTH OF THE QUANTUM THEORY

The quantum theory had a rather obscure beginning, which explains why, in spite of its great importance, it was not accepted quickly. It was put forward to explain some results on how the energy of the radiation emitted by a hot body varies with the wavelength or frequency of the radiation. Careful experiments on this matter were carried out around the turn of the century particularly by a number of physicists in Berlin, and attempts were made to explain the results. In 1894 the German physicist Wilhelm Wien (1864–1928) had put forward, on the basis of what seemed to be the best physical theory, the formula

$$\text{Energy} \quad = \quad \frac{A\nu^3}{e^{\beta\nu/T}} \qquad (8.1)$$

where $\nu$ (Greek *nu*) is the frequency of the radiation and $T$ the temperature (it is not necessary to understand this formula; I only put it in to give some idea of the kind of reasoning that went into the discovery of the quantum theory). However, the best experiments that were done in 1905, by Heinrich Rubens (1865–1922) and Ferdinand Kurlbaum (1857–1927), showed that the formula did not quite agree with the results.

This is where Max Planck, the originator of the quantum theory, entered the picture. Max Karl Ernst Planck (1858–1947; Fig. 8.3) was born in Kiel, and studied at the Universities of Munich and Berlin. Here he came under the influence of von Helmholtz, whose important work on thermodynamics we met in Chapter 2. Planck became professor of physics in Berlin in 1889, and much of his early research was in thermodynamics; as we noted in Chapter 2 he did some important work in thermodynamics, but was unlucky in that much of what he did had been done a little earlier, and slightly better, by either van't Hoff or Helmholtz.

As a student in Berlin, Planck had also been influenced by Gustav Kirchoff (1824–1887), who had done pioneering work on the radiation emitted by hot bodies. It is therefore not surprising that Planck was interested in the experimental results that were being produced in Berlin on the variation of this energy with the frequency of the radiation. Some new results obtained by Rubens and Kurlbaum were presented by Kurlbaum to the Berlin Academy on 25 October 1900, but a few days previously they had been communicated privately to Planck, who had had time to think about them. In the discussion following the presentation of the paper Planck proposed the equation

$$\text{Energy} \quad = \quad \frac{A\nu^3}{e^{\beta\nu/T} - 1} \qquad (8.2)$$

Note that this differs from equation 8.1 only by the addition of minus one in the denominator. Planck had suggested it simply because he thought it would

**Fig. 8.3**   Max Planck (1858–1947), who in 1900 explained the distribution of energy in radiation in terms of the idea that energy comes in small packets, or quanta. For this work he was awarded the Nobel Prize for physics in 1918. He was professor of physics at the University of Berlin from 1889. He strongly opposed Nazi policies, and one of his sons was executed for plotting against Adolf Hitler.

fit the data better—he had no theoretical reason at all to suggest it. At this stage, therefore, his suggestion was *empirical*.

The day after the Berlin Academy meeting Rubens, who had worked on the problem through the night, told Planck that his formula fitted all of his data within the experimental error. Much encouraged, Planck worked hard during the next few weeks to find a theoretical basis for his empirical formula.

The essence of Planck's treatment was that a solid body consists of an array of atoms each one of which is in constant oscillation, and obeys essentially the same laws as the bob of a pendulum. He considered the various energy states that these oscillators could have, and carried out a statistical treatment, making use of the methods that had been worked out by Ludwig Boltzmann. Purely as a mathematical convenience Boltzmann had often treated energy as coming in small packets, but at the end of his derivations he always required the energy packets to become of zero size.

Planck followed this procedure, and assumed that if the atoms were oscillating with frequency $\nu$, their energy came in packets of size $h\nu$, where $h$ is a constant—it is now known as the *Planck constant*. In other words, the energy of the oscillation can be 0, $h\nu$, $2h\nu$, $3h\nu$...but nothing in between. As we mentioned, when Boltzmann used this kind of mathematical procedure he always made the size of the packets become zero at the end of the work.

Planck did not, however, carry out this final step, and he did not at first appreciate the great significance of this omission. His paper giving this treatment appeared in December 1900, and 14 December 1900, when he presented the work to the German Physical Society, is often regarded as the birthday of the quantum theory.

In view of the fact that we now accept Planck's quantum theory as correct and of great importance, we would have expected his 1900 announcement to have attracted immediate attention. We would at least have expected some discussion and criticism of it. In fact, for about five years hardly any notice at all was taken of Planck's paper. The main reason for this is that the theory was presented only as applying to black-body radiation, and its wider implications were not recognized until later. Most physicists at the time were working on what appeared to be much more exciting problems than radiation, such as radioactivity, X-rays, and the electron.

When attention was finally paid to Planck's theory, it was at first usually unfavourable. The criticisms that were made generally suggested not that Planck had suggested a quantum hypothesis that was not acceptable, but rather that he had made a silly mistake in his mathematics.

## QUANTA OF RADIATION

At first Planck thought of his quantum theory as applying only to the oscillators—for example to the atoms in a solid, which could possess energy only in multiples of $h\nu$. No one thought of regarding radiation itself as being quantized, until 1905 when Albert Einstein (1879–1955; Fig. 8.4) made that important suggestion.

At the time, and for four more years, Einstein was a junior official in the patent office in Berne, Switzerland, having become a Swiss citizen in 1901. His university career had been less than promising, and not much was expected of him. However, the year 1905 was an *annus mirabilis* for him, since he published three papers each one of which was worthy of a Nobel Prize. The first of these was on the quantization of energy, which did actually get him his Nobel Prize in 1921. One of the other two was on the special theory of relativity, to be considered later.

Modern textbooks often say that Einstein's 1905 paper on quanta of radiation was based on Planck's 1900 paper on the quantization of oscillator energies. Only someone who had not looked at the paper could write that. Einstein's great contribution in his 1905 paper was his recognition that radiation itself is quantized. He made only a passing reference to Planck's work, which at the time he did not find convincing, and he made no use of Planck's constant $h$. Instead his proposal that radiation is quantized was based on a number of considerations, one of which was what is called the *photoelectric effect*.

**Fig. 8.4**   Albert Einstein (1879–1955), who was German born, but took out Swiss citizenship in 1901, and American citizenship in 1940. He revolutionized physics, particularly with his theory of relativity and with his theory that radiation can behave as a stream of particles (quanta, or photons). He won the 1921 Nobel Prize for physics, specifically for his work on the quantization of radiation (relativity theory had still not proved itself at that time). He was professor at the University of Zürich from 1914 to 1933, and later was at the Institute for Advanced Study at Princeton.

This effect had been discovered in 1887 at the University of Kiel by Heinrich Rudolf Hertz in the course of his famous experiments on radio transmission (Chapter 6). Rather paradoxically, it was a series of experiments that convincingly confirmed Maxwell's electromagnetic theory of radiation that also produced some of the first evidence for the particle nature of radiation. Hertz discovered that the length of a spark induced in a secondary circuit was greatly reduced when the spark gap was shielded from the light of the spark in the primary circuit. He showed that ultraviolet light has the effect of increasing the length of the spark that was produced. Similar effects, later realized to be due to ejection of electrons by the radiation, were reported in the same year, 1887, by Arthur Schuster in Manchester, and by Svante Arrhenius in Sweden.

Two years later, in 1889, detailed studies of the emission of electrons under the influence of light were reported by Johann Elster (1854–1920) and Hans Geitel (1855–1923). These men were often known as the Castor and Pollux of science, or as the Heavenly Twins, as they were inseparable both in their private lives and in their teaching and research. They had been school friends, and both became teachers at a Gymnasium in Wolfenbüttel near Braunschweig (Brunswick). When Elster married and moved into a house

Geitel joined the couple, and the two friends carried out research in a laboratory they established in the house. They were often confused with each other,
and a man who somewhat resembled Geitel was once addressed by a stranger
as 'Herr Elster'. His reply was 'You are wrong on two counts: first, I am not
Elster but Geitel, and secondly I am not Geitel.'

In 1898 J J Thomson in Cambridge concluded that the particles ejected by
the radiation are the same as those present in cathode-ray tubes; that is, they
are electrons (which Thomson always called 'corpuscles'). Later Phillip
Lenard (1862–1946) showed that no emission occurs if the frequency of the
radiation is less than a critical value $\nu_0$, and that the energy of the ejected
electron is equal to $h\nu - h\nu_0$ if radiation of frequency $\nu$ is employed. He found
that the intensity of the radiation has no effect on whether or not electrons
are emitted; it affects only the number of electrons emitted. (We met Lenard
earlier, in Chapter 6, as the pathologically anti-Semitic Nazi who denounced
Hertz's work, having previously admired it.)

Einstein realized that these results are inexplicable in terms of the wave
theory of radiation, according to which radiation of sufficiently high intensity,
irrespective of frequency, would be able to eject electrons. With the particle
theory, on the other hand, the explanation is straightforward (Fig. 8.5). A
particle of radiation has energy $h\nu$, and is capable of ejecting an electron provided that $h\nu$ is sufficiently large, i.e., that the frequency is sufficiently large.

Einstein's paper is often referred to today as his paper on the photoelectric
effect, but much of the paper is concerned with other matters, such as the
way in which atoms and molecules absorb and emit radiation. In other words,
it was important in leading to explanations of the detailed nature of spectra.

Einstein's suggestion that light can act as a stream of particles may seem
rather surprising in view of what we have been saying so far in this book
about radiation. The particle theory of radiation, advocated by Newton,
Laplace, and others, seemed to have been conclusively disproved by the work
of Thomas Young, Fresnel, Fizeau, and Foucault (Chapter 6), and Maxwell
had put the wave theory on a firm footing. How can the diffraction and interference of light be explained in terms of a particle theory? Also, there appeared to be a logical difficulty with Einstein's theory: the energy of radiation
$h\nu$ is expressed in terms of its frequency $\nu$, but frequency only makes sense in
terms of the wave theory.

In his 1905 paper Einstein anticipated these objections and answered them.
He began his paper by admitting that the wave theory of light was well established and was not to be displaced as far as properties such as diffraction and
interference are concerned. But, he continued, such properties, in which light
interacts with matter in bulk, relate to time averages. It is quite possible that
the wave theory will prove inadequate whenever there are one-to-one effects
such as the photoelectric effect and the emission and absorption of radiation.
In other words, what is needed is a *dual theory of radiation*. For diffraction
and interference the wave properties are relevant; for the photoelectric effect

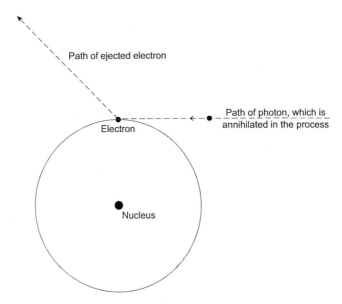

**Fig. 8.5** Explanation of the photoelectric effect. An electron is shown in its orbital motion around the nucleus, and a quantum of radiation (a photon) interacts with it. The energy of the photon is *hv*, where *v* is the frequency of the radiation. A certain minimum energy, called the *work function*, is required to remove the electron. If the photon does not have this energy, nothing can happen; increasing the intensity of the light makes no difference, since it only increases the number of photons, and at a given frequency of light they all have the same energy. If *hv* is greater than the work function, the photon will remove the electron from the atom and will be annihilated— think of the electron as just a bundle of energy. Any excess energy will be in the escaping electron.

and for emission and absorption one must regard the radiation as a beam of particles.

At this stage Einstein was not convinced that oscillator energies are quantized, but only that radiation is quantized. In fact, Einstein in this paper did not use Planck's equation (8.2), but equation 8.1 which had been given by Wien and was not satisfactory. After his paper was published this was pointed out to Einstein, and then he began to realize that Planck was quite right: oscillator energies must be quantized. He became even more convinced of this when two years later he worked on the specific heats of solids, and found that he could give a good explanation of them by using Planck's treatment of vibrations in solids.

Even Planck had difficulty in accepting Einstein's theory of the quantization of radiation—and even in accepting that his own work really required us to think about things in a different way. To Planck the idea of quanta of energy was no more than an *ad hoc* hypothesis to explain black-body radiation;

he was slow to realize that the quantum theory has a much wider importance. It was in his nature to be conservative, and having been brought up on classical physics he found it difficult to renounce the old ideas. Einstein deserves credit not only for introducing the idea of the quantization of radiation, but for persuading Planck that his theory was correct and important.

Einstein had accomplished so much in his three great publications of 1905, that it is not surprising that he did not do much research later that was as important. His subsequent career was, however, distinguished, and he inspired many younger scientists. In 1934 he accepted an appointment at the Institute for Advanced Study at Princeton, where he remained to the end of his life. His work attracted much interest among the public, and for many years he was probably the best known scientist in the world.

When I was a graduate student at Princeton just before the war I often passed him walking along the side of the golf course on his way to the Institute. He always grinned and gave an amiable wave. His costume was usually of the unconventional kind always associated with him, its particular feature being no tie and no socks.

Many stories were told about him by the Princeton students, and I do not know how many of them were true. One relates to the first arrival in Princeton of Einstein and his second wife. A professor who was a neighbour called one day when he was out, and was greeted by Mrs Einstein. The professor said that he was a chess player, and that he had heard that Professor Einstein also enjoyed playing chess; he would therefore be very pleased if they could play together some time. 'Certainly not', replied Mrs Einstein, 'the professor would never want to do anything like that', and firmly shut the door. The visitor was nonplussed until he finally learnt the reason for the misunderstanding: Mrs Einstein had thought that Professor Einstein was being invited to play jazz, which she herself pronounced 'chess'. Einstein only played classical music on his violin.

Another incident occurred during the time I was in Princeton. The secretary to the Institute received a telephone call from someone with a German accent who asked for Professor Einstein's address. The secretary said that she was sorry, but the professor had given strict instructions that his address was not to be given out. 'Now that is very difficult', said the caller, 'You see, I am Professor Einstein and I have forgotten where I live and I want to go home'.

Einstein often had difficulty going home. For some time after the Einsteins arrived in Princeton, neighbours would sometimes go into their own sitting rooms and find Einstein calmly installed reading a newspaper; he had gone into the wrong house. Consideration was given by the neighbours to painting their front doors a different colour from that of the Einsteins, so that he would more easily find his own house.

Einstein was always rather vague about trivial details like paying bills or filing income tax forms. Fortunately he always had a secretary to deal with such matters. The US income tax form ends with the questions:

1. Did anyone help you with completing this return?
2. What was the nature of the help?

To 1. Einstein answered 'My secretary', to 2. 'She told me what to write'.

It is of interest that although Einstein is most famous for his theory of relativity, his 1921 Nobel Prize was awarded to him for his work on the quantum theory. The reason is that at the time of the award there was considered to be insufficient evidence for relativity theory. One aspect of Einstein's Nobel Prize is somewhat unusual. For some years he and his first wife were not getting along well, but she was reluctant to leave him. When it became fairly certain that Einstein would win the Prize he made a bargain with his wife, which they kept, that he would give her the prize money as compensation for her leaving him.

## BOHR'S ATOMIC THEORY

Even after the quantum theory had been used to explain properties of radiation many scientists paid little attention to it. Chemists and biologists in particular thought that they could get along nicely without it. All this changed in 1913, however, when Niels Bohr showed how the quantum theory can explain the structure of atoms and many of their properties.

Niels Henrick David Bohr (1885–1962; Fig. 8.6) was born in Copenhagen and studied at the University of Copenhagen where he obtained his doctorate in 1911. In 1912, at the age of 27, he went to Cambridge with the intention of working with J J Thomson, the Cavendish professor. Somehow the association was an uncomfortable one. It appears that Bohr offered some well-meant but penetrating criticisms of a 'plum-cake' theory that Thomson had proposed in 1904, in which electrons were supposed to be embedded in a positively-charged medium. Bohr was always very courteous, but his English was not very good, and as often happens in such circumstances his criticisms did not come out as politely as he had intended. Apparently he pointed to a passage in one of Thomson's papers, and simply said 'That is wrong'. As Thomson was always remarkably good-natured it is unlikely that he resented the criticisms for very long, and was the last person to bear a grudge. Bohr, however, was unusually sensitive about such matters, and was conscious of the fact that he had committed a *faux-pas*. His relationship with Thomson therefore became strained. The situation was not helped by the fact that Thomson had suggested to Bohr an experimental problem which Bohr felt was uninspiring. Bohr's heart was always more in theoretical than in experimental work, and he became unhappy with his progress.

After a few months of this Bohr left Cambridge and went to work at Manchester University with Ernest Rutherford (1871–1937) who had already made a great reputation with his work on atomic nuclei. The collaboration between Rutherford and Bohr became a happy and fruitful one. Rutherford

**Fig. 8.6** Niels Bohr (1885–1962), who is particularly famous for his theory of the atom, developed in 1913 at the University of Manchester, where Ernest Rutherford was professor of physics. Later he became professor at the University of Copenhagen and headed an institute financed by the Carlsberg Breweries. He exerted a strong influence on atomic and nuclear physics, not only through his own work but by his support of others. He escaped from Denmark during World War II and provided vital information to the British and American governments which was an important factor in the construction of the first atomic bombs.

was himself not much of a theorist, and was initially somewhat wary of Bohr who was full of theoretical ideas; however, partly because Bohr was a good soccer player, Rutherford soon accepted him and gave him much encourage-ment and cooperation. Bohr completed most of his theory of the atom in Manchester, but put important finishing touches to it, including his work on spectral lines, after he had returned to Copenhagen.

Bohr applied the quantum theory to the hydrogen atom, which consists of a positively-charged nucleus, called a proton, and one electron. He concluded from the quantization condition that electrons are required to remain in fixed orbits round the nucleus (Fig. 8.7). In the normal hydrogen atom the orbit is close to the nucleus, but the electron can be 'excited' into certain other orbits in which it is further from the nucleus. Other orbits, in which the electron is in intermediate positions, are not allowed. Bohr worked out mathematical equations which expressed the positions and energies of these orbits.

This work had been done before Bohr left Manchester, and he then went home to Copenhagen. There he was visited by the spectroscopist Hans Marius Hansen (1886–1956) who asked him if he had interpreted the spec-trum of the hydrogen atom. In Chapter 3 we saw that spectra consist of series of lines, known as Fraunhofer lines, and that these are characteristic of the

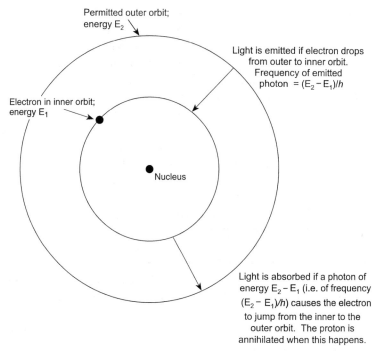

**Fig. 8.7**    Bohr's idea of a hydrogen atom, consisting of a nucleus (a proton) and one electron. The quantum condition restricts the electron to certain orbits, two of which are shown. If an electron is in an outer orbit and drops to an inner one, there is emission of radiation. The frequency $\nu$ of the emitted radiation is determined by the fact that $h\nu$ must be equal to the difference between the energies of the two levels. If light passes through the atom, a photon may be absorbed, the energy $h\nu$ released by it raising the electron from one quantized level to a higher one. The energy $h\nu$ of the photon absorbed is the difference between the energies of the two electron levels.

chemical substances present. Until Bohr did his work no one had been able to interpret the position of these lines, although some simple empirical equations had been given which accounted for them.

Bohr replied to Hansen that he had decided not to consider the spectral lines of the hydrogen atom as he thought they would be impossibly complicated. Hansen, however, persisted, and drew Bohr's attention to relatively simple empirical equations that had been put forward. Surprisingly, Bohr had forgotten about these, and after looking into them realized that his mathematical expression for the energy levels provided a complete explanation. When he put in the constants that appeared in his energy equations, he got excellent agreement with the positions of the spectral lines.

Particularly because of this good agreement, Bohr's theory was quickly accepted with some enthusiasm. Einstein, for example, spoke highly of it. In September 1913, at the 83rd meeting of the British Association for the

Advancement of Science, the theory was discussed favourably, and reports of it, as expounded at the meeting, appeared in *Nature* and in *The Times* of London. It is interesting that Bohr himself was somewhat less enthusiastic than many others, regarding his theory as 'preliminary and hypothetical.'

Besides being of great significance for the interpretation of atomic structure and spectra, Bohr's theory provided further convincing evidence for the validity of the quantum theory. The theory not only explained the distribution of energy in radiation: it related to specific heats, to the absorption of radiation, and to atomic structure, and evidently had wide implications.

At about the same time that Bohr put forward his atomic theory Johannes Stark (1874–1957), at the Technische Hochschule in Aachen, made a discovery of great importance. He found that an electric field splits spectral lines, such as the Fraunhofer lines of hydrogen, into a number of component lines. Stark's paper dealing with this appeared in November 1913, just a few months after Bohr's paper was published. Shortly afterwards the general features of the Stark effect were interpreted on the basis of Bohr's theory, and this provided further evidence for the theory. (Stark, like Lenard, later became pro-Nazi and anti-Semitic to the point of paranoia and then interpreted his results in such a way as to form the basis of an attack on the quantum ideas of Einstein.)

The Bohr theory of the atom was certainly a remarkable achievement, as it provided a general interpretation of atomic structures and spectra, which previously had seemed quite incomprehensible. However, even with many improvements that were made to the theory, serious difficulties remained, as Bohr himself freely admitted. It did not, for example, seem possible to extend the theory in such a way as to give a complete treatment of atoms containing more than one electron, or even of the simplest of molecules. Aside from these difficulties, there was a feeling that it was arbitrary just to add quantum restrictions to the old mechanics. What was needed instead was a new mechanics in which the quantum restrictions would emerge as a mathematical necessity.

Throughout his career Bohr exerted a profound influence on science in many ways—but not through his lectures. As a graduate student at Princeton in 1939 I attended a lecture he gave, and remember it vividly as the worst lecture I ever heard. I happened to be sitting on the front row, but Bohr spoke so softly that I could not hear him, and he wrote on the board in minute indecipherable letters. Of course, afterwards we all thought it a great privilege to have seen the great man, even though we could not hear him or read his equations.

## Quantum mechanics

In 1921 Max Born (1882–1970) became professor of physics at the University of Göttingen. He at once became interested in the possibility of creating a new form of mechanics which would take account of the quantum theory, and made a good start in 1924 with a paper entitled 'Zur Quanten-mechanik'; this was the first time the expression quantum mechanics had been used. In the spring of 1925 one of Born's research assistants was the 24-year-old Werner Heisenberg (1901–1976; Fig. 8.8), who had obtained his doctorate two years earlier. Like some others who later had careers of great distinction he almost failed to obtain his degree, as he had neglected his laboratory work and was unable to answer any of the questions put to him by Wilhelm Wien, the professor of experimental physics. Wien wanted to fail him, but was persuaded to give him third-class standing in his oral examination. It is amusing to note that one of the topics that caused Heisenberg trouble was the resolving power of a microscope, and he still got this wrong when he proposed his uncertainty principle in 1927.

In the spring of 1925 Heisenberg was suffering from such a severe attack of hay fever that he obtained leave to stay for some time on Heligoland, a rocky island free from vegetation. While there he developed a modification of Born's treatment in which physical quantities were represented by a particular kind of mathematical function called a complex number. In July 1925, Heisenberg sent his manuscript to Born who at once recognized its importance and submitted it for publication, in Heisenberg's name only; it appeared later in the year.

**Fig. 8.8**   Werner Heisenberg (1901–1976). With Max Born he gave the first formulation of quantum mechanics in 1925. He also suggested the uncertainty principle, according to which the more accurately we measure the position of a particle, the less accurately we can know its speed. He won the 1932 Nobel Prize for physics. He was professor of physics at Leipzig, Berlin, and Göttingen. During World War II he directed Germany's atomic energy programme, but at the same time devoted much attention to other matters, and the work did not achieve its goal.

Born did not at once recognize the importance of Heisenberg's complex numbers, but 'after eight days of concentrated thinking and testing' realized that they had the properties of functions known as *matrices*. At the time matrices were known to some mathematicians, but were rarely used by physicists; Born had learnt about them in his student days 20 years earlier, and now recalled them. Heisenberg later admitted that he knew nothing of matrices; in developing his quantum mechanics he had reinvented matrix mechanics without realizing that he was doing so. Born then further developed Heisenberg's treatment in terms of matrices. Theoretical physicists who have made a careful study of Heisenberg's work have pointed out that he by no means followed logical principles in arriving at his theory; a great deal of imagination and intuition was involved.

Another contribution of Heisenberg, made shortly after his matrix mechanics, has proved of great importance. This was his *uncertainty principle*, or principle of indeterminacy. Heisenberg arrived at the principle by carrying out a 'gedanken' (imaginary) experiment in which a beam of light is used to discover the position and momentum (mass multiplied by velocity) of an electron. Because the light disturbs the electron there is a limitation on the accuracies of the measurements. His conclusion was that it is impossible to determine both the position and the speed of a particle exactly. The more accurately one knows the position, the less accurately one knows its speed, and vice versa.

Heisenberg won the 1932 Nobel Prize in physics for his quantum mechanics. It was always a matter of regret to Heisenberg that Born had no share in this honour. The award of the prize jointly to Born and Heisenberg would indeed have been more appropriate. Not only had Born initiated the idea of a quantum mechanics: he had shown the true importance of Heisenberg's obscure mathematics by relating it to matrix mechanics. A less generous man would have delayed the submission of Heisenberg's paper until a joint paper of broader significance could be prepared. Born did receive a Nobel Prize much later, in 1954, for his work on the quantum mechanics of collision processes.

Heisenberg remained in Germany, and during World War II directed the German atomic energy programme, which is briefly discussed later in this chapter. He by no means approved of Nazi policies and for this he suffered from the vicious attacks of Stark and Lenard, who were powerful and fanatical Nazis. Born had to leave Germany, and in 1936 became professor at the University of Edinburgh, returning to Germany after his retirement in 1954.

Shortly after the appearance of Heisenberg's first paper on quantum mechanics, another important contribution was made by Paul Adrien Maurice Dirac (1902–1984; Fig. 8.9). Dirac was trained as an electrical engineer at the University of Bristol, and being unable to obtain employment secured in 1923 a research scholarship to work for his PhD degree in Cambridge under R H Fowler. In July 1925, Heisenberg lectured in Cambridge and Dirac

**Fig. 8.9**   Paul Adrien Maurice Dirac (1902–1984). In 1925 he produced a system of quantum mechanics that turned out to be mathematically identical with that of Heisenberg, but which most people find more comprehensible. He first introduced relativity theory into quantum mechanics, and predicted electron spin and the positron, a particle equivalent to the electron but having a positive instead of a negative charge. He shared with Schrödinger the 1933 Nobel Prize for physics. From 1932 to 1969 he was Lucasian professor of mathematics at Cambridge.

attended the lecture but later said that he had paid little attention to it. In September he saw the proofs of Heisenberg's paper and initially saw nothing useful in it. He soon changed his mind and quickly produced an alternative formulation, his first paper on quantum mechanics appearing late in 1925. His quantum mechanics turned out to be equivalent to Heisenberg's, but somewhat easier to understand.

Dirac followed up his first paper on quantum mechanics by many others of great importance. In 1926 he applied his methods to the hydrogen atom, and in papers that appeared in 1928 he combined his system of quantum mechanics with the theory of relativity. The resulting theory was an extremely powerful one which had far-reaching consequences. For example, Dirac was able to deduce that an electron must have a spin, and that there are two quantum numbers for electron spin; experimental evidence for this had already been obtained. We can visualize this by saying that the electron can rotate in one direction with a certain speed, and in the opposite direction with the same speed, but in no other way. Dirac also predicted from his theory the existence of an elementary particle having the mass of the electron but a positive charge instead of a negative one. This particle, the *positron*, was discovered in 1932 by the American physicist Carl David Anderson (1905–1991).

Dirac was a typical 'absent-minded professor' about whom many stories are told, some of them probably true. One of my favourites relates to a boat trip in which he had to share a cabin with a Frenchman who spoke little English. His father having been born a French-speaking Swiss, Dirac spoke French fluently, but with the Frenchman he spoke entirely in English. At the end of the trip the Frenchman overheard Dirac speaking French, and angrily asked him why he had not mentioned that he knew the language. Dirac's reply was simply, 'Well, you didn't ask me'. Once, when asked if he could play any musical instrument, he replied, after some thought, 'I don't know—I've never tried'.

Once while discussing physics with a colleague he idly watched a woman who was knitting. He finally told her that he had been working out the mathematics of knitting, and had discovered that there was an alternative procedure, which he explained. 'Yes,' she drily replied, 'it is called purling, and has already been discovered'.

When Dirac heard in 1933 that he had been selected to share the Nobel Prize for physics with Schrödinger, he wanted to refuse it, saying that he disliked publicity. Rutherford told him that he would experience more publicity if he refused it, and that convinced him to accept.

## THE WAVE PROPERTIES OF PARTICLES

We have seen that radiation has both particle properties as well as wave properties. The converse idea was also put forward, in 1923, that particles such as electrons can also have wave properties. This suggestion came from Louis Victor, Prince de Broglie (1892–1987: Fig. 8.10). His title of 'Prince' was a courtesy title conferred upon a de Broglie ancestor by Francis I of the Holy Roman Empire, and to which all descendants were entitled; after his elder brother Maurice died in 1960, Victor inherited the French title of Duc, which took precedence over the title of Prince.

Louis de Broglie had first intended to become a civil servant and in 1910 obtained a licentiate in history. Later, partly under the influence of his elder brother Maurice, Duc de Broglie (1875–1960), who was a distinguished experimental physicist, Louis became interested in physics and philosophy, obtaining a licentiate in the physical sciences at the Sorbonne in 1913. He then worked towards his doctorate, specializing in fundamental problems of space, time, and atomic structure. He was particularly interested at first in the work of the American physicist Arthur Holly Compton (1892–1962) who, by observing the scattering of X-rays, obtained in 1923 further striking confirmation of Einstein's suggestion that radiation can exhibit particle properties.

De Broglie suggested the converse hypothesis, that particles such as electrons can have wave properties. By making use of equations for radiation, he

**Fig. 8.10**   Louis Victor, Prince (later seventh Duc) de Broglie (1892–1987). In his doctoral thesis, presented to the Sorbonne in 1924, he suggested that particles such as electrons have wave properties as well as particle properties. He was awarded the Nobel prize for physics in 1929, by which time his theory had been confirmed by experiments in which electrons were diffracted. In 1960 he succeeded his elder brother Maurice as Duc de Broglie, a title that took precedence over the courtesy title of Prince.

suggested an equation that related the wavelength $\lambda$ of the wave associated with a particle to the mass of the particle and its speed. His equation is remarkably simple:

$$\lambda = h/mv \tag{8.3}$$

This just says that the wavelength $\lambda$ of a moving particle is equal to the Planck constant $h$ divided by its mass $m$ multiplied by its speed $v$.

When de Broglie presented his PhD thesis in 1924 he had some trouble with his examiners. One of them, for example, dismissed the ideas as 'far-fetched'. However, de Broglie had shrewdly sent a copy of his thesis to Einstein, who returned an enthusiastic endorsement which included the phrase 'He has lifted a corner of the great veil'. In spite of this, the examiners remained sceptical of the work but somewhat reluctantly awarded the degree. Robert Jemison Van de Graaff (1901–1967), the inventor of the electrostatic generator that bears his name, was present as a student at the defence of the thesis, and much later remarked—presumably having in mind Winston Churchill's famous remark about the RAF in the Battle of Britain—'Never has so much gone over the heads of so many'.

Other prominent physicists were also initially sceptical of de Broglie's theory, and he did not receive his Nobel Prize until 1929, by which time the theory had been confirmed experimentally. The first experiments to support the theory involved the diffraction of electrons. According to de Broglie's equations, electrons accelerated by a potential of about 100 V should have wavelengths similar to the interatomic spacing in crystals, and like X-rays (Chapter 7) are therefore suitable for diffraction experiments. Clinton Joseph Davisson (1881–1958) had been studying electron scattering in the laboratories of the American Telephone and Telegraph Company, and of the Western Electric Company in New York since 1919. In January 1927, with his colleague Lester Halbert Germer (1896–1971), he succeeded in demonstrating the diffraction of electrons by a single crystal of nickel. In May of the same year, at the University of Aberdeen, George Paget Thomson (1892–1975) and his research student Andrew Reid observed the diffraction of electrons by thin films. Ten years later Davisson and Thomson shared the Nobel Prize in physics for this work. It has been commented that J J Thomson was awarded the 1906 Nobel Prize for showing that the electron is a particle (Chapter 6), while his son G P Thomson obtained a 1937 Nobel Prize for showing that it is a wave.

De Broglie's idea played no part in the first quantum-mechanical treatments of Born, Heisenberg, and Dirac. It did, however, strongly influence the thinking of Schrödinger, who developed a wave mechanics that eventually turned out to be mathematically equivalent to the quantum-mechanical theories of Heisenberg and Dirac. This we will now discuss.

## WAVE MECHANICS

The quantum mechanics of Erwin Schrödinger (1887–1961; Fig. 8.11) has been particularly popular with chemists, as it provides a more easily visualized representation of atomic structure, in contrast to the rather formal approaches of Heisenberg and Dirac. Schrödinger, an Austrian by birth, became professor of physics at the University of Zürich in 1921, and it was there that he did his original work on wave mechanics.

Schrödinger's wave mechanics evolved directly from de Broglie's ideas, which he had first thought to be 'rubbish' until he was persuaded otherwise. In the words of the physicist Hermann Weyl (1885–1955), Schrödinger obtained his inspiration for wave mechanics while engaged in a 'late erotic outburst in his life'. His amorous activities were somewhat remarkable. His unprepossessing appearance, with his thick spectacles, hardly corresponds to the popular idea of a Lothario or Casanova, but such he was. In 1920 he married Annemarie Berthel, and although their relationship was punctuated by many stormy episodes they remained together until his death. Schrödinger had no children by his wife, but he had at least three illegitimate daughters.

His work on wave mechanics was begun during one of his amorous adventures in late 1925, when he stayed at a holiday resort with a mistress while his wife remained in Zürich, and it was perhaps not only the theory that was

**Fig. 8.11**   Erwin Schrödinger (1887–1961), who in 1927 formulated the first wave-mechanical theory, which he later showed to be mathematically equivalent to the quantum mechanics of Heisenberg and Dirac. He shared with Dirac the 1933 Nobel Prize for physics. He was Austrian born and educated at the University of Vienna. He left Germany in 1933, and after spending time in Oxford and elsewhere became director of the Institute for Advanced Study in Dublin. In 1956 he returned to Austria as professor of physics at the University of Vienna.

conceived at that time. The woman involved in that particular encounter has not been identified, but it would seem that she deserved at least a Nobel Prize nomination as Best Supporting Actress.

In Schrödinger's wave mechanics an electron in an atom is treated as a wave rather than as a particle in an orbit going round the nucleus. He did not derive his famous wave equation—that is to say, he did not arrive at it by a formal mathematical proof. Instead he proceeded by analogy with the equations used in Maxwell's electromagnetic theory for wave motion in ordinary radiation. His wave equation is a differential equation involving a wave function or eigenfuction $\Psi$ (psi), and an energy $E$ known as an eigenvalue. Quantum restrictions are not introduced arbitrarily, but appear as a direct consequence of the wave equation, for which no solution is possible unless the energy has one of a number of permitted values. The restrictions as to the orbits in which an electron may move are a consequence of the need for a wave to fit into the right amount of space.

At first Heisenberg and Dirac were critical of Schrödinger's wave mechanics. Max Born approved of it from the start, and in 1926 provided a physical interpretation of the wavefunction $\Psi$. For a problem in atomic or molecular structure, $\Psi$ according to Born is related to the probability that an electron is present in a particular small region of space. In some situations the probability is simply represented by the square of the wavefunction, $\Psi^2$.

The misgivings of Heisenberg and Dirac about Schrödinger's theory evaporated when later in 1926 Schrödinger proved that his and Heisenberg's formulations were mathematically equivalent. Starting with Heisenberg's equations, which involve matrices representing physical properties, Schrödinger showed that each physical property can be replaced by an appropriate mathematical operator, and that his wave equation was then obtained. Modern quantum-mechanical calculations are often based on this procedure.

Heisenberg's uncertainty principle, and Born's interpretation of $\Psi^2$ as representing a probability, are important components of what came to be called the *Copenhagen interpretation* of quantum mechanics, since Niels Bohr, professor in Copenhagen, had much to do with formulating this point of view. The Copenhagen interpretation implies a lack of complete determinism, in that future events do not follow inevitably from past conditions, pure chance playing some role. Most physicists have accepted this interpretation, but Einstein and Schrödinger took strong exception to it.

Einstein's objection is summarized in his often-quoted statement that 'God does not play dice.' In a number of forceful but always friendly arguments with Bohr, Einstein tried to devise ways of circumventing the uncertainty principle, but Bohr was always able to show that he was in error—sometimes by invoking Einstein's theory of relativity.

Schrödinger was an intensely emotional person, and this often led to rather bizarre personal behaviour. At the beginning of the Second World War he became director of the Institute for Advanced Studies in Dublin, but in 1945 he resigned, the reason being that he had become involved in a quarrel about the way his office was cleaned. He usually dressed unconventionally, and as a result on several occasions had difficulty gaining admission to scientific meetings, and sometimes to lectures he was to give himself.

When the Copenhagen interpretation was put forward he did not simply object to it, as a few other scientists did—he found it deeply distressing. With regard to Born's probability explanation, Schrödinger would say: 'I can't believe that an electron hops about like a flea.' In one exchange with Bohr he said that since people were giving these interpretations to his wave mechanics 'I regret having been involved in this thing.' To Schrödinger an electron has wave properties and is not to be regarded as a particle darting about. In his view electronic orbitals are to be considered in terms of wave properties, not as the movement of particles.

## THE CHEMICAL BOND

A little has been said at the beginning of Chapter 7 about the nature of the chemical bond. Not much could be done to explain the chemical bond

satisfactorily until the electron had been discovered and its properties investigated. Our ideas of the bond changed drastically with the advent of quantum
theory and especially of the new quantum mechanics. In 1916 G N Lewis,
whose important work on thermodynamics we met in Chapter 2, made a contribution of special significance. He suggested that the usual type of chemical
bond involves a pair of electrons.

Lewis had first developed a rudimentary form of his theory at Harvard in
1902, and had presented it to his students. In 1916, when he first published
his theory, he was at the University of California at Berkeley. The important
idea that Lewis introduced is that bonding can take place as the result of the
sharing of electrons between two atoms. His original suggestion, later
modified, was that groups of eight electrons ('octets') were stationed at the
corners of cubes. Bonding could occur by an overlap of the edges of two
cubes (Fig. 8.12). He recognized that in some molecules both of the bonding
electrons may originally have come from one of the atoms, an arrangement
referred to later as a coordinate link or a semi-polar bond. With suitable

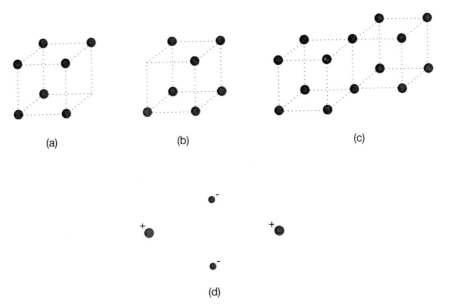

(a)                    (b)                    (c)

(d)

**Fig. 8.12**  G N Lewis's original idea of chemical bonding, as he represented it in
1916. In some atoms the electrons arrange themselves at the corners of a cube, and he
referred to them as *octets* of electrons. In a and b one electron is missing from the
octet of each atom. In c the two atoms have come together to form a molecule; by
coming together at the edge of the cube, each atom has completed its octet. The
hydrogen molecule (d) is a special case. Each atom has one electron, and the molecule
is formed by the pairing of these electrons. The two electrons, by residing between the
protons, reduce the repulsion between the protons, and a bond is possible. Lewis's
ideas became modified and refined over the years, and form the basis of the modern
ideas about the chemical bond.

modifications Lewis's model could also explain the tetrahedral arrangement of bonds emanating from a carbon atom (Fig. 7.1).

In 1919 Lewis's ideas were taken up with great enthusiasm by the American physical chemist Irving Langmuir (1881–1957), who already had a wide reputation, having done important work on solid and liquid surfaces that was later to bring him a Nobel Prize. Langmuir extended and to some extent modified Lewis's ideas, applying them to a wider range of substances. He also suggested a number of the terms that are used today, such as covalence and electrovalence. He published a dozen papers on the subject between 1919 and 1921, and he also presented the theory in a number of lectures he gave in the United States and elsewhere. Unlike Lewis, who was somewhat shy and diffident and not a good lecturer, Langmuir was always eloquent and compelling—to the extent that people sometimes complained that they found themselves convinced by things that they knew to be wrong. As a result, although Langmuir always gave due credit to Lewis, the theory began to be known as the Lewis–Langmuir theory, which did not entirely please Lewis. In Britain, in fact, the theory was sometimes known as the Langmuir theory, which produced the following reaction from Lewis in a letter in 1926: '...to persist, as they do in England, in speaking of the Langmuir theory of valence is inexcusable.' Lewis was also resentful of the indifference of most of his immediate colleagues, particularly when he was developing his theory at Harvard from 1902 onwards, and in 1929 he refused an honorary degree from Harvard.

In addition, Lewis was somewhat disappointed by the attitude of the physicists, who regarded his theory as unreasonably crude. The physicists particularly objected to the fact that Lewis seemed to be regarding the electrons as fixed in certain positions, and that the electron-pair bond that he was proposing did not seem capable of binding the atoms together. In his turn Lewis, like many other chemists, was critical of the physicists' idea of completely mobile electrons, which seemed incapable of explaining the undoubted fact, established by strong chemical evidence, that molecules have definite shapes. Over the years both Lewis's and the physicists' concepts became modified and by 1923, when Lewis's famous book *Valence and the structure of atoms and molecules* appeared, his ideas were entirely consistent with those that the physicists themselves were then putting forward.

The great importance of Lewis's theory is that it provided chemists with a valuable way of visualizing the electronic structures of atoms and molecules, and for practical purposes his ideas are still used today. In spite of that, physicists have tended to ignore his contribution. A few physicists have appreciated what Lewis did, as illustrated by the following passage from *Wave mechanics* (1945) by Walter Heitler (1904–1981), who did important quantum-mechanical work on the covalent bond:

> Long before wave mechanics was known Lewis put forward a semi-empirical theory according to which the covalent bond between atoms was effected by the formation

of pairs of electrons shared by each pair of atoms. We see now that wave mechanics affords a full justification of this picture, and, moreover, gives a precise meaning to these electron pairs: they are pairs of electrons with antiparallel spin.

To many it has seemed surprising that Lewis did not receive a Nobel Prize in view of his great contributions to the theory of bonding as well as to thermodynamics. The fact that his bonding theory in its original form seemed naive to the physicists may have been the deciding factor. In the 1920s and 1930s the nominations of Niels Bohr were given close consideration by the Nobel Prize committee, and many of his nominees received the award. He nominated a number of physicists whose work was of special interest to chemists, and Langmuir specifically for his work on surfaces, but he never nominated Lewis.

An important aspect of valency theory relates to the arrangement of the electrons in the various atoms, and this was actively investigated during the second decade of the twentieth century. In 1913 a contribution of fundamental importance was made by Henry Gwyn Jeffreys Moseley (1887–1915), who was working in Rutherford's laboratories at the University of Manchester. Moseley measured the characteristic frequencies of the X-ray lines of a number of elements, discovering a relationship from which he was able to deduce the charge on the atomic nucleus; this is the number of protons in the nucleus, and hence the number of orbiting electrons in the neutral atom. This number, the *atomic number*, is more or less the order of the elements when they are listed in order of increasing atomic weight. Moseley's brilliant career was cut short, at the age of twenty seven, when he was killed in action in the First World War at the battle of Gallipoli.

## Relativity theory

Besides proposing that radiation is quantized, Albert Einstein also formulated in 1905 the special theory of relativity. We need not say much about this here, since although of fundamental importance, the theory has little effect on our ordinary lives. The majority of research scientists can do all their work without taking the special theory into account.

The special theory of relativity is concerned with formulating the laws of physics in a way that is common to all observers under any conditions. According to the theory, the speed of light is a constant, but may appear to be different for an observer moving with a speed comparable to the speed of light. The theory predicts the results of measurement of length that are made by two observers moving at high speeds relative to one another. Certain astronomical anomalies, inexplicable according to Newton's mechanics, could be explained in terms of the theory.

Einstein's general theory of relativity was formulated in 1915. One consequence of it is of great practical importance. There is a relationship between mass $m$ and energy $E$ that is expressed by the equation

$$E = mc^2 \tag{8.4}$$

where $c$ is the speed of light. This means that under certain circumstances, energy and mass can be interconverted. For example, a process may occur in which there is a loss of mass, in which case the gain of energy is given by this equation (8.4).

For ordinary chemical reactions this can be ignored, the mass changes being too small to measure. For example, if 2 grams of hydrogen reacts with 16 grams of oxygen, to form water (the chemical equation is $H_2 + \frac{1}{2}O_2 \rightarrow H_2O$), the heat (energy) evolved is 241 750 joules. From Einstein's equation 8.4 it can be calculated that there must be a decrease in mass of 0.000 000 002 7 grams, which is much too small to measure.

When nuclear reactions occur, on the other hand, there can be an enormous production of energy. We will come back to this again later.

For a number of years after relativity theory appeared, very few scientists understood it at all. Obviously Einstein understood it, and one other who did so was the British astrophysicist Sir Arthur Eddington (1882–1944). Eddington made an important contribution by giving, in 1919, the first experimental proof of the theory. Einstein's theory had led to the conclusion that a large enough mass could exert gravitational attraction on a beam of light, and cause the beam to bend. Until 1919 it had not been possible to confirm this, but in that year there was a total eclipse of the sun, and Eddington was able to observe that light from distant stars, passing near to the huge mass of the sun, was slightly deflected in accordance with Einstein's prediction.

A nice story about Eddington dates from about that time. Someone said to him that he had been told that there were only three people in the world who understood Einstein's theory. Eddington did not reply, and was asked if he agreed with the statement. 'Well', said Eddington, 'I've been trying to think who the third person could possibly be.'

## NUCLEAR POWER AND THE ATOMIC BOMB

Although the quantum theory is of such great importance in modern science, no great technical innovation can be traced directly to it. Of course, inventions such as the transistor and the microprocessor were made by people who had to take full account of the quantization of energy and of radiation. But, unlike radio and the distribution of electric power, the transistor and the microprocessor cannot be traced back specifically to the quantum theory, or indeed to any one great advance in pure science.

The theory of relativity, on the other hand, is directly related to the harnessing of nuclear energy, including the production of the nuclear bomb, now almost always called the atomic bomb. Einstein's famous formula

$E = mc^2$ leads to the conclusion that if a process occurs with a loss of mass, an equivalent amount of energy is produced. Only processes involving atomic nuclei produce changes large enough to give rise to the enormous amounts of energy evolved when a nuclear power plant operates, or an atomic bomb is detonated.

Surprisingly, the idea of the conversion of mass into energy was considered quite early in the twentieth century. Elster and Geitel (whom we met earlier in this chapter in connection with the photoelectric effect) became interested in the source of the energy in the radiation that was emitted by a radioactive substance. After carrying out experiments to test other possibilities, they concluded in 1903 that the energy cannot come from any outside source, but must come from the atoms themselves. This was, of course, before Einstein's theory suggested any such possibility. Because of this conclusion, the house in Wolfenbüttel where Elster and Geitel lived and did their work bears a memorial plaque on which the two men are honoured as the *Entdecker der Atomenergie*, the discoverers of atomic energy.

It is also interesting to note that Lord Rutherford, who did much of the fundamental scientific work that eventually made possible the production of atomic energy, was asked in the 1930s whether energy could ever be produced on a large scale from nuclear reactions. He scoffed at the idea.

The observation that made the production of nuclear energy possible was made in December 1938, less than a year before the Second World War started. Otto Hahn (1879–1968) and Fritz Strassmann (1902–1980), working at the Kaiser Wilhelm Institute in Berlin, observed that when uranium, the heaviest element then known, was bombarded with neutrons, one of the products was barium, a relatively light element. At once Lise Meitner (1878–1968) and her nephew Otto Robert Frisch (1904–1979) interpreted Hahn and Strassmann's observation as indicating that the protons were breaking the uranium nucleus into two nuclei of roughly equal mass. At the time both Meitner and Frisch were refugees from Germany; Frisch was working in Bohr's laboratory in Copenhagen, and Meitner was working in Stockholm. They also calculated that during the process there was a substantial loss of mass and therefore, by Einstein's relationship, an enormous liberation of energy—much greater than in any nuclear process that had been previously observed. They referred to the process as nuclear fission.

In Paris in the same year, Frédéric Joliot-Curie (1900–1958) and his wife Irène Joliot-Curie (1897–1956)—the daughter of Marie Curie (1867–1934), famous for her early work on radioactivity—discovered something of great significance about the fission process. They found that there were more neutrons released in the process than had been used in the initial bombardment of the uranium. These secondary neutrons could disintegrate more uranium atoms, and thus set off a self-perpetuating chain reaction which would spread through the whole of the uranium present.

Niels Bohr learned of the nuclear-fission work in 1939 while he was on the visit to Princeton University that I mentioned earlier in this chapter. With John Archibald Wheeler (b. 1911) of Princeton, Bohr worked out a detailed theory of nuclear fission. One of the conclusions of their study was that only a tiny fraction of the uranium (it turned out to be 0.7%) is capable of undergoing nuclear fission; this was the form known as uranium-235. In this form the nucleus has 143 neutrons and 92 protons; the mass number, the sum of these two numbers, is 235. When two nuclei have the same number of protons but differ by the number of neutrons, we call them *isotopes*. The isotope present in largest proportions (about 99.3%) in natural uranium is uranium-238, which has 146 neutrons and 92 protons. It is a strange irony that Bohr and Wheeler published their important paper on 1 September 1939, the day that Germany precipitated the Second World War by invading Poland.

It is interesting to note that after the war Bohr nominated Meitner and Frisch for a Nobel Prize, which they never received. Hahn did receive the 1944 Nobel Prize for chemistry for his discovery of fission; later he was greatly depressed, and even contemplated suicide, on learning that his discovery had made possible the use of an atomic bomb against Japanese civilians. He did not hear of the Nobel award until 1945 when with a number of other German scientists he was interned in a country house near Cambridge in England; the presentation of the award to him was in December 1946.

The small international community of nuclear physicists soon became actively involved in this field of investigation. Bohr's prediction that only uranium-235 would undergo neutron-induced fission was confirmed. This meant that technologies based on uranium fission would require the isolation of uranium-235. This was an exceedingly difficult thing to do as the two isotopes behave chemically in exactly the same way, and only differ in mass by 1%.

Nuclear physicists in the major countries that would soon be at war recognized that there were two distinct but closely related technical possibilities that might decide the outcome of the war. In the first place, it might be possible to have a controlled self-sustaining chain reaction which would liberate large amounts of energy capable of producing an ample supply of electricity—in other words, nuclear power plants could be built. This alone would be a great asset, since energy is in short supply in times of war. There was also another possibility, more difficult to accomplish but more devastating in its effects. If enough uranium-235 could be isolated, it might be possible, by allowing the chain reaction to go out of control, to produce a bomb that would be vastly more powerful than any previous bomb.

On 2 August 1939, Albert Einstein, who was well aware of most of the work that was going on in the world of nuclear physics, wrote a letter to President Roosevelt in which he drew attention to the military implications of the fission experiments. At the time Einstein was at the Institute for Advanced

Study at Princeton, and he and Bohr had discussed the matter when Bohr visited Princeton in that year. Einstein's particular concern was that Germany might produce the bomb first, and would by its means be able to conquer the world. After the war he said many times that he regretted sending the letter: 'Had I known that the Germans would not produce a bomb I would not have sent it.' Einstein obviously thought that his letter had been the decisive factor in initiating the war work done by the Allies on the atomic bomb. This is hardly the case; the letter was undoubtedly an important factor, but many other influential scientists were thinking along the same lines. The person chiefly responsible for getting the American work started was Vannevar Bush (1890–1974), who in 1939 was president of the Carnegie Institute of Washington. It was he who was instrumental in persuading President Roosevelt to authorize the formation of the National Defense Research Committee (NDRC) on 27 June 1940. Einstein himself was not involved in wartime work on nuclear processes.

The American nuclear weapons programme began in 1940 with a modest grant to Enrico Fermi (1901–1954) and other refugees from Europe. Their work was done first at Columbia University in New York and later at the University of Chicago. Their first great success was to produce a self-sustaining chain reaction in natural uranium, and this they did by suitably moderating the speed of the neutrons. When natural uranium is bombarded with neutrons most of them are absorbed by the non-fissionable uranium-238, present to the extent of 99.3%. The Chicago physicists reduced the probability of this by passing them through what is called a moderator, which slows them down. The best moderator is 'heavy water', which is water in which the ordinary hydrogen atoms are replaced by the isotope called deuterium, the nucleus of which consists of one proton and one neutron. However, heavy water was not then available in sufficient quantity, and very pure graphite was used instead. Cadmium rods, which strongly absorb neutrons, were used to control the chain reaction; when the neutron level increased, showing that a chain re-action had just begun, they were thrust back to keep the process under control.

What Fermi and his colleagues had achieved, on 2 December 1942 in Chicago, was to construct the world's first nuclear reactor. This was an important step, but only a step, towards an atomic bomb, which could not be achieved with natural uranium. Among the spectators on 2 December was Arthur Holly Compton, one of the leaders of the American atomic energy project. He at once telephoned the news to his colleague James Bryant Conant, distinguished chemist and president of Harvard, with the coded and now famous message: 'The Italian navigator has landed'.

In the meantime some work done in England, by Otto Frisch and Rudolph Peierls (1907–1995) at the University of Birmingham, had been important in showing that an atomic bomb might be easier to make than many physicists had thought. It had been recognized that to make a bomb from uranium-235 it was necessary to have more than a certain critical mass of the material. If there is less than the critical mass too many of the neutrons escape into the surrounding space, and no explosion occurs. Previous estimates had led to the conclusion that many thousands of kilograms might be required, and this presented a formidable difficulty. Frisch and Peierls, however, calculated that about a kilogram would be enough (it later turned out that about five kilograms are necessary). Otto Frisch had left Bohr's laboratory and he and Peierls were then both refugees from Germany in England, where they had distinguished careers. Frisch became professor of natural philosophy (physics) in Cambridge in 1947, and Peierls, who was knighted in 1968, was professor of mathematical physics at the University of Birmingham from 1937, and Wykeham professor of physics in Oxford from 1963 to 1974.

Their work on the critical mass of uranium-235 was done unofficially, without any government support, but Frisch and Peierls kept their conclusions secret except from the British authorities. The result was that the British decided to intensify efforts towards making a bomb. They first set up a committee with the code name MAUD, the reason for this name being rather odd and due to a misunderstanding. A telegram had been received in England from Niels Bohr and it ended with the mysterious words 'and please tell Maud Ray Kent'. This was assumed to be a code, but no one could decipher it; because of it the name MAUD was adopted for the committee. After the war it was revealed that Maud Ray had been a governess in Bohr's house, and that she lived in Kent; Bohr simply wanted her to be informed that he was safe and well.

In 1942 the British government also supported President Roosevelt in the creation of the so-called Manhattan Project, a collaborative American–British project devoted to constructing an atomic bomb. As a result, thousands of physicists and engineers, in Britain, Canada, and the United States, gave their attention to the problem, and were ultimately successful. The main centre for the construction of the atomic bomb was Los Alamos, New Mexico, the work being directed by J Robert Oppenheimer (1904–1967).

Physicists had also become aware of another possible route to an atomic bomb. They had predicted, and it was later confirmed, that when the abundant isotope uranium-238 captured a neutron it was transmuted into another element which is now called plutonium. The nucleus of this element underwent fission when bombarded by neutrons, in much the same way as uranium-235 did, and again vast amounts of energy were liberated. Since plutonium is a distinctly different element, rather than an isotope of uranium, it can easily be separated from uranium. The nuclear reactor, made possible by Fermi's work, was thus a producer of plutonium, from which another type of atomic bomb could be made. The chemical separation of plutonium was first achieved in 1941 by Edwin Mattison McMillan (b. 1907) and Glenn Theodore Seaborg (b. 1912) at the University of California. Their work was voluntarily withheld from publication until the end of the war.

The Allied decision was to proceed along both paths. The laborious efforts to separate uranium-235 from natural uranium were continued, mainly at Oak Ridge, Tennessee. There are several possible methods, the one most used being gaseous diffusion. The method involves converting uranium, which is a solid chemical element, into a chemical compound that is a gas. The molecules of the compound containing uranium-235 move just a little faster than those of the heavier uranium-238, and diffuse a little faster through a thin porous membrane. The compound chosen was uranium hexafluoride, $UF_6$. However, the ratio of the molecular masses of the two molecules is only 1.0086, so that the difference between the diffusion rates is extremely small. A cascade of many diffusions was necessary, and it took almost two years for the first batch to be produced.

At the same time that the uranium separation was being carried out, reactors were being operated for the production of plutonium, and this work was carried out mainly at a plant built at Hanford, Washington. The Oak Ridge plant also produced plutonium, and in fact delivered the first batch of it.

By the beginning of 1945 more than $2 billion had been spent on the atomic bomb projects, and it was still not certain that success would ever be achieved. Later in the year both projects had succeeded. Enough uranium-235 had been produced, and the technique used with it was to make a bomb containing two masses of uranium-235 that were subcritical, that is, they were too small to detonate by themselves. To detonate the bomb, the two masses were shot together suddenly by means of ordinary explosives, producing a critical mass which at once produced a stupendous explosion. This is the type of bomb (called 'Little Boy') that devastated Hiroshima, Japan, on 6 August 1945. It was used without any prior test; only small amounts of uranium-235 had been separated, and it was not thought desirable to waste any on a test to see if it worked.

For technical reasons (the tendency of plutonium to undergo fission spontaneously, before being bombarded with neutrons) a somewhat different procedure had to be used for the plutonium bomb. A sphere of plutonium of subcritical size was surrounded by ordinary explosives. To produce the detonation, a chemical detonation was produced which compressed the plutonium to a much higher density than it normally has. Since an adequate supply of plutonium was available, a test explosion was carried out at Trinity, near Los Alamos, on 16 July 1945. The test was successful, and a plutonium bomb (called 'Fat Man') was dropped on Nagasaki in 9 August 1945. In this way the Second World War came to a swift conclusion.

Niels Bohr played an important part in the war, some of his experiences being so remarkable that if described in a book of spy fiction they would seem implausible; they are, however, well documented, and show that scientists do not always lead dull lives. After the German occupation of Denmark in 1940, Bohr received many invitations to move to Britain or the United States. At first he declined, feeling that his place should be in Denmark where he was able to maintain the research of his institute. In 1941 he received a visit from Heisenberg under rather mysterious circumstances. By this time Heisenberg had been ordered by the German government to work on the military use of nuclear energy. In October 1941 he was sent to Copenhagen to give a public lecture, which Bohr and his colleagues at the institute declined to attend. Heisenberg, who was on friendly personal terms with Bohr having worked with him, then arranged a private meeting. As to what was said at this meeting, Heisenberg's uncertainty principle applies. According to Heisenberg's account at the end of the war, he had some ethical objections to the use of nuclear energy for waging war, and wanted to discuss the matter with Bohr. According to Bohr, on the other hand, Heisenberg was anxious to work on the atomic bomb, as he believed that with its use Germany might win the war. At the time Bohr himself was not aware of the extent of the technical work that was going on in Britain and the United States, and did not believe that a bomb could be made.

Early in 1943 Bohr received a visit from Captain (later Colonel) Volmer Gyth (1902–1965), an officer of the Danish army general staff who was heavily involved in the resistance movement. Gyth wanted to know if Bohr was willing to receive an important message from England which would be on a piece of ultramicrofilm hidden in a hollowed-out section of a key, which would be one of a bunch of keys. Bohr agreed, and the transcribed message turned out to be a letter from Professor James Chadwick (1891–1974) of the University of Liverpool; Chadwick had discovered the neutron in 1932, for which he received the 1935 Nobel Prize for physics. The letter told Bohr that arrangements could be made to transport him secretly to England if Bohr were willing to undertake the adventure.

At that time life for Bohr and his colleagues in the institute was still tolerable, and he declined the invitation for the time being. His letter of reply was reduced to ultramicrofilm and given to a courier who had a dentist seal it into the filling of a tooth; in England another dentist removed the filling, and the film went to Chadwick.

In mid-September 1943 the Germans greatly increased their control over Denmark, and Bohr learnt that his life and that of his family were at risk. He at once got in touch with the Danish underground, and as a result a dangerous passage to Sweden was arranged for them. Bohr and his wife, and a group of others, walked to a beach near Copenhagen, crawling on all fours during the last part of the trip. They travelled through the night, in a small fishing boat and then in a trawler, and arrived in Malmø, Sweden, where they first stayed in the police station. Bohr was then taken by train to Stockholm. A few days later he was flown to London's Croydon airport in the empty bomb rack of an unarmed Mosquito bomber. When the plane approached occupied Denmark the pilot took the plane to a high altitude and instructed Bohr through the intercom to don his oxygen mask. For some reason Bohr did not do so, and when the pilot tried to communicate with him he received no answer. Fearing (as it turned out correctly) that Bohr might have lost consciousness through lack of oxygen, the pilot decided to take the risk of descending to a lower level; much of the rest of the trip was at near sea level. This action probably saved Bohr's life; he was conscious on arrival at Croydon, but had been unconscious for a period, and might not have survived.

This operation was supposed to be highly secret, but unfortunately the *New York Times* carried a news item on 9 October 1943 about Bohr's arrival in England. That was the last news item about Bohr until the end of the war.

During the remainder of the war, Bohr spent time at some of the nuclear establishments in both Britain and the United States. He made several extended visits to the establishment at Los Alamos, New Mexico, and engaged in fruitful discussions.

Although willing to cooperate in the construction of an atomic bomb for use in the Second World War, Bohr was greatly concerned that the post-war period might lead to an arms race that would have disastrous consequences. He therefore devoted much effort to persuading those responsible for the atomic bomb programme to introduce suitable international safeguards. In particular, he wanted complete cooperation with the Russians, who had their own nuclear energy programme but were being kept in the dark about what the British, Americans, and Canadians were doing.

With this object in mind, Bohr was able to arrange a half-hour interview with Prime Minister Winston Churchill, on 16 May 1944. Present also was Churchill's scientific adviser Lord Cherwell (1886–1957), who as F A Lindemann was professor of physics at Oxford. Apart from the fact that Bohr convinced Cherwell, the meeting was a disaster. Bohr was never successful as a communicator; he spoke in a scarcely audible voice, and his argument in any case was hard to follow. Churchill, never a patient or polite man, soon lost interest and interrupted Bohr, taking up most of the time arguing with Cherwell. Bohr was most upset; he commented to a colleague 'It was terrible: he scolded us [Bohr and Cherwell] like two schoolboys.'

Worse still, Churchill was left with the impression that Bohr was not to be trusted, and feared that he would communicate atomic secrets to the Russians. In a memorandum he even commented that Bohr 'ought to be confined or at any rate made to see that he is near the edge of mortal crimes.' Fortunately nothing came of this suggestion. Later Bohr was able to have a much lengthier and uninterupted conversation with President Roosevelt. Roosevelt listened carefully and politely, but it does not appear that Bohr's arguments had much effect on him either.

We have seen that Heisenberg was ordered by the German government to work on an atomic bomb. Before the outbreak of war Heisenberg had been invited to accept scientific appointments outside Germany, but had always declined, saying that his duty was to stay in his country. Heisenberg was a decent apolitical man who tried to be a loyal German while at the same time greatly disliking Nazi policies and actions. In spite of his low opinion of his government, he wanted his country to win the war, believing that political problems would somehow sort themselves out. As a result of the position he took, he came under harsh criticism after the war. In particular it was felt by some that Heisenberg could have used his great scientific prestige to oppose some of the Nazi policies, particularly the persecution of the Jews. No doubt Heisenberg felt that as a result of such action he would lose whatever influence he had, and would be rendered completely ineffective.

He was a theoretical physicist with not much aptitude for engineering, and he and his colleagues did not solve any of the crucial practical problems that were necessary for producing an atomic bomb. They may have known how to make a reactor, but had not made one; nor had they begun to take the additional long step towards making a bomb. After the surrender of Germany but before the dropping of the atomic bombs on Japan, Heisenberg and other German scientists were interned for a time in a country house called Farm Hall in the tiny English village of Godmanchester, not far from Cambridge. The house had been used as a 'safe house' by the British intelligence agency MI6 for parachuting agents into German-occupied territory. There the Germans lived in some comfort, but their rooms were 'bugged'. Their reactions when they heard over the radio of the the bombing of the Japanese cities, and realized that the Allies had been successful with their atomic bomb development, were revealing. From their conversations with each other it was clear that they themselves had made little progress, even towards building a reactor. Later they tried to justify their failure by saying that for moral reasons they had not really tried to make a bomb; they had merely convinced Hitler that they were doing so. This was perhaps true for some of them, but not all. After the war Albert Speer (1905–1981), who had been the German minister of armaments, reported that towards the end of 1942 his country's atomic bomb project was dropped as impractical, on the advice of nuclear physicists.

Was it justifiable for the Allies to drop atomic bombs on two Japanese cities in August 1945? That is, of course, an ethical question, and this is a book about science—should we be discussing it here? I am sure we should. Scientists are just as concerned with the results of their work as are any other members of the public. As they can assess the scientific information a little better than the average person, it is important that they express their views.

The ultimate responsibility for dropping the two bombs lay with US President Truman, who fully accepted that responsibility—'The buck stops here' as the sign said on his desk. We may be sure that his decision had the support of many of his American advisers, and no doubt of Prime Minister Churchill and his advisers. The Russians were presumably not consulted, as they were kept in the dark about war work done by their allies, and particularly about work on the atomic bombs.

The view is often expressed today, particularly by people who were not around at the time, that the dropping of the bombs on civilians was an inhumane act, and should never have happened. I respect that view, but cannot agree with it. I think that on balance human suffering was reduced as a result of the action.

Many factors have to be taken into account. Perhaps most important were the large number of prisoners, many of them civilians, who had been held by the Japanese for up to three and a half years and subjected to barbarous treatment. After the attack on Pearl Harbour in December 1941 the Japanese armies swept through Thailand, Burma, Penang, Hong Kong, and most of the Malay Peninsula. Atrocities were committed on a massive scale against both combatants and non-combatants, and many millions of people lost their lives. The survival rate among those who were held in captivity was low; those who survived their treatment—and I knew a few of them personally—suffered daily for the rest of their lives, and most of them died before their time.

When the Germans surrendered on 8 May 1945 most of the world was tired of war, and it was hoped that the Japanese, who could not possibly win in the end, would soon give up. They insisted that they would never surrender. There was thus a choice between waging a conventional war against them, or dropping atomic bombs.

The experience had been that in a conventional war which involved crossing a body of water, the odds were heavily loaded against the invaders. The Germans had found this with the Battle of Britain of 1940. The Allies had found, at Dieppe and later at Dunkerque, that an invasion, even if eventually successful, entails heavy casualties. An invasion of Japan, because of the greater distances to be travelled over the water, would have been much more costly in lives and suffering. These losses would have been sustained more by the US than by any of their allies, who were exhausted by their war with Germany.

It seems to me to be highly likely that the suffering resulting from not dropping the atomic bombs on Japan would have been much greater that the suffering inflicted by the two bombs. It might be argued that at least the first bomb should have been dropped on some remote part of Japan. However, the defiance of the Japanese government was such that they probably would have paid no attention to such an action. After all, they refused to surrender after the first bomb was dropped. If, as seems to be the case, only two bombs were in existence, and one was dropped on a remote region and the other on a city, there might well have been no surrender. The war might have dragged on for years, with little success for the Allies but extensive casualties.

It is surely significant that half a century after the two bombs were dropped on Japan, no other atomic weapons have been used in anger. Nor has there been another world war. Perhaps, paradoxically, the use of the weapons on Japan has ensured that it would be unthinkable to use them again.

# NINE

# *Scientists, science, and society*

When I started to write this book I thought I understood fairly well how scientific discoveries are made, and how they are transformed into technologies. As I proceeded, however, I realized that things were a good deal more complicated than I had suspected. In particular, I came to realize how easily one can go wrong if one tries to generalize. With the topics dealt with in the previous seven chapters, it is hard to discern any common pattern. In the present chapter I shall discuss a few themes that have become apparent in the earlier chapters, but to a great extent I shall be emphasizing divergencies rather than similarities.

We will first discuss scientists—what sort of people they are, and whether they differ in any essential way from people who do intellectual work of other kinds. Then we will consider something about the way—or I should say, the many ways—in which scientific discoveries are made. Then we will say something about the relationship between science and technology. We will briefly consider how the public sees the effects of science, and whether it seems to be judging them fairly.

## THE MAKE-UP OF A SCIENTIST

Non-scientists often think that scientists are in some way a special breed of people. Indeed, some scientists seem to share that point of view. My own opinion, both from having had the good fortune to know personally a considerable number of scientists (and many of them extremely good ones), and also from having read many biographies of scientists, is that this is not the case.

In the first place, I feel sure that many of the good scientists I have known, or have read about, would have been successful in anything they undertook. Many scientists did not originally intend to be scientists, and we have met some examples of this in this book. J J Thomson, Paul Dirac, and Henry Eyring, for example, originally intended to be engineers, and it is hard to believe that they would not have been good ones. Joseph Black took a medical degree and practiced medicine as well as conducting classes in chemistry. Thomas Young, Hippolyte Fizeau, Léon Foucault and many others also began their careers as medical students. Several scientists, including William

Grove and Joseph Plateau, became lawyers before becoming scientists. Grove finally went back into law, and became a judge.

Several great scientists have had success in fields other than science, and some have won great renown. Some of them are listed in Table 9.1. Sir Christopher Wren, after being professor of geometry at Oxford and making important contributions to mathematics, went to even greater fame as an architect. The buildings he designed include St Paul's Cathedral in London and the Sheldonian Theatre in Oxford. Robert Boyle, the 'father of chemistry' as he is sometimes called, wrote books on a variety of subjects, including religion and sociology. In his scientific work he was helped by a large group of assistants, and he was left with time to devote to his other interests as well as to many philanthropic causes.

One of the most versatile of all scientists was Benjamin Franklin. He was first successful as a printer and publisher, and was—in the best sense of the word—an amateur scientist. Later he was highly successful as a politician and diplomat, being one of the authors and signers of the American Declaration of Independence.

Alexander Porphyrvich Borodin (1833–1887) considered his main work to be that of professor of chemistry at the University of St Petersburg. He made significant contributions to chemistry, but he is unusual in being better known to the general public as a composer than, even to modern scientists, as a scientist. His musical compositions, however, were created in his spare time,

**Table 9.1**    Scientists distinguished in other fields

| Scientist | Field of science | Other field |
| --- | --- | --- |
| Robert Boyle (1627–1691) | Chemistry | Philosophy, sociology |
| Christopher Wren (1632–1723) | Mathematics | Architecture |
| Benjamin Franklin (1706–1790) | Physics | Statesmanship |
| Mikhail Lomonosov (1711–1765) | Physics | Literature |
| Peter Paul Roget (1779–1869) | Physiology | Philology |
| Alexander Borodin (1833–1887) | Chemistry | Musical composition |
| Revd Charles Dodgson ('Lewis Carroll'; 1832–1898) | Mathematics | Imaginative fiction |
| Bertrand Russell (1872–1970) | Mathematics | Philosophy |
| Chaim Weizmann (1874–1952) | Chemistry | Statesmanship (he became the first President of Israel in 1948) |
| John Maynard Keynes (1883–1946) | Mathematics | Economics |
| Michael Polanyi (1891–1976) | Chemistry | Philosophy, sociology |

or when he was ill. It is said that his musical friends sometime greeted him with the comment 'I hope you are not feeling too well today'.

Scientists who became famous in other fields are, of course, the exception. There are many other scientists who, while not attracting wide attention for their work, did competent work in other fields (Table 9.2). Archaeology has been of special interest to several scientists. Thomas Young, whose important work on the nature and properties of light was considered in Chapter 3, did work of fundamental importance on the deciphering of the hieroglyphic inscriptions on the Rosetta Stone. William Henry Fox Talbot, who introduced the photographic technique that is essentially the same as the one we use today, devoted much of his time to archaeology, and helped to decipher the cuneiform inscriptions at Nineveh. Both these men were so enthusiastic about their other interests that they sometimes put aside their excellent scientific innovations. Young has something of a reputation for never quite finishing anything. Talbot, having discovered his photographic method, did nothing about announcing it for over a decade; he only did so when he realized that he was being 'scooped' by Daguerre.

Scientists, then, seem to be much like other people who are interested in intellectual pursuits. Many scientists have told me that they had trouble deciding whether to go into science or some other field. In their general behaviour, too, scientists seem just like other people. Some are generous, and the proportion of generous scientists is not obviously different from the proportion of generous people as a whole. Religious belief does not seem to be affected by whether one is a scientist or not. We might think that a scientist would be less likely than the average person to be a religious fundamentalist, but even this seems not to be the case. Michael Faraday was a Sandemanian, with beliefs similar to that of the Quakers, Katherine Lonsdale became a Quaker, and Henry Eyring was born a Mormon and rose to high office in

**Table 9.2**  Scientists with important other interests

| Scientist | Field of science | Other interests |
|---|---|---|
| Joseph Black (1728–1799) | Chemistry | Medicine |
| Thomas Young (1773–1829) | Physics | Archaeology |
| Henry Fox Talbot (1800–1877) | Mathematics, chemistry | Archaeology |
| William Robert Grove (1811–1896) | Chemistry | Law |
| James Clerk Maxwell (1831–1879) | Physics | Poetry |
| Robert W Wood (1868–1855) | Physics | Humorous verse |
| Alfred W Stewart (1880–1947) | Chemistry | Detective fiction (as J J Connington) |

that church. Some scientists are agnostics, but their proportion seems about the same as among the public.

Some scientists are highly gregarious, some are hermits, and most are somewhere in between. Most scientists are enthusiastic about discussing their ideas with others, but some fear that their ideas will be stolen by others and are secretive. Isaac Newton is the supreme example of that; rather than publish something that might turn out to be wrong, he would encode his ideas; if they were wrong he could forget them, if they were right he would claim credit for them. Röntgen, although sociable in some other ways, is believed never to have discussed his scientific work with anyone; even his assistant was kept in the dark about what he was doing. Röntgen published his results promptly, but never attended scientific meetings. Oliver Heaviside lived like a hermit, but corresponded freely with others, including Hertz.

## SCIENTISTS AND THEIR SOCIAL BACKGROUND

We tend to think today that ours and ours alone is the age of opportunity, when anyone can rise to great heights, and that this was not the case in earlier centuries. The fact is, however, that in earlier times a surprising number of people of humble background and poor education were able through sheer ability to overcome their early disadvantages and rise to positions of responsibility. Humphry Davy (1778–1829) was the son of a Cornish woodcarver and his schooling was haphazard. At the age of 16 he was apprenticed to an apothecary-surgeon, but at 19 was considered by Thomas Beddoes to be competent enough to be superintendent of the Pneumatic Institute at Clifton; his rise to fame was then rapid. He became lecturer at the Royal Institution at the age of 23, and its professor of chemistry at 24; he was knighted at the age of 34. John Dalton (1766–1844), the son of a Cumberland weaver, attended school only until he was 12 but he rose to a position of great scientific prestige—although he received no appointment of any distinction. William Whewell (1794–1866), the son of a Lancashire carpenter, became Fellow of Trinity College, Cambridge, at an early age, held two Cambridge professorships (of mineralogy and moral theology), and became Master of Trinity in 1841. The great German mathematician Carl Friedrich Gauss (1777–1855) was the son of an illiterate farm steward but obtained his doctorate at the age of 22, and was the leader in various branches of mathematics for the rest of his life. Of course, we can have no idea of how many men and women of great ability were denied the possibility of living productive lives.

Michael Faraday's background was as disadvantaged as can be imagined. His father, James Faraday, was a blacksmith who for a time worked near Outhgill in North Yorkshire. His smithy was next door to an inn and was on a drovers' route for moving cattle or sheep to market. Early in 1791 a harsh winter and a poor economy were making it impossible for James Faraday to sustain his wife and the two children they then had, and they decided to try their luck in London. We know that they arrived there shortly before 20 February 1791, and Michael was born on 22 September, so that it is possible that Mrs Faraday was pregnant when they made the trip. The family—the parents and their three, and later four, children—first lived in a single squalid room near London Bridge, and Michael's education was extremely rudimentary.

Another scientist whose upbringing was far from advantaged was Oliver Heaviside, who in 1902 deduced the existence of what came to be called the Heaviside layer. His father was an artist who made a precarious living and Oliver apparently had a wretched boyhood in London. He was unable to attend a university, but nevertheless managed to become a highly accomplished mathematician. He was trained as a telegrapher, and lived a reclusive life in Devon. In 1908 the Institution of Electrical Engineers made him an honorary member after striking him off their rolls for failure (due to his poverty) to pay his dues.

It is an interesting paradox that it was easier in earlier times than it is now for a person of limited education to achieve success in science. When many of the fundamental principles of science remained to be discovered, a scientific education was not essential; men of natural genius like Davy and Faraday could pick up the necessary science for themselves. Today, because science has become much more complex, it would be quite impossible for anyone to make significant advances in science without a lengthy specialized education. Of course, Davy's and Faraday's success depended on a good deal of luck as well as on their great ability. One wonders if Faraday would have achieved success if Davy had not found himself without an assistant at a critical time.

A few eminent scientists have come from the other end of the social and educational scale. One of these was the Honourable Robert Boyle, who was born in Ireland, the fourteenth child of the Earl of Cork, who at one time was probably the richest man in England. Boyle was privately educated for the most part, and picked up the science he needed by taking lodgings in Oxford, which at the time was the main centre of scientific work in England. Later he resided with his sister in London and established a research laboratory in their house.

The Honourable Henry Cavendish (1731–1810) was a kinsman of the enormously wealthy Dukes of Devonshire. His father was Lord Charles Cavendish, who was the son of the second Duke of Devonshire and the younger brother of the third Duke. Henry Cavendish was educated at Cambridge but left without taking a degree. He was intensely reclusive by nature, and established a private laboratory in London where he devoted himself to scientific research. He made a number of discoveries of great importance; some of it, since it was unpublished, was not revealed until many years after his death. He was probably the first to prove that water is not an element.

John William Strutt (1842–1919), who won the 1904 Nobel Prize for physics, inherited his father's title and became the third Baron Rayleigh in

1873. Although of considerable wealth he devoted much energy to his scientific studies and research. At Cambridge he was Senior Wrangler and Smith's Prizeman, and he succeeded Maxwell as professor of experimental physics at Cambridge, holding the post from 1879 to 1884; later he was professor at the Royal Institution. His preference, however, was to carry out his research at his country seat at Terling Place, Essex.

One of the more interesting characters in the history of science was Randal Thomas Mowbray Rawdon Berkeley, the eighth (and last) Earl of Berkeley (1875–1947). He was not as distinguished a scientist as Lord Rayleigh, but he did a good deal of highly competent and careful work. He was perhaps the last of the amateur physical scientists, in the sense that he had no formal education in science. He went into the Royal Navy in 1878, at the age of 13, but resigned his commission nine years later. He bought a house at Foxcombe, just outside Oxford, and applied to the university to take a degree while still residing in his house. The rules of the University, however, stipulated that a student must reside within a mile and a half from Carfax, the hub of the University, and the rules were found to be inflexible. This decision somewhat embittered Berkeley, but he established good relations with Balliol College, which allowed him to work in some cellars they had converted into a laboratory. (These cellar laboratories later housed, for a short time, Frederick Soddy, who in them did some of the work on isotopes for which he won the 1921 Nobel Prize for chemistry).

In these laboratories, in collaboration with Ernald George Justinian Hartley (1875–1947), a distinguished member of Oxford's Inorganic Chemistry Laboratory, Berkeley carried out a series of investigations on osmotic pressure. They introduced important new experimental techniques, and were able to obtain results of high precision which were useful in the development of theories of solutions. Some of the Earl's later work was done at his house at Foxcombe. In about 1930 laboratory equipment from the house was transferred to Balliol College, and a set of false teeth was included; these were used in the cellar laboratory to demonstrate that they do not fluoresce under ultraviolet radiation, in contrast to natural teeth.

A question that is sometimes asked is whether scientific ability is inherited. What often prompts the question is the fact that there are a number of interesting cases of close family relationships between eminent scientists. For example the distinguished astronomers Sir William Herschel (1738–1822) and Caroline Lucretia Herschel (1750–1848) were brother and sister. Sir William's son was Sir John Herschel (1792–1871) who was distinguished as an astronomer and also as a chemist; as we saw in Chapter 3, he made pioneering contributions to the science of photography, and during his lifetime had great popular fame as a scientist.

Marie Curie (née Marya Sklodowska, 1867–1934) and Irène Joliot-Curie (1897–1956) were mother and daughter. Marie Curie shared the 1903 Nobel Prize for physics with her husband Pierre and with Antoine Henri Becquerel for their pioneering work on radioactivity, and she won the 1911 Nobel Prize for chemistry for the discovery of radium and polonium. Her daughter Irène shared with her husband Frédéric the 1935 Nobel Prize for physics for producing the first artificial isotope.

The father-and-son Bragg team have been mentioned already (Chapter 7); they are unique in being the only such team to share the same Nobel Prize, and Lawrence was the youngest person to win the prize. J J Thomson (1856–1940) and his son G P Thomson (1892–1975) both won Nobel Prizes for physics, but at different times (1906 and 1937). William Thomson (Lord Kelvin) had no children, but he was the son of a distinguished mathematician, James Thomson (1786–1849) and the brother of James Thomson (1822–1892) who was an extremely able physicist and engineer. Michael Polanyi (1891–1976) was a distinguished physical chemist some of whose work merited a Nobel Prize; his son John Charles Polanyi (b. 1929) shared the 1986 Nobel Prize for chemistry for his pioneering studies in chemical dynamics.

The 3rd Baron Rayleigh was mentioned earlier in this chapter. His son, Robert John Strutt, the 4th Baron (1875–1947) was also a distinguished physicist, being for a time professor of physics at Imperial College, London.

The Becquerel family provide an example of a remarkable scientific dynasty, in that four generations of them were of outstanding ability:

| | |
|---|---|
| Antoine César Becquerel (1788–1878) | 1838–1878 |
| Alexandre Edmond Becquerel (1820–1891) | 1878–1891 |
| Antoine Henri Becquerel (1852–1908) | 1891–1908 |
| Jean Becquerel (1878–1953) | 1908–1948 |

(The dates given to the right are the dates that the men were Directors of the Musée d'Histoire Naturelle in Paris.)

The best known of the Becquerels is Antoine Henri, the discoverer of radio-activity, but the others were of considerable distinction, especially Edmond, some of whose work was mentioned in Chapter 3. We see that when Jean Becquerel retired from that position in 1948, the Musée for the first time in 110 years did not have a Becquerel as Director. And when he died in 1953 the French Académie des Sciences did not have a Becquerel as a member for the first time in well over a hundred years.

Of course, these relationships, interesting in themselves, tell us nothing about whether scientific ability, or any other kind of ability, is inherited. Environment is obviously of great importance; a child brought up in a family of intellectuals has a greater chance of becoming interested in the same kinds of activities as other family members. From our present knowledge of evolution it appears that the probability of a genetic component in intellectual ability is small.

Another question of interest with regard to scientists is the age at which they tend to do their best work. Scientists themselves tend to think that the best work is done when a scientist is fairly young, at any rate under 40. It is true that much great work has been done by young people; Michael Faraday, Werner Heisenberg, and Dirac are obvious examples. The youngest person to win a Nobel Prize was Lawrence Bragg, who was 25 when he won it and 22 when he did the work.

On the other hand there are examples of much older people doing work of outstanding originality. W H Bragg, who shared the Nobel Prize with his son, was well in his fourties when he did the work, and previously he had hardly done any research at all. Röntgen was 50 when he discovered X-rays, and Becquerel was 42 when he discovered radioactivity. J J Thomson was 41 when he did his great work of measuring the mass and charge of the electron. Max Planck was 42 when he first formulated his quantum theory. Schrödinger was 39 when he formulated his system of quantum mechanics, and Max Born was in his mid-forties when he gave his statistical interpretation of the Schrödinger equation. When Born did the work for which he received the 1954 Nobel Prize for physics, he was in his sixties.

Besides William Bragg, another late starter in research was the British chemist Nevil Vincent Sidgwick (1873–1952); he also illustrates the point that scientists are sometimes highly proficient in other fields. He belonged to a family that was distinguished in the classics, and at Rugby School he studied both classics and science until he was 18. He first applied for a classical scholarship at Oxford, but failed; he was then successful in winning a science scholarship to Christ Church. In 1895 he obtained a first-class degree in chemistry, but finding that his family rather turned up their noses at a science degree, he decided that he would satisfy them by taking honours in literae humaniores ('Greats'), which involves Latin and Greek history and philosophy studied in the original languages. In 1887 he got first class honours in Greats, greatly helped by a brilliant *viva voce* (oral) examination in philosophy.

He took a PhD degree from Tübingen University in 1901, and returned to Oxford where he lived at Lincoln College until the end of his life; he never married. At first he did little research, all of rather mundane quality. Later, mainly as a result of meeting Rutherford on a ship to Australia in 1914, when he was 41, he decided to devote himself to the electronic theory of atoms and molecules. In 1922, when he was nearly 50, he published his *The electronic theory of valency*, which made him famous in scientific circles. For many years he devoted himself to understanding the vast body of hitherto disorganized facts of inorganic chemistry in terms of atomic theory, and this work culminated in his 750 000-word classic *The chemical elements and their compounds*, published in 1950 when he was 77. Until his death two years later Sidgwick retained his extraordinary memory for facts, his mental keenness, and the clarity, freshness, and liveliness of his writing.

In addition, there are many examples of people who did outstanding work when they were young, and continued it into old age. Joel Henry Hildebrand (1881–1983), whose career was mainly in the chemistry department of the University of California, continued to carry out research well into his nineties.

Gerhard Herzberg (b. 1904) was doing research at least until his ninetieth birthday. Friedrich Hund, who did pioneering work on molecular structure and who celebrated his hundredth birthday on 4 February 1996, continued to give scientific lectures at least until he was 96.

It may be true that work of great originality is more likely to be done by a younger person than an older one. However, outstanding work, particularly of an inductive kind where scientific knowledge is systematized, is often done by older people.

## IS THERE A SCIENTIFIC METHOD?

There is no clear pattern of scientific discovery, or of technical advance. Some writers on this topic, even scientists themselves, tend to oversimplify the matter. Active research scientists have often commented to me that they have read something that was supposed to explain the scientific method, but that they could not relate the way they themselves worked to what had been written. We are sometimes told that a scientist assembles data and then for-mulates a theory or explanation that best fits the data. This may well be the way in which a detective solves a crime, or a lawyer prepares a case. It may be the way in which some routine and unimportant science is carried out. Every active scientist of any competence, however, would agree that this is *not* the way good scientific work is done.

The truth is that there is no one scientific method. As a scientist or en-gineer struggles through the mass of experimental data, there is no well-marked path to a satisfactory theory. There are not even any clear-cut rules of the road, but only a few general guidelines. A successful scientist feels free to use any methods, and must not be afraid to use imagination and intuition, and to break any rules that might be thought to apply. When the great chemist van't Hoff delivered his inaugural lecture to the University of Amsterdam in 1878, he used as his title 'Imagination in science'. He said that he had made a special study of the way in which the great scientific advances of the past had been made, and stressed that imagination and observation were both of great importance.

The reader may here protest that surely there are some things that one must not do. Surely, for example, one must not invent data. It would indeed be wrong to do so, but sometimes one does better by ignoring some of the data. This point, which some people may find disturbing, requires a little discussion.

Isaac Newton provides us with good examples of what I mean. It is well known that Newton was the first to formulate a universal law of gravitation, and to apply it to calculating the orbits of planets round the sun. In doing so he sometimes found that there were discrepancies between his calculations and the observed data. Newton thought nothing of brushing under the carpet data that did not agree with his calculations, and for this he has sometimes been criticized as being dishonest. I do not really think that this criticism is justified. We now know that the discrepancies were due to the attractions of planets not yet discovered. As it has turned out, his theory was essentially correct (aside from Einstein's small relativity corrections which do not come into the present discussion). It would have been unfortunate if Newton had failed to put forward his great theory because of the discrepancies that he knew of.

Newton's whole scientific career is an object lesson on the importance of not sticking to a so-called scientific method. In many ways Newton's methods were quite 'unscientific'. Besides keeping quiet about facts that did not fit his theories, he was much guided by metaphysical and religious arguments, which any modern scientist would condemn. In spite of all these weaknesses, he surely was one of the greatest scientists who ever lived, and that was because of his insight and imagination.

There are several other interesting examples where great progress has been made by people who conveniently overlooked data that did not fit in with their ideas. One famous example relates to the Arrhenius theory of ionic dissociation; this has been referred to briefly in Chapter 7, but deserves some further discussion. When Michael Faraday was doing his work on electrolysis (Chapter 4) he realized that charged particles were moving in solution, but did not think that they ever became completely free. When, for example, a solution of copper sulphate, $CuSO_4$, is electrolysed, copper is deposited at the negative electrode, and Faraday realized that some charged species (which we now know to be $Cu^{++}$, a copper atom from which two electrons have been removed) have moved through the solution. Faraday decided to give a name to these charged species. He put the matter to the Master of Trinity College, Cambridge, the Revd William Whewell (1794–1866), whose interests were very broad, covering the classics as well as the sciences. For a positively-charged species like $Cu^{++}$ Whewell suggested the word *cation*, which comes from the Greek κατα (kata), downwards, and ιον (ion), which is the Greek present participle of the verb 'go'. For a negatively charged species such as $Cl^-$, Whewell suggested the word *anion*. The general word *ion* could be used to refer to either a cation or an anion. It is important to understand that these charged species were not considered by Whewell or Faraday to exist by themselves in solution. Instead these ions were believed to slip from one molecule to another, always being held to one of them. A helpful analogy is provided by a country dance, in which a person is passed along a line of people but at no time is completely free, always holding the hand of one person or another.

Some time later it was suggested by several people, including Rudolf Clausius whose thermodynamic work we met in Chapter 2, that perhaps small amounts of these ions could exist by themselves in solution. Later, in 1887, Svante Arrhenius made the bold suggestion that ions can exist in substantial amounts in solution, and that an equilibrium exists between the undissociated molecules and the ions. The story of how Arrhenius's suggestion came to be made is an interesting one in itself, and in brief is as follows. In 1884 Arrhenius, then aged 25, presented his PhD thesis at the University of Stockholm. In it he described something like this theory of dissociation into ions, although his explanation was not as explicit as it might have been. His examiners were not at all impressed, and gave him such a low rating that in the ordinary way an academic career would have been denied to him. He sent copies of his thesis to several people, including Wilhelm Ostwald of the University of Leipzig, who although still in his thirties had a considerable reputation as a physical chemist. Ostwald was so impressed that he at once travelled to Stockholm to discuss the matter with Arrhenius. He also derived a simple equation, the Ostwald dilution law, which provided a simple test of Arrhenius's theory. This equation contains a constant called a *dissociation constant*, which can be calculated from experimental data of various kinds, particularly the conductivities of solutions. If it were found that the 'constant' varied, i.e., it was not really a constant, the conclusion should be that the theory was wrong.

As well as Ostwald, van't Hoff also became an enthusiastic supporter of Arrhenius's ideas, and the three men became known as 'Die Ioner'—the Ionists. Many other scientists were sceptical of the idea, and the Ionists set to work to obtain further evidence for the theory. Trouble arose almost at once. With some substances, such as acetic acid and ammonia, the conductivity data were quite consistent with Ostwald's dilution law—the constants were really constant. These substances were recognized to be examples of what are called *weak electrolytes*. They are not much dissociated into ions; if we write the dissociation as

$$AB \rightleftharpoons A^+ + B^-$$

the equilibrium lies over to the left—that is, there are not many ions compared with the undissociated AB molecules. Some substances, on the other hand, are what are called *strong electrolytes*, and we now know that they are completely dissociated into ions. Sodium chloride, NaCl, which is common salt, is a good example. When it is dissolved in water it occurs entirely as $Na^+$ and $Cl^-$ ions; there are no NaCl molecules at all. We have, already seen in Chapter 7 that Lawrence Bragg showed in 1912 that no NaCl molecules exist in solid sodium chloride; the crystal is a lattice of $Na^+$ and $Cl^-$ ions (Fig. 7.9).

These strong electrolytes thus presented an embarrassing problem for the Ionists. In fact, even before Ostwald's paper on his dilution law appeared in print in 1888, it was discovered that with strong electrolytes the calculated 'constants' varied greatly when the concentration of the solution was varied. They could be more than a thousand times greater at one concentration than at another. This was not the case for just a few substances. There were a great many strong electrolytes that did not agree with the theory; all inorganic salts, and many acids and bases, are of this type.

These widespread discrepancies between the theory and the experimental data could have been regarded as a disaster for the theory. But when the matter was pointed out to Arrhenius he calmly suggested that the results for the strong electrolytes should be ignored, and that further work should be done with the weak electrolytes, which did obey the Ostwald equation. He justified this rejection by saying (correctly) that with the strong electrolytes the conductivities themselves did not vary very much when the concentration was changed, although the 'constants' changed enormously; a better test of the theory, he argued, was therefore provided by the weak electrolytes, which gave a larger change of conductivity.

Walter Nernst, who was a littler younger than the other three men, also became an enthusiastic Ionist, and did all he could to persuade people that the Arrhenius point of view was correct. In 1891 he published an important book on physical chemistry, called *Theoretische Chemie*, which was translated into English and exerted a wide influence for many years. In connection with the theory of ionic dissociation Nernst wrote that 'the Ostwald formula does not exactly fit the observed facts in the case of highly dissociated acids and bases'. When he wrote that, Nernst knew perfectly well that the formula simply did not fit the data at all, and that the 'constants' were varying by factors of a thousand or more. His expression 'does not exactly fit' must be the record understatement of the century, or perhaps of all time.

At first sight these comments of Arrhenius, Nernst, and others might seem to be unscientific, if not dishonest. How can one justify ignoring all results that did not agree with the theory? I do not think, however, that we should judge the Ionists harshly. In the end it turned out that they were right to consider only the weak electrolytes, although they could not have justified their position at the time. It later became clear that the behaviour of strong electrolytes is really to be explained in quite a different way. For strong electrolytes a different theory was needed, involving the electrical forces between the ions in solution. Such a theory was not perfected until 1924, when it was worked out by the Dutch physical chemist Peter Debye (1884–1966) in collaboration with the German physicist Erich Hückel (1896–1980). Debye ended his distinguished career at Cornell University in Ithaca, New York.

But what if the Ionists had turned out to be barking up the wrong tree altogether? Suppose it had turned out that ions do not really exist in solution, and that there was an entirely different explanation for the variations in the electrical conductivities. The Ionists would then have been in an embarrassing position. They would have been roundly condemned for not reporting the evidence that did not support their position, and they would have had little to say in their defence.

I am not sure just what the moral of this is. We should surely teach students to be honest in reporting results—and then perhaps admit that sometimes it is justifiable to ignore certain results if it really seems that they are due to extraneous factors. The Ionists were, of course, men of outstanding intellect, who also had an instinct for arriving at the truth. Their theory of ionic dissociation was of profound importance, and they were quite sure it was essentially right. They were not going to be put off by discrepancies, even though they were large and numerous.

An interesting postscript to the ionic dissociation story is that Arrhenius himself never accepted the new Debye-Hückel theory of strong electrolytes. He would not even discuss the matter; if it was raised at a meeting, he would walk out. He was, of course, not the only eminent scientist who was stubborn in not accepting new ideas; Newton had been just as bad over some matters. Arrhenius's colleagues van't Hoff, Nernst, and Ostwald readily agreed that a new theory was required for strong electrolytes. Ostwald, incidentally, was extremely stubborn on another matter; he was not even convinced of the real existence of atoms until 1909—more than a hundred years after Dalton had proposed his atomic theory.

## THE VIRTUE OF INNOCENCE

That the truth can sometimes be reached by neglecting some of the experimental data can be shown by another example. This example teaches us a further lesson; sometimes the people who have done painstaking work on a problem are unable to reach the truth because, paradoxically, they are too familiar with the facts! The truth is instead reached by someone coming in from outside, who sees the problem in a broad way, and is not too concerned with explaining all the details.

Beginning in the 1890s, and for many decades, extensive experimental work was done on the reaction between hydrogen and chlorine gases, with the formation of hydrogen chloride (which can be called hydrochloric acid). The equation for the reaction is

$$H_2 + Cl_2 \rightarrow 2HCl$$

The reaction will occur if the mixture of gases is heated, and we then call it a *thermal* reaction. It will also occur if the mixture is exposed to light, and we then say that we have a *photochemical* reaction. Many readers will have seen a demonstration in which a mixture of the two gases, confined in thin plastic (the kind used for packaging), is brought up to a flood lamp; the mixture explodes with a spectacular bang, and one can smell the hydrochloric acid produced.

Two scientists did most of the experimental work on both the thermal and photochemical reactions; they were the German physical chemist Max Bodenstein (1871–1942) and the English physical chemist David Leonard Chapman (1859–1958). Both were remarkably careful and skilful investigators. Bodenstein's work was done mainly at the University of Hannover. Chapman did his early research at the University of Manchester, but in 1907 he was appointed a Fellow of Jesus College, Oxford, and director of the college laboratories; these were used for both teaching and research from 1907 until 1946.

The objective of Bodenstein and Chapman was to study the rate of the reaction under various conditions, so as to discover the mechanics of the reaction—the details of how it occurs. There were many experimental difficulties, particularly an exasperating lack of reproducibility in the results. For example, the investigators would measure the rates of reaction one day, and then find on the following day that the results were quite different, even though the conditions seemed exactly the same. This lack of reproducibility was traced to the fact that the rates were extremely sensitive to the surface of the vessel in which the gases were enclosed, and to the presence of minute traces of impurities. Oxygen gas, in particular, had a remarkable effect on the reaction rate. If the oxygen was completely removed the rate became extremely high and difficult to measure. Addition of a minute trace of oxygen would then greatly reduce the rate.

This kind of behaviour is unusual. By contrast, the similar reaction between hydrogen and iodine,

$$H_2 + I_2 \rightarrow 2HI$$

behaves quite respectably. The rates measured are the same from day to day, the nature of the vessel in which the gases are contained does not matter, and impurities like oxygen have no effect. The conclusion is that by and large the reaction occurs by a collision between $H_2$ and $I_2$ molecules, the atoms reorganizing themselves and forming two HI molecules.

No such simple mechanism can apply to the hydrogen–chlorine reaction. Such a mechanism does not explain all the complications that occur, such as the effects of surface and impurities. In 1913 Bodenstein made a general suggestion of great importance, although he was unable to develop the details of his suggestion. He suggested the idea of a *chain reaction*, an idea that was later realized to be important for nuclear explosions. His idea was that the reaction does not go in one stage, but that certain intermediates are formed, and undergo further reactions. Many years afterwards Bodenstein's student Walter Dux (1889–1987)—who was forced out of Germany by the Nazis and resided for the rest of his life in England—reported that Bodenstein had liked to demonstrate the idea of a chain reaction with the use of his gold watch chain. He would take this out of his pocket and show that if he shook one end of the chain the impulse would travel along the chain. After the Second World War Dux tried to find this historic chain, but found that Bodenstein had patriotically turned it in to the German government. Dux had a replica made and presented it to the University of Hannover, where it still is.

Bodenstein was not, however, successful in suggesting what these intermediates could be. He thought that ions, positive or negative, might be formed, but later realized that there was not enough energy to produce them.

He also suggested that the intermediates might be 'excited' molecules, perhaps molecules vibrating more vigorously and therefore having extra energy.

The key to the solution of the problem was provided in 1918 by Walther Nernst. We have seen, particularly in Chapter 2, that Nernst did important work in thermodynamics, and was responsible for what is sometimes known as the third law of thermodynamics. Nernst also made contributions of the greatest importance to the understanding of the behaviour of ions in solution. Until 1918, however, he had paid little attention to the rates of chemical reactions. He had made no study of the hydrogen–chlorine reaction, and there is no reason to believe that he knew much about the peculiarities of the reaction, such as the influence of the surface of the vessel and the effects of traces of impurities.

Because of this unawareness he had no hesitation in suggesting a mechanism for the reaction. He knew of Bodenstein's idea of a chain mechanism, and he suggested that the chains occur as follows. First a few $Cl_2$ molecules are dissociated into chlorine atoms:

$$Cl_2 \rightarrow 2Cl$$

Then a chlorine atom reacts with a hydrogen molecule:

$$Cl + H_2 \rightarrow HCl + H$$

The hydrogen atom produced in this reaction can then react with a chlorine molecule, pulling off a chlorine atom:

$$H + Cl_2 \rightarrow HCl + Cl$$

The important point to be noted is that in this pair of reactions a chlorine atom has produced two molecules of HCl, and that at the end a chlorine atom has been regenerated. The pair of reactions can thus occur again and again, so that a small number of chlorine atoms can bring about a considerable amount of reaction. This is the essential feature of a *chain reaction*.

There is no doubt that Nernst's idea was basically correct, and we usually give him the credit for first suggesting a chain mechanism for a chemical reaction. It is important to recognize, however, that his mechanism did not explain the experimental results that Chapman and Bodenstein had so meticulously collected. The facts that the reaction is so sensitive to the nature of the surface of the vessel, and to traces of oxygen and other impurities, were not explained at all. For this reason Chapman and Bodenstein would never have suggested such a mechanism, and they were at first reluctant to accept it. The mechanism had to come from an outsider like Nernst, ignorant of the experimental details. So much for the idea of a scientific method in which experimental data lead to the right explanation.

The point is that Nernst's idea was correct only in its essential detail—that it is the atoms of hydrogen and chlorine that carry the chains. To explain the results more fully, some important details had to be added to Nernst's simple scheme. It had to be recognised that removal of chlorine atoms at the surface of the vessel must be taken into account, and that impurities such as oxygen react with the hydrogen atoms. It took more than a decade for these details to be filled in.

Both of these examples, the ionic dissociation theory and the hydrogen–chlorine reaction, have shown how an excellent theory in its original form may be quite inconsistent with experiment. The second example has the additional feature that it took someone from outside the field to show the way to a correct theory.

## Where angels fear to tread

Nernst's suggested chain mechanism is not, of course, an example of Alexander Pope's famous statement that 'Fools rush in where angels fear to tread'. Nernst was far from being a fool; he was simply ignorant of the details of the experimental behaviour that he wanted to explain—and yet he, and not the experts, was led towards the truth. Sometimes the situation is more extreme, in that a person quite ignorant of the fundamentals of science achieves success by taking an action that the experts considered absurd. An excellent example was provided by Marconi when he sent a radio signal across the Atlantic Ocean.

It is going too far to call Marconi a fool, but he was close to being one as far as radio transmission was concerned. He knew hardly any physics, and had failed to gain admission to the Italian naval academy and to any university. As we saw in Chapter 7, his experiments with radio were all made with equipment invented by others. He was working with electromagnetic radiation, but on many occasions insisted that he was not. His were Marconi-waves, he said, and they had different properties from electromagnetic waves; unlike them, his waves did not travel in straight lines, and he could direct them as he wished. This was certainly foolishness, but paradoxically it led him to do something that the experts would not try, and he was un-expectedly successful.

To the people who knew what they were doing, such as Oliver Lodge and Lord Rayleigh, the idea of sending the waves across the Atlantic was an absurd one, because of the curvature of the earth. Marconi, in his ignorance, tried to do so, and succeeded in 1901. The physicists were stunned, and at first could provide no explanation. Then in 1902 Oliver Heaviside speculated that the waves must be reflected by a layer in the upper atmosphere, and this came to be called the Heaviside layer and later the ionosphere. At the time there was no direct evidence for the layer; Heaviside postulated it as the only way out of an embarrassing difficulty. It was not for more than twenty years that direct evidence for the layer was obtained.

The moral of this is that one should always be guided by theory, but that sometimes predictions may be wrong because of unknown factors—in this example, the ionizing layer.


## THE RIPENESS OF TIME

The reader will have been struck several times by the fact that discoveries have often been made ahead of their time, so that there had to be a delay before use could be made of them. A noteworthy example relates to the persistence of vision (Chapter 3). Reference to this was made in the second century AD by Ptolemy, who noticed that a disc painted in different colours appeared uniformly coloured if spun rapidly. Over the centuries many other people must have noticed the same kind of thing; a lamp swung rapidly appears as a continuous arc of light, for example. Not for many centuries, however, did anyone do anything about the effect. It was Peter Mark Roget who started things going, in 1824. Drawing ingenious conclusions from the fact that the spokes of a moving wheel appeared curved when seen through a vertical Venetian blind, Roget rediscovered the persistence of vision, and saw how it could be applied. Michael Faraday and Sir John Herschel were immediately interested, and both constructed discs (sometimes called 'thaumatropes') to demonstrate the effect. In Belgium, J A F Plateau rediscovered the effect, apparently independently of Roget, in 1832. The Austrian scientist van Stamper took out a patent on 'stroboscopic discs' in 1833.

With the introduction in 1839 of photographic techniques, the interest in the persistence of vision became even more widespread. The discs previously constructed were not very impressive, as the pictures on them had to be drawn by hand. With improvements in photographic speeds it became possible to photograph rapidly moving animals at short intervals of time, and in this way to create much more convincing discs. As we saw in Chapter 3, it was these discs that led to the development of cinematography. In this example we see the recognition of a physical effect (persistence of vision) and the development of techniques finally going hand in hand. Previously it had not seemed possible to do anything practical based on the persistence of vision.

The Fraunhofer lines—those black lines that are seen in many coloured spectra, such as the solar spectrum—provide another example of a discovery being ignored because the time was not ripe. When Newton did his experiments on the solar spectrum at Trinity College, Cambridge, in 1672 he must have noticed the black lines, but he made no mention of them. Perhaps he just thought that they were boundaries between the different colours, as William Hyde Wollaston did in 1802. It was not for another two decades or so that proper attention was paid to these lines by the Bavarian lens manufacturer Joseph von Fraunhofer. By studying them carefully and systematically he was able to show that they are not unimportant blemishes resulting from faulty instruments, but were essential features of certain spectra. Soon their great significance was recognized, and they became of great value in the identification of chemical elements, and in other ways. Again, instrumentation, in the capable hands of Fraunhofer, contributed to the appreciation of an important discovery.

Becquerel discovered radioactivity in 1897 (Chapter 6) and Rutherford, in his Silliman Lectures at Yale University (published in 1906 as *Radioactive transformations*), made an interesting comment. He said that someone might well have made the discovery a century earlier, since anyone who placed a piece of uranium ore (discovered in 1789) near to a charged gold leaf electroscope would have observed the discharge of the electroscope. If the observation had been made earlier, however, it would probably not have been reported; it might have been put down to charge leakage of unknown cause, as no possible explanation could have been suggested. By 1887, however, other radiations were known, so that explanations were possible—although it did take quite a number of years before the details could be sorted out. Again we see a link between scientific understanding and technology.

In this connection it is interesting to note that when Rutherford gave his Silliman Lectures the time was still not ripe for an interpretation of the lifetimes of radioactive substances, a matter that Rutherford had studied deeply. Why do some nuclei undergo radioactive transformation and others fail to do so, when all are present under exactly the same conditions? That question could be answered only later, when the implication of the quantum theory had been appreciated.

There are many striking examples of advances in the field of computers that were ahead of their time. In the early years of the nineteenth century Charles Babbage (1792–1871) carried out some quite remarkable work on calculating machines, but it could not be applied for about a century because related techniques were not available. One person who worked with him was Ada, Countess of Lovelace (1815–1852), who was the daughter of the poet Lord Byron. Her ideas were completely forgotten until over a century after her death. When in the late 1940s certain computer patents were taken out it was discovered that they were invalid, as the principles on which they were based were clearly laid out in Lady Lovelace's book *Sketch of the analytical engine*, which was published in 1843. The computer programming language ADA was named in her honour, as it incorporates many of her ideas on the functioning of a computer system.

## WRONG TURNS INTO DEAD ENDS

One point that is to be stressed about scientific research is that mistakes are often made, but that the progress of science does not really suffer as a result.

We have seen many examples of this in earlier chapters of this book. Sometimes an incorrect theory persisted for centuries, but worthwhile work was nevertheless carried out. Until the middle of the nineteenth century, much work on heat was based on the incorrect theory that heat is a substance (Chapter 2); the great work of Antoine Lavoisier, Joseph Black, and Sadi Carnot was done with that theory accepted. When it was finally realized that heat is instead a form of energy, their work was not invalidated; instead it was quickly reinterpreted in terms of the new theory.

Two more specialized examples are, I think, instructive. One is concerned with the theory of the rates of chemical reactions. Today we usually relate the rates of reactions to what we call the *Arrhenius equation*, according to which we plot the logarithm of a rate against the reciprocal of the absolute temperature (the Celsius temperature plus 273), and usually get a straight line; we saw an example in Fig. 2.30. From the slope of the line we then calculate what we call the activation energy, which is the height of the energy barrier for the reaction; it is the amount of energy the colliding molecules must have in order to react together.

This approach to chemical reactions was not reached without considerable controversy, which most modern chemists have forgotten all about. In 1895 a rival theory was put forward by the chemist Vernon Harcourt (1834–1919) and the mathematician William Esson (1839–1916), both of Oxford University. Much earlier the two had done some pioneering work on the rates of chemical reactions. What they claimed in 1895 (and again later in 1912, after they had obtained more evidence), was that a much better straight line is obtained if one plots the logarithm of the rate against the logarithm of the absolute temperature—instead of against the reciprocal of the temperature. There is no doubt that what they said was correct; the plotted lines were almost always straighter.

If we accepted the simple idea that the best theory is the one that best fits the experimental results, we would have to agree that Harcourt and Esson had proved that the Arrhenius equation was not the best equation, and that theories of reaction rates should be modified. This is not, however, what happened. The Arrhenius theory survived, and Harcourt and Esson's work remains a scientific curiosity that only a few people interested in the history of chemistry know about.

It was remarked earlier that scientific theories do not arise on the basis of fitting data to a formula, and this is an excellent example. There are two reasons why the Harcourt–Esson formula was quietly forgotten. The first is a purely mathematical one. Taking a logarithm has the effect of narrowing the range of data, so that one is intrinsically more likely to get a good straight line after one has taken a logarithm. Secondly, the Harcourt–Esson empirical relationship was theoretically barren; one simply could not relate it to other ideas about how chemical reactions proceed. The Arrhenius equation, on the other hand, gave a clear theoretical picture of a reaction, in terms of an energy barrier to reaction. The fact that the lines obtained in the Arrhenius plots were not always straight is explained in terms of complexities in the reactions—for example, more than one process may be occurring at the same time.

The second example relates to the way in which chemical reactions come about. In about 1913 the French scientist Jean Baptiste Perrin (1870–1942), who won the 1926 Nobel Prize in physics for his work on Brownian movement, suggested that all chemical reactions occur as a result of the absorption of radiation from the surroundings. It was already known that some reactions, called photochemical reactions, are brought about by radiation, but Perrin went further than this; he believed it to be true of all reactions. We need not go into the evidence he put forward for what became known as the radiation hypothesis; it was convincing to some, but after a decade or so of controversy the hypothesis was overturned.

The important point to be made here is that this wrong turn into a dead end really did no harm to the progress of science. On the contrary, it stimulated much research done with the object of proving it wrong. It was finally shown that radiation is only important in certain reactions. Others proceed as a result of the energy the molecules receive by chance, as a result of collisions with other molecules.

## QUALITATIVE AND QUANTITATIVE SCIENCE

Our discussion of scientific method in the last few sections has been rather negative, in that we have emphasized the fact that there seem to be no simple guidelines; one must follow one's nose (in chemistry, sometimes quite literally). There is, however, one point that should be stressed, because people who are not scientists often go wrong about it when they think that they are following the scientific method. Scientists distinguish clearly between quantitative and qualitative aspects. In other words, they recognize that some properties can be expressed in terms of numbers, and that others cannot. A scientist who looks at a painting, or listens to music, will judge the work by aesthetic standards, and will not be deluded into thinking that the quality of the art can be judged in terms of numbers.

Of course a scientist can examine a painting and measure the wavelengths of the colours in it, and for some purposes that would be a useful thing to do—it might help in deciding on its authenticity, for example. But no scientist would think that the quality of the painting could be established from such an investigation. Nor would a scientist think that the quality of a piece of music could be related to the frequencies of the notes.

It is true that scientists try to make quantitative measurements as far as they can, but when they do this they recognize the limited importance of the measurements. James Clerk Maxwell, and later Edwin Land, made a quantitative study of colour (Chapter 3), but they were well aware that their measurements could only in an indirect way relate to how one sees a painting or photograph. Sometimes quantitative measurements are used to help to establish quality, but only when there seems to be no more satisfactory way. Schools and universities evaluate students to a considerable extent on the basis of marks. If the quality of students were assessed entirely from the impressions of teachers, serious mistakes and injustices might result. The present method seems the best possible, provided that examinations are set in an appropriate way. For example, it is obviously unsatisfactory to judge students entirely by how they solve numerical problems, as seems to be done in some science courses. Science students should also be tested for their ability to express their ideas.

Sometimes, however, assigning numbers to make a qualitatative assessment is taken to quite absurd lengths. For example, a well-known Canadian periodical, excellent in many ways, has started to evaluate the quality of universities by assigning numbers to certain aspects of them. For example, a number is assigned on the basis of the number of books in a university's library. A moment's reflection shows us the absurdity of this; a smaller library of well-chosen books can be much better that a large library of ill-chosen books. The quality of a library cannot be defined in terms of numbers, because quantity and quality are two completely different things. Similarly, the proportion of students who come long distances to attend a university has no bearing on its quality. Conclusions drawn on the basis of numbers relate only to quantity, and can have no bearing on quality.

## SCIENTIFIC STRATEGY

As there seems to be no such thing as a scientific method, it follows that one cannot devise a general strategy for solving scientific problems. This opinion is by no means universally accepted. J D Bernal, in a penetrating account of the work of William Astbury published in 1963, remarked that it is 'ideally possible to construct a rational strategy for any scientific research, even for such a complicated investigation of protein structure'. As we saw in Chapter 7, Astbury worked in this field, but did not arrive at a satisfactory kind of structure. Bernal went on to say that, by hindsight, we can see the mistakes that Astbury made. I do not myself believe that any such strategy really exists, since research problems are so different from one another that a different strategy is needed for each, and we do not know the correct strategy until the work is done.

Bernal was right in suggesting that there is a spectrum of scientific approaches, ranging from 'dashing at a problem the moment it presents itself', to sitting down in front of it and 'taking it by slow siege'. He was right in saying that Astbury's approach was of the first type, and that as a result he made strategic mistakes. Astbury did show impulsiveness in assuming the folding of proteins to be in two dimensions, and stubbornness in not accepting the later correct suggestion that the folding is in three dimensions. But it does not seem to me that anyone could formulate a general strategy that should have been followed in that type of investigation. Instead, the successful investigator is the one who instinctively formulates the best strategy, and continuously modifies it, as the work proceeds.

## THE ORGANIZATION OF SCIENCE

We should now address a question of great importance: to what extent should scientific research be organized? This is a matter that has been extensively debated among scientists and others. Extreme positions have been taken, and the best answer is somewhere in between—but where, exactly?

Not surprisingly, the extreme position that science should be completely under the control of the state was put forward and implemented by the Soviets. This point of view was brought into sharp focus in 1931 at the International Congress of the History of Science and Technology, held in London. There a group of Soviet scientists explained their position in great detail. Present at the Congress was J D Bernal, an ardent communist, and he was most impressed. In an article published in the *Spectator* in 1931 Bernal discussed science as it was done outside the USSR, and referred to 'the appalling inefficiency of science at the present time, tied to academic and impoverished universities and to secretive industries and national governments'. By contrast, he went on, 'the USSR had 850 linked research institutes and 40 000 research workers. Is it better to be intellectually free but socially totally ineffective or to become a component part of a system where knowledge and action are joined for one common social purpose?'

This makes somewhat strange reading today, now that we know that the achievements of the Soviets in science and technology have been vastly inferior to those of capitalist countries, particularly the US and Britain. Bernal, however, was too intelligent and practical a man to think that there was any likelihood that he could persuade countries like Britain and the US to adopt the Soviet methods in their entirety. He was convinced, however, that some degree of organization was necessary for science and technology to proceed more efficiently. This was in fact becoming evident to many scientists, particularly since scientific work was becoming much more expensive. The pioneering work of people like Faraday and Maxwell was all done with cheap equipment, most of it already available in a simple university or government laboratory. Because of the great advances in science, however, the work that was being undertaken by the middle of the twentieth century was for the most part much more expensive, and special sources of funding were necessary. Competition between different groups of scientists made it necessary for governments to have scientific policies. Until the Second World War few governments had given any consideration at all to this problem.

In 1939 Bernal published an influential book, *The social function of science*. The basic idea emphasized in it was that science has an important function in society, and that proper scientific planning could greatly improve human conditions. Bernal emphasized that science was already capable of solving the world's material problems, and that the factors hindering their solution were not scientific, but social, psychological, and political. He estimated that the efficiency of research was then only about 2%. By that he meant that with the benefit of hindsight he concluded that scientific and technical goals would have been realized fifty times more effectively if they had been done in the most efficient way possible. He realized, of course, that we never do have the advantage of hindsight, and that it is difficult to see one's way ahead through a scientific problem. He thought, however, that with suitable forethought the efficiency of research and development could be improved a little.

One who strongly supported Bernal's ideas was Patrick Maynard Stuart Blackett (1897–1974), a physicist of great distinction. He did pioneering work on nuclear transformations, and was the first to confirm experimentally the existence of the positron (the electron's 'antiparticle', having the same mass but a positive charge). He was awarded the 1948 Nobel Prize for Physics, and was created a Life Peer.

Bernal's views did not go unchallenged. One of his first opponents was John Randal Baker (1900–1984) of the Department of Zoology at Oxford (he never became a professor, but was later Reader in Cytology and a Professorial Fellow of New College). In 1939 he published, in the *New Statesman*, a 'Counterblast to Bernalism'. This article attracted the attention of Michael Polanyi, then at the University of Manchester, where he was a colleague of Blackett; the two had had many arguments on the subject. The botanist and ecologist Sir Arthur Tansley (1871–1955) agreed with Baker and Polanyi, and in 1940 the three founded the Society for Freedom of Science. The opposition to Bernal's ideas was mainly on the grounds that any attempt to lead science into socially desirable channels would stifle creativity.

By that time, of course, the Second World War had started, and science had become organized in a particular way for the purpose of winning the war. It is of interest that Bernal played an important part in the application of science to the war effort. Although a communist, he was appointed to a key scientific position under Lord Louis Mountbatten, who had a high regard for Bernal's abilities. This led to an amusing incident. Bernal was anxious to obtain the services of a former colleague, but found that his efforts were frustrated. On asking why, he was told that the only thing against the man was that he was known to have associated with a notorious communist. It soon emerged that the communist was Bernal himself, and the appointment was then made at once.

One of the ideas that developed from Bernal's *The social function of science* was operational research, or operations research. This is the scientific analysis of a process or operation. During the war, under the leadership of Blackett, it was applied with great success to bombing and other military operations. Since the war the idea has been applied successfully to many governmental and industrial processes.

During subsequent years, a dual pattern has developed in the democratic countries. A certain amount of organization has been introduced, but many scientists are still able to pursue research in their own way. Most scientists today agree that there is room for improvement. The main weakness is that those who administer research funds place too much emphasis on the possible practical applications of the work.

I hope that this book has shown how unreasonable this attitude is. We have seen many examples where the consequences of a scientific discovery could not possibly have been recognized at the time the work was done. Michael Faraday never dreamed of the large-scale production and distribution of electrical power. Maxwell never thought of the possibility of radio transmission, and even Hertz, who first transmitted a radio signal, expressed the opinion that one could never broadcast over more than a few metres. Lord Rutherford, whose work directly made possible the application of nuclear energy, ridiculed the idea that it would be a practical possibility.

Some of the scientific work of the chemist Chaim Weizmann provides a particularly instructive example. Born in Russia in 1874, he first worked in Germany and Switzerland and obtained his doctorate at the University of Fribourg. He had become active in the Zionist movement, and in 1904 moved to Manchester, partly because he thought that Britain was the most likely country to establish a Jewish national homeland in Palestine. He worked at Manchester University with the professor of chemistry, William Henry Perkin, Jr (1860–1929), who was particularly interested in dyestuffs.

Weizmann decided to work on the general problem of the synthesis of organic compounds by bacteria. He had no practical aim in view; he just thought that there should be more basic knowledge on the subject. He worked with a bacterium now called *Clostridium acetobutylium*, and discovered that it would convert sugar into an alcohol called butyl alcohol. When, however, he told Perkin about this, the reply was 'Butyl alcohol is a futile alcohol; throw it down the sink'. This was a surprising comment in two ways; Perkin had no sense of humour, but apparently even he could not resist this little joke, and also it showed a lack of wisdom on his part. In fact, as we shall see, Weizmann's discovery had widespread applications. (Perkin was not very bright in other ways. He was a good synthetic organic chemist, but could not appreciate anything outside his field. He became professor of chemistry at Oxford in 1912, and was fond of making the incredibly stupid remark that 'physical chemistry is all very well, but of course it does not apply to organic compounds'.)

Weizmann found that the bacterium also produces acetone. At the time this was of little use, since acetone could be produced in sufficient quantities from wood. However, when the First World War broke out there was an urgent need for acetone for the manufacture of gun propellants, and not enough could be obtained from wood. In the end, through the intervention of Winston Churchill, a gin factory was taken over, Weizmann was appointed its chief chemist, and large amounts of acetone (instead of gin) were made from alcohol by use of the bacterium. Some of this work was done in Toronto, using maize for the production of the alcohol. Weizmann's technical success had much to do with the help given by the British government towards the establishment of a Jewish homeland, of which he became the first president in 1948.

Weizmann's bacterium had yet another success. Butyl alcohol turned out to be far from futile. It is an excellent fast-drying paint solvent, and is now extensively used in the automobile industry. *Clostridium acetobutylium* is still hard at work.

This saga should convince even the most hard-bitten bureaucrat that it is folly to expect pure scientists to justify their intended work on the basis of utility. Almost invariably, practical consequences can be identified only after, not before, the research is done. The only criterion for approving a scientific project should be its quality.

## SCIENCE AND TECHNOLOGY SINCE THE SECOND WORLD WAR

The World War of 1939 to 1945 brought about a great change in public attitudes to science. The scientific successes of the war, such as the harnessing of nuclear energy, the application of radar, and the improvements in computer technology, produced at first a general public euphoria. It was thought that nuclear power would bring about a complete solution to all problems of energy and food supply, and that communications techniques would greatly enhance dissemination of information. Science seemed to offer a general panacea, and there was a dramatic increase in the number of students wanting to enter the field of science. One result was the founding of many new universities; another was an increase in government support of science.

From the scientific and technical points of view these hopes have been realized. Nuclear power plants have been installed in a number of countries, and in some regions a substantial fraction of the energy used is provided by nuclear power. Particularly spectacular have been the improvements in communications technology resulting from the introduction of the transistor in 1947 and the microprocessor in 1971. In combination these two technical innovations, by allowing extensive miniaturization, have completely transformed communications. They have also transformed other areas, such as the auto-mobile industry and the manufacture of household devices.

At the same time chemists and biochemists have brought about less spec-tacular but nevertheless important advances. New plastics, some with quite remarkable properties, have become available. Many medical needs have been satisfied by a variety of new pharmaceutical products, and some diseases, such as smallpox, have been completely eradicated. Treatments for many other conditions have been discovered, and there is much promise for the future.

## PUBLIC ATTITUDES TO SCIENCE AND TECHNOLOGY

In spite of these scientific and technical successes, the public euphoria about science and technology has gone into a recession. This is not surprising in view of the deep chasm between what science and technology have achieved, and the general state of the world today. The scientific advances are by no means reflected in the welfare of the earth's inhabitants. The lives of millions are profoundly affected by war and the threat of war, and vast numbers of people are undernourished and living in misery. In the time that it takes to read this paragraph, it is likely that a few people in this world will have starved to death. Even in the so-called developed countries there is much unemploy-ment. The public cannot fail to see that some of these problems are directly related to advances in technology.

The trouble is that society has not adjusted properly to scientific advances. Much of the responsibility for this lies with politicians and their advisers, who for the most part are completely ignorant of science and its relationship with technology and the needs of society. They seem, in fact, to be unaware of the relationship, and to make no effort to find out about it. Rarely do politicians make any mention of the obvious fact that the advances of science and tech-nology are completely transforming society. They do not seem to have recog-nized that the new technologies have reduced the need for manual labour, and have greatly increased the need for persons with advanced technological skills. Many politicians have a background in law, and few know much about science. It is not necessary for a politician to have a detailed knowledge of science, but in the modern world some appreciation of science, and its rela-tionship to technology and society, is essential. Politicians spend much of their time on legalistic problems related to the criminal code and the consti-tution. These matters are important, but they should take second place to reforming the economic structure so that it will react to the changed techno-logical environment.

The deterioration in public attitudes towards science and technology is also based to some extent on misunderstanding and misinterpretation. Some ill effects of science-based technology have become apparent, but in judging them the public has not taken a balanced view of the situation. There has been a reaction against science itself, which has been judged in terms of the ill effects that can arise from it, without consideration of the good effects.

In many places our environment has been harmed by technologies that had not been kept properly under control. Improvements in nuclear technology led to a proliferation of nuclear weapons, and to real and potential pollution. Scientists are even more acutely aware of these problems than the general public, because they can understand and predict them better. It is important, in coming to conclusions, not to focus entirely on the negative aspects of science and technology. Part of the trouble is that members of the public so often misunderstand the relationship between science and technology, and fail to appreciate how science and technology advance. The main aim of this book has been to help to dispel misconceptions. We have seen that science and technology advance on a broad front, and that the advances are inter-related in a rather complex and variable way. In particular, the practical applications of pure science are usually not immediately recognizable.

Two factors of fundamental importance must be recognized. One is that once scientific knowledge has been gained, there is no way that it can be hidden or suppressed. The second is that science cannot advance and make only socially desirable innovations. It is for society at large to make respons-ible and well-informed decisions as to the consequences of scientific and tech-nological advances. It is inappropriate for science to be blamed for many things for which society as a whole is responsible.

Today there are some public campaigns which exert a strong influence on public policies. These campaigns sometimes do good, but too often they are based more on emotion than on accurate scientific information and sober decision-making. It was noted in Chapter 1 that there is good and bad science, but that the bad science tends to be quickly eliminated. There is also good and bad environmentalism, but unfortunately the bad tends to survive for long periods of time. It is undeniable that there must be standards and safeguards to protect the health and welfare of the earth's inhabitants. They should, however, be imposed with all factors taken into account, not just some undesirable consequences; these must be balanced against the advant-ages. Environmentalists must take the responsibility of getting their facts straight, and must base their conclusions on all of the evidence, not just the evidence that supports their case.

Physics has come to have a bad name in some people's minds, largely because it has given us the atomic bomb. Nuclear energy plants, using a related technology, have some disadvantages, and in some places there has certainly been lack of adequate control of them. However, it is wrong to condemn nuclear power plants on the sole basis of problems that have arisen with some of them. This is one field in which a great deal of nonsense is uttered by environmentalists. It is not uncommon to read statements to the effect that the amount of energy required to mine and purify uranium for use in a nuclear plant is so great that nuclear energy offers no advantages over conventional fuels like coal. It is true that it takes about a hundred times more energy to produce a ton of suitable uranium than a ton of coal, but a ton of uranium in a nuclear reactor produces 20 000 times more energy than a ton of coal in a furnace. Aside from the energy advantage, the production of uranium is much safer; compare the number of uranium miners killed with the number of coal miners killed. Also, the nuclear industry produces much less pollution than energy technology based on fossil fuels.

One matter that has to be faced today is that the world population is increasing at an alarming rate, so that the conventional sources of energy—water power, coal, and oil—will soon prove inadequate. New technologies, such as solar energy, are being developed, but it seems unlikely that they will satisfy future needs. It may well be that nuclear energy will be necessary to prevent starvation, which in the future may be even more widespread than at present. As in all decisions, the evidence as a whole must be considered.

In the minds of some people, chemistry has become an ugly word, associated with the pollution of the atmosphere, the drug culture, and the proliferation of the means of waging war. Some people talk as if they would like to suppress chemistry altogether. Such people would be indignant if they found themselves denied the use of gasoline, anaesthetics, and pharmaceutical drugs. They forget that the work of chemists is essential to the production of these materials. There is no possible way in which chemistry can advance and make only socially desirable materials. It is for society to determine what use is made of chemicals.

A widespread misunderstanding about chemistry relates to whether substances have been extracted from materials found in nature, or have been synthesized by chemists in the laboratory. Members of the public holding extreme views on such matters have even invented a confusing and misleading terminology. They apply the words 'natural' or 'organic' when direct use is made of substances present in nature, while materials made in the laboratory are called 'artificial'. The situation is in reality quite different. Consider first what chemists call the chemical elements, of which carbon, iron, and nickel are examples. There are two ways in which we can obtain these elements. The ordinary and by far the most common method is to get them from the earth or the earth's atmosphere. Iron and nickel, for example, are found in ores, which are treated by chemical methods to obtain the pure elements. In a few special cases we obtain elements by means of nuclear devices, which produce them from other elements, usually by bombarding them with neutrons. The radioactive isotopes that are extensively used in therapy are produced in this way.

An element is just the same however we obtain it, and it makes no scientific sense to suggest, as is sometimes done, that an element obtained in one way is superior to the element obtained in another way. Advertisers have sometimes tried to make the public believe that calcium produced from seashells is better than that produced from limestone. This is scientific nonsense. Calcium obtained from any source and then purified is exactly the same as that obtained from another source, and this is true for all of the chemical elements.

It is equally true that a chemical compound is the same however we have obtained it, and it makes no difference whether we have obtained it directly from a natural source (from a plant, for example) or have synthesized it in the laboratory. There is one important difference between 'natural' and synthetic, however. Of all the millions of chemical compounds that chemists have identified, only a tiny proportion exist in nature. If, then, we insist on 'natural' compounds, we are unnecessarily depriving ourselves of an enormous number of compounds that have been found to be important for our welfare. The vast majority of the substances used as anaesthetics and as pharmaceutical drugs cannot be extracted from materials in our environment. Most of them could not possibly be produced in adequate amounts by biological means. They must be made in a chemical laboratory or an industrial plant. They are none the worse for being 'artificial', and one does not do oneself any favour by refusing to use them.

It thus makes no sense to argue, as is so often done, that 'natural' or 'organic' methods are necessarily more socially desirable than 'artificial' ones. In fact, materials obtained from 'natural' products, as well as 'artificial' substances, may have some undesirable impurities. The important point to realize is that if synthetic methods of producing many materials were abandoned, their supply would be quite inadequate—if it were possible to find them at all.

Campaigns are often directed against agricultural practices based on the use of fertilizers made in the laboratory. There is a lack of realization that if these methods were suppressed, a vast number of people would sink below the starvation level. People should not argue against such practices unless they can suggest a more desirable way of avoiding extensive food shortages.

A serious weakness of some of the public campaigns, such as the environment movement, is that they focus on obvious problems, such as polluted air and water, and overlook less evident difficulties which may be more important in the long run. Scientists are acutely aware that today's rate of growth of the world population will soon lead to much greater food shortages than now exist. More fundamental scientific work is needed as well as more technology. It is possible that disaster will only be averted by the use of nuclear and agricultural technologies that are now popularly held to be undesirable. The answers can be reached only through well-informed decisions in which the scientific as well as the social aspects are taken into proper account.

This discussion leads us inevitably to the question of population control. The experts tell us that it seems unlikely that future energy technologies (including nuclear technologies) will provide enough food to feed the world if the population continues to grow without check. Admittedly, an important theme of this book has been that scientific advances have often led to technical improvements that would have been ridiculed before they appeared. But we cannot be absolutely sure that new discoveries will allow us to cope with a rapidly increasing population. Are we justified in gambling on that possibility, when widespread human misery and starvation will result if the necessary technical results are not achieved? I think not; we must surely control the increase of population.

Population control must be considered in relation to the welfare states that are being established throughout the world. I define a welfare state as one that at least as a matter of policy accepts responsibility for the welfare of its citizens, and at the very least does everything in its power to prevent them from starving. Some conservative people seem to dislike the expression welfare state, thinking (correctly) that it was brought about largely by the efforts of persons of liberal views. (In some quarters 'liberal' seems to have become a term of abuse!) But does any reasonable person think that society should return to the policies of the nineteenth century, when countries that considered themselves civilized did allow many of their citizens to starve to death? Surely some form of the welfare state—call it something else if you prefer—must be our aim for the whole world.

The dilemma we face is that the welfare state on the one hand, and failure to control population on the other, are incompatible, in that they will inevitably lead to a population explosion. If people are not encouraged to adopt birth-control procedures, and at the same time welfare states are maintained, the result will be worse misery than has ever existed. It is unrealistic to put the blame on individuals who produce more children than they are capable of providing for. Greater blame must lie with powerful organizations which in the name of a narrow-minded morality condemn birth-control methods. If they take a broader view they will see that they are violating a higher morality by leading to a society in which there will be suffering of a magnitude never yet encountered.

Finally, a general comment about communication, capable of extensive further discussion, should be made. Users of the 'information superhighway' have become aware of some of the more obvious difficulties, but are sometimes less aware of more fundamental and important ones. They are alert to the fact that some of the information they receive is incorrect. They are less aware of the danger of thinking that the effective transmission of information solves all our intellectual problems. The point cannot be made better than was done over sixty years ago by T S Eliot:

> Where is the wisdom we have lost in knowledge?
> Where is the knowledge we have lost in information?

*     *     *     *     *     *

It may be useful to repeat the conclusions (already in the Preface) that arise inevitably from the discussions in the present book:

1. Pure research should be judged entirely on the basis of its quality, and not in terms of possible practical applications.
2. Technology and engineering must be based on pure science; the time for empirical invention is long past.
3. Decisions about science and technology must be based on a careful consideration of all the factors involved.

# Bibliography and notes

There is a vast literature dealing with matters relating to the subject matter of this book. Only a few of the books and articles that seem especially helpful can be mentioned.

## CHAPTER 1   SCIENCE AND TECHNOLOGY

Many of the books and articles mentioned in this section are referred to again later; the subsequent reference will give only a brief identification.

Not many histories include both science and technology. Two that do, and are well worth reading, are:

R J Forbes and E J Dijksterhuis, *A history of science and technology*, 2 vols., Penguin Books, London, 1963.

Donald Caldwell, *The Fontana history of technology*, Fontana Press, London, 1994 (in North America, *The Norton history of technology*, Norton, New York, 1994). Although the word science does not appear in the title of this book, a good deal of pure science is covered. The discussion of the relationship between science and technology is particularly well done.

There are many histories of science, some dealing with special aspects; the following are authoritative and of great interest:

H Butterfield, *The origins of modern science*, 1300–1800, G Bell and Sons, London, 1965. This book, written by a distinguished historian rather than a scientist, gives an accurate account of early science and its relationship with the development of the modern world.

Sir William Dampier, *History of science, and its relations with philosophy and religion*, Cambridge University Press, 1938.

A Rupert Hall, *From Galileo to Newton*, Harper & Row, New York, 1963; Dover Publications, New York, 1983.

J R Partington, *A history of chemistry*, 4 vols., Macmillan, London 1961. This book makes somewhat dry reading, but is a remarkably accurate and useful reference book.

William H Brock, *Fontana history of chemistry*, Fontana Press, London, 1992 (US title *Norton history of science*, Norton & Co, New York, 1993). This book is particularly interesting to read.

H T Pledge, *Science since 1500: a short history of mathematics, physics, chemistry, biology*, H M Stationery Office, London, 1939, 2nd edition 1966; Harper & Brothers, New York, 1939; reprint with Prefatory Note, 1959.

Mary Jo Nye, *From chemical philosophy to theoretical chemistry: dynamics of matter and dynamics of disciplines*, University of California Press, 1993.

The history of technology is well treated in Cardwell's book, already mentioned, and in the following books:

R A Buchanan, *History and industrial civilization*, Macmillan, London, 1979.

R A Buchanan, *The engineers: a history of the engineering profession in Britain,* 1750–1914, Kingsley, London, 1989.

T K Derry and Trevor I Williams, *A short history of technology*, Clarendon Press, Oxford, 1960.

Trevor I Williams, *The history of invention*, Facts on File Publications, New York & Oxford, 1987.

Trevor I Williams, *Science. A history of discovery in the twentieth century*, Oxford University Press, 1990.

Of special and fascinating interest is

*A bedside **Nature**: genius and eccentricity in science* 1865–1963 (Edited by Walter Gratzer, Macmillan, London, 1996). This volume contains facsimile reproductions of articles and portions of articles published in *Nature*, from its founding by Norman Lockyer in 1865 to its publication in 1963 of the famous Watson and Crick article on DNA (Chapter 7). Gratzer's selection of items is admirable, and his comments on them are apt and pithy. Besides good science, there is much humour and much of interest outside the field of science. The word bedside is not to be taken as meaning that it is a good book to help one to go to sleep; in my case, it kept me awake and delighted until the small hours. In what follows it will be much quoted, with the brief title *Bedside **Nature***.

Some encyclopaedias, in many volumes, kept up to date every few years, are useful to consult on special topics. Most of the articles are clearly written, and often include some of the historical aspects:

*McGraw-Hill encyclopedia of science and technology*, McGraw-Hill, New York, 1960; 7th edition, 1992.

*Encyclopedia of physical science and technology*, Academic Press, New York, 1987; 2nd edition, 1992.

The following reference books cover both science and technology:

A Hellermans and B Bunch, *The timetables of science. A chronology of the most important people and events in the history of science*, Simon and Schuster, New York, 1988; updated softcover edition, 1991.

R M Gascoigne, *A historical catalogue of scientists and scientific books from the earliest times to the close of the nineteenth century*, Garland, New York, 1984.

C C Gillispie (Ed.), *Dictionary of scientific biography*, Charles Scribner, New York, 1970–1990. In what follows, the dictionary will be referred to as *DSB*.

For biographies of most of the scientists mentioned in this book see, for example,

Magnus Magnusson (Ed.), *Cambridge biographical dictionary*, Cambridge University Press, 1990.

Hazel Muir (Ed.), *Larousse dictionary of scientists*, Larousse, Edinburgh and New York, 1994.

David, Ian, John, and Margaret Millar (Eds.) *The Cambridge dictionary of scientists*, Cambridge University Press, 1996.

The British Association for the Advancement of Science is mentioned a number of times in this book. For an excellent account of its early years, and indeed of the scientific work done at the time, see Jack Morrell and Arnold Thackray, *Gentlemen of science*, Clarendon Press, Oxford, 1981. A brief but interesting account also appears in an article by John Mills, 'The History of the British Association', in the *Strand Magazine*, Vol. 22 (1901).

For a fascinating discussion of many of the important problems of modern physics and technology see:

Roger Penrose, *The emperor's new mind: concerning computers, minds, and the laws of physics*, Oxford University Press, 1989; paperback edition, Vintage Press, 1990.

The history of the zip-fastener is a complicated but fascinating one. For an interesting summary of the history see Robert Friedel, *Invention and Technology*, **10**, No. 1, 8–16, Summer 1994. Further details are in Friedel's book *Zipper: an exploration in novelty*, W W Norton, New York, 1994.

CHAPTER 2    JAMES WATT AND THE SCIENCE
OF THERMODYNAMICS

Much has been written on steam engines. One interesting account is Howard Jones, *Steam engines*, Ernest Benn, London, 1973. For a particularly authoritative account of the history of steam engines see H W Dickinson, *A short history of the steam engine*, 1st edition, Cambridge University Press, 1938, 2nd edition, Frank Cass, London, 1963. This book contains many excellent illustrations, including portraits of many of the pioneers. Dickinson was a practical engineer with much knowledge of the history of science and engineering, having been for many years at the Science Museum in South Kensington, London. For the history of the Newcomen engine see L T C Rolt and J S Allen, *The steam engine of Thomas Newcomen*, Moorland Publishing Co., Hartington, UK, and Science History Publications, USA, 1977. This book also has many illustrations, excellently reproduced.

An old book that is well worth consulting is John Harris, *Lexicon technicum: or, an universal English dictionary of arts and sciences: explaining not only the terms of art, but the arts themselves*, London, 1704; a facsimile edition of this book was published in 1966 by Johnson Reprint Corporation, New York. In writing this book the Revd John Harris (1666–1719), who held a number of church appointments, obtained some of his information from the greatest authorities of his time, including Robert Boyle, Isaac Newton, Edmond Halley, and many others. This book, published over 40 years before Dr Johnson's famous dictionary, was the first general scientific encyclopaedia, and was written with exceptional competence. The article 'Engines' gives an excellent account of the subject, as does the one on 'Hydrostatics'.

For biographies of James Watt see H Dorn's article in *DSB*, **14**, 96–199 (1976), which gives references to other biographical accounts. Besides the books there mentioned, a biography by Andrew Carnegie, *James Watt*, Doubleday, Page & Co., New York, 1913, is of special interest, particularly on account of its authorship; Andrew Carnegie (1835–1918) was the famous Scottish-born American industrialist, author, and philanthropist, whose name is widely remembered from the many libraries he endowed. His biography of Watt is written with great enthusiasm, and includes many interesting details. In his Preface, Carnegie tells us that when he was first invited by the publisher to write the biography he declined. Then he reflected that he had made a fortune from steam power, and changed his mind. He admitted that at first he knew nothing of Watt or of steam engines, but realized that the best way to learn about them was to write the book. He added that after completing the book he realized that he had written about 'one of the finest characters that ever graced the earth'. He also describes Watt with a quotation from Shakespeare's Julius Caesar:

His life was gentle, and the elements
So mix'd in him that Nature might stand up
And say to all the world, 'This was a man'.

For a detailed assessment of James Watt's work, from various aspects including thermodynamic efficiency, see R V Jones, 'The "Plain Story" of James Watt', *Notes and Records of the Royal Society*, **24**, 194–220 (1970).

For the history of railroads with special reference to steam engines see, for example, Hamilton Ellis, *British railway history: an outline from the accession of William IV to the nationalization of the railways*, George Allen & Unwin, 1954, 1959; and A W Bruce, *Steam locomotion in America*, W W Norton, New York, 1952.

For an interesting and clear general account of the history of air engines, such as the Stirling engine, see T Finkelstein, 'Air engines'. *The Engineer*, **207**, 492–497, 522–527, 568–571, 720–723 (1959). A detailed account of the Stirling engines, with some biographical information about Robert Stirling and other members of his family, is in G Walker, *Stirling engines*, Clarendon Press, Oxford, 1980; this book includes much about the development of Stirling engines since the 1930s. A useful more recent account of automobile engines of all kinds, at not too advanced and technical a level, is M L Poulton, *Alternative engines for road vehicles*, Computational Mechanics Publications, Southampton, UK, and Boston, USA, 1994; Chapter 9 (pp. 117–124) deals with Stirling engines, and conveniently summarizes their advantages and disadvantages. For a more advanced technical treatment of Stirling engines see M J Collie, *Stirling engine design and feasibility for automotive use*, Noyes Data Corporation, Park Ridge, NJ, 1979. For an account of a recent Stirling engine see 'Energy: Stirling engine comes full cycle', *The Engineer*, 15/22 August, 1991, pp. 22–23.

The thermodynamics of the Stirling (and other) engines were first treated by the Scottish engineer William John Macquorn Rankine (1820–1872). Rankine had a profound knowledge of thermodynamics, but he was by no means a lucid writer. His first paper on engines was entitled 'On the geometrical representation of the expansive action of heat, and the theory of thermo-dynamic engines', *Philosophical Transactions*, **144**, 115–176 (1854) and is reproduced in *Miscellaneous scientific papers of W J M Rankine* (Ed. P G Tait), London, 1881. Rankine also presented a thermodynamic analysis of engines at the September 1854 meeting of the British Association for the Advancement of Science; it is published in the *Reports of the Association* with the title 'On the means of realizing the advantages of the air engine'. An account of the Stirling and Ericsson engines and the role of the regenerator is to be found in Rankine's important book *A manual of the steam engine*, C. Griffith, London & Glasgow, 1859 (17th edition, 1908). For a thermodynamic discussion of the Stirling and Ericsson engines, based in part on Rankine's analysis, see E Daub, 'The regenerator principle in the Stirling and Ericsson hot air engines', *British Journal for the History of Science*, **7**, 259–277 (1974); to appreciate this article one must have a good understanding of thermodynamics.

The transition from steam engineering to thermodynamics is authoritatively treated in D S L Cardwell, *From Watt to Clausius*, Manchester University Press and Cornell University Press, Ithaca, New York, 1971. An interesting but less detailed discussion is also to be found in Cardwell's *Fontana (or Norton) history of technology*. Another publication dealing with the same topic is R Fox, 'Watt's expansive principle in the work of Sadi Carnot and Nicholas Clément', *Notes and Records of the Royal Society*, **24**, 233–253 (1969–70).

For biographies of Sadi Carnot see J F Challey's article in *DSB*, **2**, 79–84 (1971), M Kerker, 'Sadi Carnot', *Scientific Monthly*, **85**, 145–149 (1957), and S S Wilson, 'Sadi Carnot', *Scientific American*, **245**, 134–145 (August 1981). Evidence that Carnot died in a hospital for the mentally disturbed is contained in an article by A Birembaut, 'Sadi Carnot et l'essor de la thermodynamique', published by the Centre National de Recherche Scientifique, Paris, 1976. More on the relationship between Carnot's work and the second law of thermodynamics is to be found in M Barnett, 'Sadi Carnot and the second law of thermodynamics', *Osiris*, **13**, 327–357 (1958), and in the article by R Fox mentioned in the previous paragraph.

The fascinating life of Benjamin Thompson (Count Rumford) is summarized in S C Brown's article in *DSB*, **13**, 350–352 (1976). Some amusing details about Rumford, particularly with reference to his part in the founding of the Royal Institution, are to be found in John Meurig Thomas, *Michael Faraday and the Royal Institution; the genius of man and place*, Adam Hilger, Bristol, Philadelphia & New York, 1991: Sir John Meurig was Director of the Royal Institution for five years. For an account of J R Mayer's work see R S Turner, *DSB*, **9**, 235–240 (1974). An excellent account of the development of the first and second laws of thermodynamics is to be found in S G Brush, *The kind of motion that we call heat*, North-Holland Publishing Co., Amsterdam, 1976.

Much has been written about James Prescott Joule. For a summary of his work see L Rosenfeld, *DSB*, **6**, 180–182 (1972). Cardwell's *From Watt to Clausius* and his *History of technology* have much of interest about Joule, and for a more detailed discussion of his work see Cardwell's chapter 'The origin and consequences of certain of J P Joule's scientific ideas', in *Springs of scientific creativity* (Ed. R Aris, H T Davis, and R H Strewer), University of Minneapolis Press, 1983.

There is also much biographical material about William Thomson, Lord Kelvin. The *DSB* article by J D Buchwald, **13**, 374–388 (1976) is very helpful. J G Crowther's *British scientists of the nineteenth century*, Kegan Paul, Trench & Trubner. London, (1935), includes an excellent article on Kelvin. Another valuable account is in D K C Macdonald, *Faraday, Maxwell, and Kelvin*, Doubleday & Co., New York, 1954. The most recent biography, and an excellent and detailed one, is by C W Smith and M N Wise, *Energy and empire: a biographical study of Lord Kelvin*, Cambridge University Press, 1989.

More about Sir Robert Grove is to be found in E L Scott, *DSB*, **5**, 559–561 (1972), and in an article by K R Webb, 'Sir Robert Grove (1811–1896) and the origins of the fuel cell', *Journal of the Royal Institute of Chemistry*, **85**, 291–293 (1961). Entertaining aspects of Grove's career are included in Sir John Meurig Thomas's book on Michael Faraday, quoted earlier with reference to Rumford. Those interested in the part Grove played as one of the defence counsel in the trial of William Palmer will find a detailed account, including a transcript of the trial, in G H Knott, *William Palmer* (in the *Notable British trials* series), William Hodge & Company, Edinburgh, 1912. Palmer was tried and convicted, in the Central Criminal Court of the Old Bailey, only for the murder of John Parsons Cook. For information about his numerous other alleged murders see J H H Gaute and Robin Odell, *The murderers' who's who*, Harrap, London, 1979, reprinted by Pan Books, London, 1980; this book gives many further references.

A great deal has been written about von Helmholtz. The *DSB* article is by R S Turner, **5**, 241–253 (1972); it gives an excellent summary of his work, and includes many references. An article by James Clerk Maxwell, 'Scientific worthies: Hermann Ludwig Ferdinand Helmholtz', *Nature*, **15**, 389–391 (1877) is well worth reading; it is reproduced in *The scientific papers of James Clerk Maxwell* (Ed. W D Niven), Cambridge University Press, 1890, reprinted by Dover, New York, 1965.

The book by S G Brush, mentioned earlier, covers the second law of thermodynamics and its implications in excellent detail. There are also a number of useful articles about the second law, such as: M Baron, 'With Clausius from energy to entropy', *Journal of Chemical Education*, **66**, 1001–1004 (1989); E E Daub, 'Atomism and thermodynamics', *Isis*, **58**, 293–303 (1967); E E Daub, 'Entropy and dissipation', *Historical Studies in the Physical Sciences*, **2**, 321–354 (1970); R Fox, 'Watt's expansive principle in the work of Sadi Carnot and Nicholas Clément', *Notes and Records of the Royal Society*, **24**, 233–253 (1969–1970); Elizabeth Garber, 'James Clerk Maxwell and thermodynamics', *American Journal of Physics*, **37**, 146–155 (1969), and

'Clausius and Maxwell's kinetic theory of gases', *Historical Studies in the Physical Sciences*, **2**, 299–319 (1972); M J Klein, 'Gibbs on Clausius:, *Historical Studies in the Physical Sciences*, **1**, 127–149 (1969); T S Kuhn, 'Carnot's version of "Carnot's cycle"', *American Journal of Physics*, **23**, 91–95 (1955); C Truesdell, *The Tragicomedy of Classical Thermodynamics*, Springer-Verlag, Vienna and New York, 1971. For a discussion of the thermodynamics of the Ericsson engines, see the article by Daub cited above in connection with the Stirling engines.

For information about Clausius the *DSB* article by E E Daub, **3**, 303–311 (1971) gives a helpful account, as does Brush's book and the article by Klein cited in the previous paragraph.

For accounts of the life and work of Maxwell and Boltzmann, see under Chapter 5 (pp. 365–6). For more on Maxwell's demon see Martin J Klein, 'Maxwell, his demon, and the second law of thermodynamics', *American Scientist*, **58**, 84–87 (1970). A valuable collection of articles on the demon and related matters is *Maxwell's demon: entropy, information, computing* (Ed. Harvey S Leff and Andrew F Lex), Princeton University Press, 1990. This book includes the article by Klein, and contains a valuable summarizing introduction by the editors, including several amusing artistic representation of the demon.

For more on Willard Gibbs, see the *DSB* article by M J Klein, **5**, 386–392 (1973), and also M J Klein, 'The scientific style of Josiah Willard Gibbs', in *Springs of scientific creativity* (Ed. R Aris, H T Davis, and R H Stuewer), University of Minnesota Press, 1983. An excellent biography is by Muriel Ruckeyser, *Willard Gibbs*, Doubleday, Doran, Garden City, NY, 1942, reprinted 1964; it is valuable for its coverage of the background of Gibbs's work.

For accounts of the history of Nernst's heat theorem (the third law) see E N Hiebert, 'Hermann Walther Nernst', *DSB*, **10**, 432–453 (1974), and Sir Francis Simon, 'The third law of thermodynamics: an historical survey', *Yearbook of the Physical Society*, 1956, pp. 1–22.

For more on the career of G N Lewis, see R E Kohler, *DSB*, **8**, 289–294 (1973), and various issues of Volume 61 (1984) of the *Journal of Chemical Education,* where a number of articles are to be found. See also the excellent book by J W Servos, *Physical chemistry from Ostwald to Pauling: the making of a science in America*, Princeton University Press, 1990. An article by Michael Kasha, 'Four great personalities of science: G N Lewis, J Franck, R S Mulliken and A Szent-Györgyi', *Pure & Applied Chemistry*, **62**, 1615–1630 (1990), contains many interesting personal comments about Lewis by one who knew him well.

The foundations of thermodynamics and statistical mechanics are discussed in detail in S G Brush's book, cited previously. Other articles dealing with the subject are S G Brush, 'Foundations of statistical mechanics', *Archive for History Exact Sciences*, **4**, 145–183 (1967–8); J Mehra, 'Einstein and the foundations of statistical mechanics', *Physica*, **79A**, 447–477 (1975); L Rosenfeld, 'On the foundations of statistical mechanics' (in English), *Acta Physica Polonica*, **14**, 3–39 (1955).

A very interesting account of the relationship between the second law of thermodynamics and chemical kinetics is to be found in Frank L Lambert, 'Shakespeare and thermodynamics: dam the second law: the human importance of activation energies', *The Chemical Intelligencer*, April, 1996, pp. 20–25. The dam referred to is, of course, the energy barrier to reaction. This article discusses effectively C P Snow's remarkable statement that lack of knowledge of the second law on the part of a nonscientist is to be compared to failure of a scientist ever to read a word of Shakespeare. I agree entirely with Professor Lambert's conclusion, that understanding the second law is a lot harder than being familiar with Shakespeare's works. After all, Lord Kelvin, the co-discoverer of the second law, never really understood it properly, as he could not understand entropy!

More about kinetics is to be found in my article 'Chemical kinetics and the origins of physical chemistry', *Archive for History of Exact Sciences*, **32**, 43–75 (1985). Some special applications of the Arrhenius equation, for example to the creeping of ants, are further discussed in my 'Unconventional applications of the Arrhenius law', *Journal of Chemical Education*, **49**, 343–344 (1972); this paper gives references to the original articles. For even more on the creeping of ants see R Thomas Myers, 'Ants and chemical kinetics', *Journal of Chemical Education*, **67**, 761–762 (1990).

## CHAPTER 3    DAGUERRE, TALBOT, AND THE LEGACY OF PHOTOGRAPHY

There are many general accounts of the history of photography, and among them may be mentioned:

Beaumont Newhall, *The history of photography*, The Museum of Modern Art, New York, 1982.

Naomi Rosenblum, *A world history of photography*, Abbeville Press, New York, 1981.

Both of these books deal comprehensively with the scientific aspects of photography, including its application to scientific problems. The Rosenblum book, for example, has a section 'New ways of seeing: images in aid of science', which includes high-speed photographs, photographs taken with scanning electron microscopes, colour-enhanced X-ray photographs, and thermographs.

Michel Auer, *The illustrated history of the camera*, New York Graphic Society, Boston, 1975, is also of considerable interest. An excellent short review of nearly two centuries of investigation is A V Simcock, 'Essay Review: 195 Years of Photochemical Imaging', *Annals of Science*, **48**, 69–86 (1991). This article begins with Elizabeth Fulhame's work of 1794.

Although it is particularly concerned with the Eastman Kodak Company, an excellent general account of photography, including some of its early history, is to be found in Douglas Collins, *The story of Kodak*, Harry N Abrams, New York, 1990.

A number of books deal particularly with early work in photography. The following are especially recommended:

Brian Coe, *The birth of photography*, Ash and Grant, London, 1976.

Michael Pritchard (Ed.), *Technology and art: the birth and early years of photography*, Royal Photographic Society Historical Group, Bath, 1990. The article in this publication by Larry J Schaaf discusses the early work of Mrs Fulhame and others on photoimaging.

Helmut Gernsheim, *The rise of photography, 1850–1880: the age of collodion*, Thames & Hudson, London, 1988.

Larry J Schaaf, *Out of the shadows: Herschel, Talbot, and the invention of photography*, Yale University Press, 1992, is an excellent account which includes much material not available elsewhere.

Larry J Schaaf, *Records of the dawn of photography: Talbot's notebooks P and Q*, Cambridge University Press, 1996. Talbot's notebooks P and Q span the period from the first public announcement of photography in 1839 to 1843.

A commemorative publication, *John Herschel, 1792–1871*, The Royal Society, London, 1992, includes articles on Herschel's photographic investigations by Chris Roberts and Larry J Schaaf. A V Simcock, *Photography 150*, Museum of the History of Science, Oxford, 1989, is a short pamphlet accompanying a small exhibition of early photographs at the Museum; it gives a succinct and accurate account of the early work.

More on Edwin Land and his work is to be found in F W Campbell, *Biographical memoirs of the Fellows of the Royal Society*, **40**, 197–214 (1994); Ernest V Heyn, *Fire of genius*, (1976); and Peter C Wensberg, *Land's Polaroid: a company and the man who invented it*, Houghton Mifflin, Boston, 1987. A clear account of Land's colour theory, entitled 'Experiments in color vision' is given by Land himself in *Scientific American*, **200**, No 5, 84–99 (May 1959). Other good accounts of the work are to be found in K McLaren, 'Edwin H Land's contributions to colour science', *Journal of the Society of Dyers and Colourists*, **102**, 378–383 (1986) and S Zeki, 'The visual image in mind and brain', *Scientific American*, **267**, No. 3, 42–50 (March, 1992).

There are many excellent biographies of Newton, which give details of his work on colour; for example

Richard S Westfall, *Never at rest: a biography of Isaac Newton*, Cambridge University Press, 1980.

A Rupert Hall, *Isaac Newton: adventurer in thought*, Cambridge University Press, 1992.

I Bernard Cohen's account in the *DSB*, **10**, 20–101 (1974), is also excellent.

For the ingenious detective work on the rooms in which Newton carried out his experiments on light see Lord Adrian, 'Newton's Rooms in Trinity', *Notes and Records of the Royal Society*, **18**, 17–24 (1963).

For the history of colour photography, with clear accounts of the different processes, with excellent diagrams, see Brian Coe, *Colour photography: the first hundred years*, Ash and Grant, London, 1978. This book lists a number of other books dealing with the same topic.

Examples of Edmond Becquerel's coloured daguerreotypes still survive at the Conservatoire des Arts et Métiers in Paris and at the Science Museum in London, and hand-tinted engravings made by Becquerel from his daguerreotype plates are also reproduced in his *La lumière, ses causes et ses effets*, which appeared in 1867. The work of Louis Ducos du Hauron is summarized in his book *Les couleurs en photographie*, published in 1869; in it he made a specific proposal for producing a coloured print by a subtractive process.

There are many books that deal with other specialized aspects of photography. Janet E Buerger, *French daguerreotypes*, University of Chicago Press, 1989, gives an excellent account of the subject, and includes some of the early applications of photography to scientific problems, such as spectroscopy, with numerous illustrations. Also of interest is *Photography in nineteenth century America* (ed. Martha A Sandweiss), Harry N Abrams, New York, 1991.

Applications of photography to spectroscopy are to be found in the books by Janet Buerger and Larry Schaff, mentioned earlier. For the spectroscopic work of Lockyer see A J Meadows, *Science and controversy: a biography of Sir Norman Lockyer*, Macmillan, London, 1972. Besides being an excellent and interesting biography, this book gives a valuable account of spectroscopy in the second half of the nineteenth century.

For an account of Roget's work on the persistence of vision see the only biography that appears to have been written of him: D L Emblen, *Peter Mark Roget; the word and the man*, Thomas Y Crowell, New York, 1970. Surprisingly, there is no biography of Roget in the *DSB*.

For the history of cinematography see, for example, the remarkably detailed *Chronicle of the cinema* (Editor in Chief: Robyn Karney), Dorling Kindersley, London and New York, 1995. The first part of this book gives the historical and technical details of the technique in some detail, with many interesting photographs (for example, of Edison's Black Maria, his Kinetoscope, Fred Ott's sneeze, and the Lumières' first cinematograph).

For more on the remarkable Eadweard Muybridge (including details of his trial for murder) see Kevin MacDonnell, *Eadweard Muybridge: the man who invented the moving picture*, Little, Brown and Company, Boston, 1972.

There are many biographies of Thomas Edison, for example Ronald W Clarke, *Edison: the man who made the future*, 1977; and Neil Baldwin, *Edison: inventing the century*, Hyperion, New York, 1995. The last book is particularly detailed and deals with the technical work clearly.

Work on high speed photography is included in most of the general accounts of photography. Harold E Edgerton wrote several books on his own work, and a good account of his work is to be found in *Stopping time: the photographs of Harold Edgerton* (text by Estelle Jussim, edited by Gus Kayafas), Harry N Abrams, New York, 1987.

A useful and interesting reference book dealing with many modern photoimaging techniques is John C Russ, *The image processing handbook*, CRC Press, Boca Raton, FL, 2nd edition, 1994. To appreciate it requires some background in science, but little mathematics is included, and there are many illustrations of considerable interest.

To understand lasers and holography requires a sound basic knowledge of physics. Readers who lack such a knowledge but are prepared to go to some trouble to acquire it are recommended to read Jeff Hecht, *Understanding lasers: an entry-level guide*, The Institute of Electrical and Electronics Engineers, New York, 1992. This book gives a remarkable clear account of the basic physics required, and of the many aspects and applications of lasers. There are many books on lasers that assume more knowledge of physics; for example, M Bertolotti, *Masers and lasers: an historical approach*, Adam Hilger, Bristol, 1983; and M J Beezley, *Lasers and their applications*, Taylor and Francis, London, 1971; Chapter 1 (pp. 1–19) gives a good historical account of the earlier work, with full references.

A clear account of the basic principles of holography has been given by Emmett N Leith and Juris Upatnieks who, by using lasers, were the first to be successful with the technique: 'Photography by laser', *Scientific American*, **212** (no. 6), 24–35 (June 1965). For more recent work on lasers it is probably best for the general reader to consult scientific encyclopaedias, such as those mentioned on p. 356. Readers with knowledge of physics could consult Nils H Abramson, *The making and evaluation of holograms*, Academic Press, New York, 1981; and P Hariharan, *Optical holography: principles, techniques and applications*, Cambridge University Press, 1984.

CHAPTER 4    MICHAEL FARADAY AND ELECTRIC POWER

For accounts of early work on electricity and electrochemistry see J R Partington, *A history of chemistry*, Macmillan, London, 1964: Vol. 4, Chapter 21, 'Electrochemistry'; G Dubpernel and J H Westbrook (Eds.), *Selected topics in the history of electrochemistry*, The Electrochemical Society, Princeton, 1978; and J T Stock and M V Orna (Eds.), *Electrochemistry, past and present*, American Chemical Society, Washington, DC, 1989. The introductory paper, by J T Stock, 'Electrochemistry in retrospect: an overview', is a useful review of electrochemistry.

For an excellent account of Benjamin Franklin's life and his work on electricity see Carl Van Doren's *Benjamin Franklin*, Viking, New York, 1938; Professor I Bernard Cohen has referred to this as 'possibly the best biography of a scientist in English'. Other excellent accounts of Franklin and his work are I Bernard Cohen, *Benjamin Franklin's experiments*, Harvard University Press, Cambridge, MA, 1941, and *Franklin and Newton*, American Philosophical Society, Philadelphia, 1956. R J Seeger, *Benjamin Franklin: New World physicist*, Pergamon Press, Oxford, 1973, gives an interesting short account of Franklin's life, and reproduces a number of his letters on scientific problems. Oddly, it does not include any of his published papers, or even give a list of them; many of Franklin's papers are based heavily on his letters.

For more details on Benjamin Franklin's motor run by static electricity, see M B Schiffer, 'Franklin invents the motor', *Invention and Technology*, Summer 1994, p. 64. For Thomas Davenport's motor see M B Schiffer, 'The blacksmith's motor', *Invention and Technology*, Winter 1994, p. 64.

A clear account, with full mathematical details, of Ampère's theory of electromagnetism is to be found in R A R Tricker, *Early electromagnetism*, Pergamon Press, Oxford, 1965.

An excellent detailed biography of Faraday is L Pearce Williams, *Michael Faraday*, Basic Books, New York, 1965; softcover reprint, Da Capo Press, 1987. A less detailed but authoritative and very enjoyable account of Faraday and the Royal Institution is to be found in John Meurig Thomas, *Michael Faraday and the Royal Institution; the genius of man and place*, Adam Hilger, Bristol, Philadelphia & New York, 1991. Another interesting account is to be found in D K C Macdonald, *Faraday, Maxwell, and Kelvin*, Doubleday, New York, 1964. Some interesting reminiscences of Faraday appeared in *Nature*, **6**, 412 (1877) and are reproduced in Walter Gratzer's *Bedside Nature* on p. 31.

A book by G Cantor, *Michael Faraday: Sandemanian and scientist*, Macmillan, Basingstoke, 1991, is particularly concerned with the relationship between Faraday's science and his religion. Evidence can be adduced for a whole spectrum of opinion on this matter. Some have argued that Faraday kept his religion and science completely apart—that he was a religious man on Sundays and a scientist on the other days of the week. At the other extreme it has been argued that Faraday's scientific opinions, many of which were unconventional, were largely a result of his unconventional religion. Cantor's book tends to take the latter point of view, and some readers will disagree with him; it is useful, however, to have this point of view laid out. It cannot be doubted that religion influenced Faraday's mode of life and his always kindly treatment of others.

Today we remember the Sandemanian sect mainly because Faraday belonged to it. The name Sandeman is now well known for an entirely different reason—for a fine brand of port wine. The firm of port-wine merchants was founded by George Sandeman, who was the nephew of Robert Sandeman, after whom the sect was named.

Jane Marcet (1785–1858), whose book *Conversations on Chemistry* (2 vols., 1809) so greatly influenced Faraday, was born Jane Haldimand in Geneva. In 1799 she married Alexandre John Gaspard Marcet (1770–1822), who had an MD degree. Both were chemists, and lived in England for some time; he was elected FRS in 1808.

Important accounts and discussions of Faraday's work are to be found in David Gooding and F A J L James (Eds.), *Faraday rediscovered; essays on the life and work of Michael Faraday*, Macmillan, Basingstoke and New York, 1985.

For accounts of Faraday's work on electrolysis see F A J L James, 'Michael Faraday's first law of electrochemistry', in *Electrochemistry, past and present*, pp. 32–49; and L Pearce Williams's biography. For further details about Faraday's electrochemical terms see Sydney Ross, 'Faraday consults the scholars: the origins of the terms of electrochemistry', *Notes and Records of the Royal Society*, **16**, 187–220 (1961) and 'Scientist: the story of a word', *Annals of Science*, **18**, 65–85 (1964). The second of these articles is reproduced, with some additions, in Sydney Ross, *Nineteenth century attitudes: men of science*, Kluwer Academic Publishers, Dortrecht, 1991.

A good outline of Joseph Henry's life and career is to be found in the *DSB* article by N Reingold, **6**, 277–281 (1972). Another excellent account of Henry's scientific work is W F Magie, 'Joseph Henry', *Reviews of Modern Physics*, **3**, 465–495 (1931). For a longer and interesting biography of Henry, with clear explanations of his scientific work, see T Coulson, *Joseph Henry, his life and work*, Princeton University Press, 1950. This book includes an account of Smithson and the founding of the Smithsonian Institution, and of the scientific frauds perpetrated by Robert Keeley in the late nineteenth century. Henry is also well treated in J G Crowther, *Famous American men of science*, Books for Libraries, Freeport, New York, 1937, reprinted 1969. Joseph Henry's first paper in which he described the production of motion from an electromagnet (see Fig. 4.9) is 'On a reciprocating motion produced by magnetic attraction and repulsion', *American Journal of Science* (usually known as *Silliman's Journal*), **20**, 340–343 (1831). Another interesting paper on an early electric motor is M H Jacobi, 'On the application of electromagnetism to the moving of machines', *Annals of Electricity*, **1**, 408–415, 419–444 (1836–1837).

For an interesting account and critical evaluation of the early work on electromagnetic induction, including that of Ampère as well as of Faraday, see Sydney Ross, 'The search for electromagnetic induction', *Notes and Records of the Royal Society*, **20**, 184–219 (1965).

A useful and accurate account of the early development of electromagnets and electric motors, and of the telegraph and telephone, is to be found in W A Atherton, *From compass to computer: a history of electrical and electronic engineering*, Macmillan London and San Francisco, 1984.

Walter Gratzer's *Bedside Nature* contains many interesting references to the telegraph and telephone. The story about Lord Rayleigh's belief that the telephone would not come to much practical use, told by J J Thomson, is on p. 198; Thomson emphasized that Rayleigh was usually better at predicting the uses of inventions. The use of the telephone in surgery for the detection of bullets is mentioned on p. 153 (from a 1915 article). Reference is made to the fact that in 1881· Graham Bell attempted to detect the bullet in President Garfield's body, and appeared to find a series of bullets in a regular pattern. They turned out to be the nodes in the bedspring, and the President died before the matter was sorted out. A surprising use of the telephone in the entertainment field was reported in *Nature*, **22**, 329 (1880) and is on p. 59 of *Bedside Nature*. A choral concert in Zürich was transmitted about 80 kilometres to the telegraph office in Bâle, where a large audience heard the singing 'with perfect fidelity'.

For a discussion of Johnstone Stoney's suggestions about the electron see J G O'Hara, 'George Johnstone Stoney FRS and the concept of the electron', *Notes and Records of the Royal Society*, **29**, 265–276 (1974–1975). This article includes a photograph of Stoney as does the obituary by J Joly, *Proceedings of the Royal Society*, **A, 86**, xx–xxxv (1912).

For a biography of Ferranti and an account of his pioneering work on the distribution of electricity, see Noel Currier-Briggs, *Doctor Ferranti: The life and work of Sebastian Ziani Ferranti*, London, 1970.

CHAPTER 5   JAMES CLERK MAXWELL AND RADIO
TRANSMISSION

The first biography of Maxwell was by Lewis Campbell and William Garnett, *The life of James Clerk Maxwell*, London, 1882; it was reprinted in 1969, with a preface by Robert H Kargon and some additions, by the Johnson Reprint Corporation, New York and London. This book is a goldmine of valuable and interesting information.

Lewis Campbell (1830–1908), a distinguished classical scholar, was a boyhood friend of Maxwell, and was chiefly responsible for the more personal aspects of the book. Garnett (1850–1932) had been a student of Maxwell at Cambridge, and his first research asistant; in the biography he dealt with Maxwell's scientific work, and he collected documentary material for the book. An interesting feature of this book is that although the authors were admirers of Maxwell, they did not fully appreciate his greatness as a scientist. However, the anonymous person who wrote his obituary for the Royal Society (*Proceedings of the Royal Society*, **33**, i–xvi (1882)), who may well have been his friend and admirer P G Tait, did clearly recognize his genius.

Garnett also published a long obituary notice of Maxwell in *Nature*, **21**, 45 (1879); part of it is reproduced on p 57 of Walter Gratzer's *Bedside* **Nature**.

More recent biographies of Maxwell are C W F Everitt, *DSB*, **9**, 198–230 (1974); C W F Everitt, *James Clerk Maxwell, physicist and natural philosopher*, New York, 1975; Ivan Tolstoy, *James Clerk Maxwell: a biography*, University of Chicago Press, 1981; M Goldman, *The demon in the aether: the story of James Clerk Maxwell*, Paul Harris, Edinburgh, 1982; D K C Macdonald, *Faraday, Maxwell and Kelvin*, Doubleday & Co., New York, 1954.

For more on Maxwell's work on photoimaging in colour see Ralph M Evans, 'Maxwell's color photography', *Scientific American*, **205**, 117–128 (November 1961). This article includes a coloured facsimile image of the ribbon. Maxwell's work on kinetic theory, including the investigation of the radiometer, is covered in detail in S G Brush, *The kind of motion that we call heat*, North-Holland Publishing Co., Amsterdam, 1976. This book also deals with Maxwell's work on thermodynamics, as does Elizabeth Garber, 'James Clerk Maxwell and thermodynamics', *American Journal of Physics*, **37**, 146–155 (1969).

Martin J Klein, 'Maxwell, his demon, and the second law of thermodynamics', *American Scientist*, **58**, 84–87 (1970) explains, in an Appendix, the significance of $dp/dt$, which Maxwell sometimes used as his signature. By a somewhat contrived argument Maxwell and Tait used the equation $dp/dt = JCM$ as a way of expressing the second law of thermodynamics; $t$ is temperature, not time. See also two papers by Elizabeth W Garber. 'Clausius and Maxwell's kinetic theory of gases', *Historical Studies in the Physical Sciences*, **2**, 299–319 (1972) and 'Aspects of the introduction of probability into physics', *Centaurus*, **17**, 11–39 (1972). Articles on the demon and related matters are in *Maxwell's demon: entropy, information, computing* (Ed. Harvey S Leff and Andrew F Lex), Princeton University Press, 1990.

Some of Maxwell's original papers are lucid and can be very helpful to the modern reader. Many of them are reproduced in *The scientific papers of James Clerk Maxwell* (Ed. W D Niven, 1890), which was reprinted by Dover Publications in 1965. This publication has two unfortunate weaknesses. Several of Maxwell's quite important papers are omitted, and the reference given to each paper does not give the year of publication. There is an irony here, as Maxwell himself would not have approved at all of omitting dates. In 1931, the centenary of his birth, *Nature* reproduced a letter of his (reproduced in *Bedside* **Nature**, p 215), which began 'How vile are they who quote newspapers, journals and translations by number and vol. of page, instead of the year of grace.... Lockyer [the founder of *Nature* and its Editor for half a century] always alters a letter to NATURE for Sept. 7, 1876 into vol. ?. p ?, as if all promoters of natural knowledge counted everything from the epoch when NATURE first began'. Maxwell was on the whole a remarkably good tempered man, but he seems to have disliked Lockyer, as is suggested by the tone of this letter, his use of the word vile, and the ditty quoted on p. 97.

The biographies of Maxwell mentioned above, by Everitt, Tolstoy, and Goldman, give good non-mathematical accounts of electromagnetic theory. Those wishing a mathematical treatment should consult modern textbooks of physics and not Maxwell's *Treatise*, which is difficult reading. Some of his papers on the theory are quite easy to follow; see, for example, Paper XXXVI of *The scientific papers of James Clerk Maxwell* for a comparison of electrostatic and electromagnetic units. Maxwell's correspondence and unpublished scientific papers, with many plates and diagrams and a detailed commentary, are to be found in P M Harman (Ed.), *The scientific letters and papers of James Clerk Maxwell*, Cambridge University Press, Vol. 1 (1846–1862), 1992; Vol. 2 (1862–1873).

Accounts of the establishment of the wave theory of light are to be found in the histories of science mentioned under Chapter 1. A good summary, with additional references, is to be found in the article on Thomas Young in *DSB*: Edgar W Morse, *DSB*, **14**, 562–572 (1976). For excellent accounts of the work of Fizeau and Foucault see the *DSB* articles on them: J B Gough, *DSB*, **5**, 18–21 (1972) and Harold L Burstyn, *DSB*, **5**, 84–87 (1972).

For an excellent brief account of Hertz's life and work see Russell McCormmach, *DSB*, **6**, 340–350 (1973), The biography–autobiography edited by Hertz's mother (J Hertz) is entitled *Heinrich Hertz: Erinnerungen Briefe, Tagebüche*, and was published in Leipzig in 1927. There are a number of articles dealing with Hertz's work on electrodynamics, such as Peter Heimann, 'Maxwell, Hertz, and the nature of electricity', *Isis*, **62**, 149–157 (1970). The book by P M Harman (Peter Heimann), *Energy, force, and matter: the conceptual development of nineteenth-century physics*, Cambridge University Press, 1982, contains much about the work on electromagnetic theory by both Maxwell and Hertz.

For further details about how Hertz came to adopt Maxwell's point of view on electromagnetic theory see J G O'Haha and W Pricha, *Hertz and the Maxwellians*, Peter Peregrinus Science Museum, London, 1987. The book contains much interesting correspondence between Hertz, Heaviside, Fitzgerald, and Lodge.

For general accounts of the development of radio see W A Atherton, *From compass to computer: a history of electrical and electronic engineering*, Macmillan and San Francisco Press, London and San Francisco, 1984, and Harold I Sharlin, *The making of the electrical age: from the telegraph to automation*, Abelard Schuman, London, New York, Toronto, 1963. Early work on radio is covered in detail in H G J Aitken, *Syntony and spark: the origins of radio*, Wiley, New York, 1976.

It is hard to understand why Sir Oliver Lodge and his great work are so unrecognized and underestimated today. The brief article about him in *DSB*, **8**, 443–444 (1973) does not do him justice; one gains no appreciation of the fact that for many years Lodge was recognized by physicists as a leading figure, and that he was well known to the public particularly for his radio broadcasts. A helpful account of him is Sir Richard Gregory and A Ferguson, 'Oliver Joseph Lodge', *Biographical Memoirs of the Fellows of the Royal Society*, **3**, 551–574 (1941). For a detailed biography see W P Jolly, *Sir Oliver Lodge: a biography*, Constable, London, 1974. Of considerable interest is Lodge's autobiography: *Sir Oliver Lodge, past years: an autobiography*, Hodder and Stoughton, 1931. Also of great interest is his book *Advancing science: being personal reminiscences of the British Association in the nineteenth century*, Ernest Benn, London, 1931; this book includes Lodge's account of his demonstration of radio transmission at the 1894 meeting at Oxford, including the objection raised by Boltzmann to his calling Hertz 'no ordinary German'. Detailed accounts of Lodge's radio work, from different aspects, are in Peter Rowlands and J Patrick Wilson (Eds.), *Oliver Lodge and the invention of radio*, PD Publications, Liverpool, 1994. This is a valuable account of the subject.

Heaviside's work is outlined in Charles Süsskind, *DSB*, **6**, 211–212 (1972), and in 'A R', *Proceedings of the Royal Society*, **A, 110**, xiv-xv (1926). For more on Heaviside see Paul J Nahin, *Oliver Heaviside, sage in solitude*, IEEE Press, New York, 1988.

For accounts of the work of FitzGerald see Oliver Lodge's obituary in *Proceedings of the Royal Society*, **75**, 152–160 (1905), and Joseph Larmor (a fellow Irishman), *Nature*, **63**, 446 (1901); the latter obituary quotes a letter from Oliver Heaviside which is reproduced in *Bedside Nature*, p. 108.

Marconi's work is well covered in Aitken's *Syntony and spark*, Wiley, New York, 1976; and in W P Jolly, *Marconi*, Stein and Day, New York, 1972.

An extremely valuable source of information about the British Association for the Advancement of Science, particularly in its earlier years, is Jack Morrell and Arnold Thackray, *Gentlemen of science: early years of the British Association for the Advancement of Science*, Clarendon Press, Oxford, 1981. This book gives an excellent overview of the state of science in Britain in the nineteenth century.

## CHAPTER 6   J J THOMSON AND THE ELECTRONIC AGE

References to Faraday's work on electrolysis, and Johnstone Stoney's suggestions about the electron, are to be found under Chapter 4.

Otto von Guericke's account of his experiments on the production of a vacuum were published in his book *Experimenta nova (ut vocantur) Magdeburgica de vacuo spatio*, 1672. A translation of relevant parts of this book appear in Martha Ornstein, *The role of scientific societies in the seventeenth century*, University of Chicago Press, 1938. An account of von Guericke's work had appeared as early as 1657, and was probably known to Hooke and Boyle.

The basis of the statement that 16 horses should have been able to separate the Magdeburg hemispheres is as follows. The diameter of the sphere seems from the engraving to be less than 1 metre, but we will take it to be 1 m so that we will be sure to calculate too high a force. A diameter of 1 m gives an area of

$$\pi(0.5)^2 \approx 0.8 \text{ m}^2$$

Atmospheric pressure is roughly $10^5$ Pa $= 10^5$ kg m$^{-1}$ s$^{-2}$. The force corresponding to this is $8 \times 10^4$ m$^2$ s$^{-2}$. The acceleration of gravity is about 10 m s$^{-2}$, so that this force corresponds to a mass of $8 \times 10^3$ kg. Each of the 16 horses thus had to exert a force equivalent to a mass of 500 kg. Since a jerk, rather than a sustained pull, is all that is required, this should have been easy for horses; human weightlifters have accomplished much more under similar circumstances.

The work of Boyle is described in R E W Maddison, *The life of the honourable Robert Boyle, FRS*, Taylor and Francis, London, 1969. This includes diagrams and descriptions of the air pumps designed by Robert Hooke. The discovery of 'Boyle's law' is described in I Bernard Cohen, 'Newton, Hooke, and "Boyle's law" (discovered by Power and Towneley)', *Nature*, **204**, 618–621 (1964). The real discoverers were Richard Towneley (1629–1668) and Henry Power (1623–1668); they communicated their result to Boyle, who asked Hooke to confirm it, and Hooke replied that he had already discovered the law. The book in which Boyle described Hooke's air pump, with a diagram of it and descriptions of their experiments appeared in 1660 with the title *New experiments physico-mechanicall, touching the spring of the air, and its effects: made, for the most part, in a new pneumatical engine*. Boyle did not claim to have discovered Boyle's law himself, but he did much to publicize it.

Maxwell's work on Crookes' radiometer is covered in some detail in S G Brush, *The kind of motion that we call heat*, North-Holland Publishing Co., Amsterdam, 1976.

For an interesting general account of early work relating to the discovery of the electron see B A Morrow, 'On the discovery of the electron', *Journal of Chemical Education*, **46**, 584–588 (1969).

For the work of J J Thomson see J B Morrell's excellent biography, with a number of key references, in *DSB*, **13**, 362–372 (1976). There are several authoritative biographies of Thomson, including Lord Rayleigh, *The life of Sir J J Thomson, OM*, Cambridge University Press, 1943; G P Thomsom (J J's son, and himself a Nobel Prize Winner in physics), *J J Thomson and the Cavendish laboratory in his day*, Cambridge University Press, 1965. For interesting details about J J Thomson see the reminiscences of his son and daughter, George and Joan Thomson, 'J J Thomson as we remember him', *Notes and Records of the Royal Society*, **12**, 201–210 (1957).

Investigations of positively charged particles are described in J H Beynon and R P Morgan, 'The development of mass spectrometry: an historical account', *International Journal of Mass Spectrometry and Ion Physics*, **27**, 1–30 (1978).

A helpful survey of the development of electronics is to be found in W A Atherton, *From compass to computer: a history of electrical and electronics engineering*, Macmillan, London, and San Francisco Press, California 1984. See particularly the section on electronic radio, pp. 194–212. The work of J A Fleming is outlned by Charles Süsskind in *DSB*, **5**, 32–33 (1972), and described in more detail in J T MacGregor-Morris, 'Sir Ambrose Fleming (jubilee of the valve)', *Notes and Records of the Royal Society*, **10**, 134–144 (1955). Fleming's work is also well described in G Shiers, 'The first electron tube', *Scientific American*, **220**, 104–112 (March 1969).

For more on Lee De Forest see Charles Süsskind's item in *DSB*, **4**, 6–7 (1971), and De Forest's autobiography, outrageously and immodestly titled *Father of radio*, Chicago, 1950.

An excellent account of the later developments in electronics, with particular reference to computers, is to be found in *Understanding computers: computer basics*, Time-Life Books, Alexandria, Virginia, 1989.

## CHAPTER 7    THE BRAGGS AND MOLECULAR ARCHITECTURE

Much of the general structural material in the early part of this chapter is to be found in high-school textbooks of chemistry, and in other introductory books on the subject. A particularly accurate and interesting source of information is W H Brock's *Fontana* (or *Norton*) *history of chemistry*, Fontana Press, London, and W W Norton, New York, 1993.

The fact that Archibald Scott Couper rather than Kékulé was first responsible for the correct structures of a number of organic compounds was discovered in 1929 by Richard Anschütz. While writing his biography of Kékulé, *August Kékulé*, Verlag Chemie, Berlin, 1929, Anschütz carried out some detective work which led him to an important 1858 publication by Couper. For further details see *Chemistry in Britain*, February 1993, p. 126, and Alfred Bader, 'A chemist turns detective', *Chemistry in Britain*, September 1996, p. 41.

For a readable and accurate account of Röntgen and the discovery of X-rays see Graham Farmelo, 'The discovery of X rays', *Scientific American*, November 1995, pp. 86–91. Abraham Pais also gives a good account of the man and his work, particularly on pp. 35–42 of his *Inward bound: of matter and forces in the physical world*, Oxford University Press, 1986. For a full-length biography see Otto Glasser, *Wilhelm Conrad Röntgen and the early history of the Röntgen rays*, C C Thomas, Springfield, IL, 1945, reprinted by Norman Publishing Co., 1993.

The X-ray photograph supplied by Walther Nernst, possibly of his own hand, was published in *Nature*, **53**, 324 (1896), and is reproduced in *Bedside Nature*, p. 101. On the same page of *Bedside Nature* is a note which mentions Edison and points out that X-rays may be injurious to the eye. The detective novel mentioned in the text is R Austin Freeman, *The eye of Osiris*, Hodder and Stoughton, London, 1911. The US edition is entitled *The vanishing man*, Dodd, Mead & Co., New York, 1912; it was later reprinted in the S S Van Dine Detective Library, Scribner's Sons, New York, 1929. Chapter 18 of the book gives a detailed account of how an X-ray photograph was taken in those days.

An excellent source of information on X-ray diffraction is *Fifty years of X-ray diffraction*, International Union of Crystallography, Utrecht, The Netherlands 1962. This book was edited by Paul Peter Ewald (1888–1985), a distinguished crystallographer, and was issued on the occasion of the Commemoration Meeting in Munich of the International Union, held in July 1962. Although written primarily for experts, this book contains much that is easily understood by the non-scientist, everything being explained in non-mathematical terms. There are biographies of many distinguished crystallographers, and articles by many others on their early work. There are some excellent diagrams.

Sir David Phillips's biography of William Lawrence Bragg (*Biographical Memoirs of Fellows of the Royal Society*, **25**, 75–143 (1979)), besides being an excellent biography, gives an admirable and authoritative summary of the development of X-ray crystallography. Much the same is true of Dorothy Hodgkin's biography of J D Bernal, *Biographical Memoirs of Fellows of the Royal Society*, **26**, 17–84 (1980). Other biographies that are well worth reading for the insight they give into the X-ray work are Henry S Lipson, 'The fifth Bragg lecture: W L Bragg—scientific revolutionary', *Journal of Chemical Education*, **60**, 405–407 (1983); Dorothy Hodgkin, 'Kathleen Lonsdale', *Biographical Memoirs of Fellows of the Royal Society*, **21**, 447–489 (1975); S Arnott, 'John Monteath Robertson', *Biographical Memoirs of Fellows of the Royal Society*, **39**, 349–362 (1994); and J D Bernal, 'William Thomas Astbury', *Biographical Memoirs of Fellows of the Royal Society*, **9**, 1–35 (1963). Being written by experts, these accounts are of outstanding scientific quality. The summaries of the work of the Braggs, Astbury, Kathleen Lonsdale, and J D Bernal, in the *DSB*, are also useful and reliable, and give references to additional material.

An excellent account of early work on the structure of proteins, as established by X-ray crystallography, is to be found in the earlier part of John C Kendrew's *The thread of life: an introduction to molecular biology*, Harvard University Press, 1966; this book is based on a series of lectures given by Kendrew and organized by the British Broadcasting Corporation. More specialized accounts are to be found in Sir Lawrence Bragg, 'First stages in the X-ray analysis of proteins', *Reports on Progress in Physics*, **28**, 1–14 (1965) and Max Perutz, 'Early days of protein crystallography,' *Methods in Enzymology*, **114** , 3–18 (1985). An excellent but advanced account of the structure of proteins and of their applications in medicine is Max Perutz, *Protein structure: new approaches to disease and therapy*, W H Freeman, New York, 1992.

Henry Armstrong's letter 'Poor common salt', deriding Bragg's structure, is in *Nature*, **120**, 478 (1927), and is reproduced in *Bedside **Nature***, p. 202.

Several biographies of Linus Pauling have been published. An excellent one is Thomas Hager, *Forces of nature: the life of Linus Pauling*, Simon and Schuster, 1995; this is comprehensive (over 700 pages), and Pauling's scientific work is clearly and accurately presented. A much shorter book (136 pages), in which the scientific aspects are explained very clearly and accurately, is David E. Newton, *Linus Pauling: Scientist and Advocate*, Facts on File, New York, 1994. A readable account of Pauling's life and work is Ted and Ben Goetzel (with the assistance of Mildred and Victor Goetzel), *Linus Pauling: a life in science and politics*, Basic Books, New York, 1995. Unfortunately the book contains a few obvious scientific errors.

Watson and Crick's paper on the double-helical structure of DNA was in *Nature*, **171**, 737 (1953), and is reproduced in *Bedside Nature* (the last item) on p. 249. Reference is there given to the Pauling and Corey paper suggesting the triple helix.

The first book on the double helix structure of DNA was J D Watson, *The double helix: a personal account of the discovery of the structure of DNA*, Atheneum, New York, 1968. Comments on this book have been made in the text. It is worth reading if only for its entertainment value (I admit to having read it several times), but the reader must take it with a considerable grain of salt; in order to create a humorous effect (and to sell more copies of the book) the author takes liberties with the facts. The book must not be regarded as a true account of how that particular piece of scientific work was done, or as giving a reliable impression of how any scientific work is done. The book is based largely on Watson's first impressions, and some of its statements are quite outrageous. His comment on Sir Lawrence Bragg was 'I quietly concluded that the white-mustached figure of Bragg spent most of its days sitting in London clubs like the Athenaeum'. This refers to a man who was doing, and continued to do for many more years, highly original research, while at the same time running the Cavendish Laboratory with great success. It says much for Bragg's sense of humour and his generosity that after reading the book he wrote a pleasant Foreword in which he expressed some approval—adding, however, that those who figure in it must read it in a forgiving spirit.

A much more accurate, but brief, account of the Watson–Crick work is included in Francis Crick's *What mad pursuit: a personal view of scientific discovery*, Basic Books, New York, 1988.

For authoritative historical accounts of the DNA work see Robert Olby, *The path to the double helix*, Macmillan, London, 1974, H F Judson, *The eighth day of creation*, Simon and Schuster, New York, 1979, and D A Chambers (Ed.), *DNA: The double helix: perspective and prospective at forty years*, New York Academy of Sciences, New York, 1995. There is also a film dealing with the subject, entitled *Life story* in the UK and *Double helix* in the US. It is enjoyable to watch, and in Crick's opinion it is reasonably accurate as to the facts.

Rosalind Franklin's role in the DNA research is well presented, in an easily understood way, by Sharon Bertsch McGrayne in *Nobel Prize women in science*, Birch Lane Press, New York, 1993; the book includes a few distinguished women who for one reason or another were not successful in receiving the award. A biography of Rosalind Franklin, not improved by its taking a rather blatantly and quite unnecessary feminist position, has been written by Anne Sayre, *Rosalind Franklin and DNA*, W W Norton, New York, 1975. It is hard to believe, in view of the successes of Kathleen Lonsdale and Dorothy Hodgkin, that Rosalind Franklin suffered in any way from the fact that she was a woman. From all accounts she was acerbic in manner, and her colleagues found her difficult to work with. Her research is discussed in some scientific detail by Aaron Klug, 'Rosalind Franklin and the discovery of the structure of DNA', *Nature*, **219**, 808–810, 843–845 (August 24, 1968), and 'Rosalind Franklin and the double helix', *Nature*, **248**, 787–788 (April 26, 1974). A balanced account of Rosalind Franklin and her work has been given by her sister Jenifer Glynn, 'Rosalind Franklin, 1920–1958', in *Cambridge women: twelve portraits* (Ed. Edward Shils and Carmen Blacker), Cambridge University Press, 1966.

There are several good biographies of Charles Darwin which put his work into perspective. The biography often regarded as definitive is by Adrian Desmond and James Moore, *Darwin: the life of a tormented evolutionist*, Michael Joseph, London, 1991; Norton Paperback 1994. This book is detailed (over 700 pages), accurate, and interesting to read. A shorter account is by Michael White and John Gribbin, *Darwin: a life in science*, Simon and Schuster, London, 1995; this again is readable, and has the excellent feature of relating Darwin's work to the recent developments in molecular genetics. Also interesting and well worth reading for an understanding of evolution is Jonathan Weiner, *The beak of the finch*, Jonathan Cape, London, 1994.

For an excellent and highly readable account of Huxley's life and work (up to 1870) see Adrian Desmond, *Huxley: the devil's disciple*, Michael Joseph, London, 1994.

There are a number of accounts of the famous Huxley–Wilberforce debate, with some inconsistencies between them as far as details are concerned. An interesting account is in Chapter 1 of William Irving, *Apes, angels, and Victorians: the story of Darwin, Huxley, and evolution*, McGraw-Hill, New York, 1955. This book is well worth reading for its overview of the early history of Darwinian theory and its general insights; it does, however, go wrong on a few details (one of them is mentioned below). For other commentaries on the debate see J R Lucas, 'Wilberforce and Huxley: a legendary encounter', *Historical Journal*, **22**, 313–340 (1977), and J V Jensen, 'Return to the Wilberforce–Huxley debate', *British Journal for the History of Science*, **21**, 161–179 (1988).

The wording of the remarks made by Wilberforce and Huxley that is given in this chapter is taken from *The autobiography of Charles Darwin and selected letters*, Ed. Francis Darwin, John Murray, London, 1887. The account is second hand, as Darwin was not present, but he was well informed by several of his friends. One who was present at the meeting was Michael Foster (1836–1907), who later built up the Cambridge School of Physiology. In a memoir after Huxley's death in 1895 Foster has interesting comments about the mood of the meeting: *Nature*, **52**, 318 (1895), re-produced in *Bedside **Nature***, p. 99. Another account, by a woman who was present, published in *Macmillan's Magazine*, 1898, with the title 'Reminiscences of a grand-mother', is reproduced in *The Oxford book of Oxford* (Ed. Jan Morris), Oxford University Press, 1978, and ends with the passage

> 'He [Huxley] was not ashamed to have a monkey for his ancestor; but he would be ashamed to be connected with a man who used great gifts to obscure the truth. No one doubted his meaning, and the effect was tremendous. One lady fainted and had to be carried out; I, for one, jumped out of my seat.'

The first account of the debate was written on 2 July 1860 (two days after the Saturday meeting) by the famous botanist Joseph Dalton Hooker (1817–1911) in a letter to Darwin; it is reproduced in *The correspondence of Charles Darwin*, Volume 8, 1860, Cambridge University Press, 1993. The main interest today in this rather intemperately-worded account is that it is so unreliable, prejudiced, and self-serving. Hooker's reference to the distinguished American scientist John Draper is rather regrettable: 'A paper of a yankee donkey called Draper on 'Civilization according to the Darwinian hypothesis' or some such title was being read & it did not mend my temper; for of all the flatulent stuff and all the self sufficient stuffers, it was all a pie of Herbt Spencer & Buckle without the seasoning of either....'. Hooker's account of what happened after 'Sam Oxon's' [the Bishop's] statement was that 'Huxley answered admirably & turned the tables but he could not throw his voice over so large an assembly, nor command the audience; & he did not allude to Sam's weak points nor put the matter in a form or way that carried the audience. The battle waxed hot. Lady Brewster fainted...' Hooker then says that he himself then 'hit him [Sam the Bishop] in the wind at the first shot in 10 words taken from his own ugly mouth..... He continues in similar vein, and leaves the impression that he and not Huxley had won the day for Darwin. At first Darwin was taken in by all this, for in a letter dated 3 July 1860 he says that the Bishop had ridiculed him '& Hooker answered him'; there was no mention in this letter of Huxley. On the same day

Darwin also wrote to Huxley saying 'I fancy from what Hooker says he must have answered the Bishop well', but not mentioning Huxley's own reply; this must have surprised Huxley. All other accounts differ from Hooker's, who was obviously playing up to Darwin in a way that is not very creditable to him. Hooker undoubtedly did make a lengthy and reasoned reply to the Bishop, but it obviously did not have the same impact on the audience as Huxley's reply.

Huxley's letter to Darwin 'giving so long an account of the Oxford doings' (replied to by Darwin on 5 July) has unfortunately not survived. Darwin wrote in his reply 'How durst you attack a live Bishop in that fashion? I am quite ashamed of you! Have you no reverence for fine lawn sleeves? By Jove, you seem to have done it well...God Bless you...get well, be idle & always reverence a Bishop.' There is no mention of Hooker in Darwin's reply; perhaps this was tact, or perhaps he had heard from others that Huxley had made the more important attack.

Draper also fares badly in more recent accounts of the debate. In White and Gribbin's *Darwin*, Draper is not mentioned by name, but is referred to as a 'little-known American'! They also say that Huxley and Hooker regarded Draper's ideas as quackery. In Cyril Bibby, *T H Huxley*, Watts, London, 1959, we read simply that 'Draper droned on...'. In Irvine's *Apes, angels and Victorians* the reference to Draper is quite scathing; although an American himself, Irvine even makes fun of Draper's American accent, and his pronunciation of 'mawnkey'. What is odd about this is that Draper (then aged 49) was born and brought up in St Helens, Lancashire, and educated at London University, emigrating at the age of 21. It is unlikely that he had such a strong American accent; could it possibly be that 'mawnkey' related to a Lancashire accent? The definitive biography of Darwin by Desmond and Moore does get this right; they too comment on the fact that Draper was English born and educated.

Admittedly Draper's book *History of the conflict between religion and science*, Appleton, New York, 1874, is rather boring and has little relevance today. Draper's intention at the 1860 meeting, according to the Athenaeum report (reproduced in *The correspondence of Charles Darwin*), was partly to defend science against the attacks from religious people. It may well be that he was boring on that subject, but he was surely entitled to more respect than he has received as far as the 1860 meeting is concerned.

An interesting account and commentary on the Huxley–Wilberforce confrontation, by an Anglican clergyman, is on pp. 161–163 of Desmond Bowen, *The idea of the Victorian church: a study of the Church of England, 1833–1889*, McGill University Press, Montreal, 1968. A helpful discussion of one aspect of the general problem of the relationship between science and religion is by Frank M Turner, 'The Victorian conflict between science and religion: a professional dimension', *Isis*, **69**, 356–367 (1978).

Trollope's comment about women not fainting is in *Phineas redux* (1874): 'Women do not faint under such shocks'. He was referring to the shock received by Lady Laura Kennedy on hearing that her dear friend Phineas Finn had been arrested for a particularly brutal murder. Surely that was a greater shock than hearing a bishop criticized; Lady Laura in fact collapsed, but Trollope assures us that she was merely pretending to faint.

The sermon by the Revd Frederick Temple, preached in St Mary's Church, Oxford, on Sunday 1 July 1860, was published in 1860 in *Christian Remembrancer*, Vol. **38**, p. 244, and as a small book with the title *The present relations of science to religion: a sermon*.

At the time of the Huxley-Wilberforce debate Dodgson ('Lewis Carroll') was not yet ordained. Like Harcourt he was a Student (i.e., Fellow) of Christ Church, Oxford, and on being appointed to that position he had assumed that he was not required to be ordained, the university reforms having been instituted. However the Dean of Christ Church, Dr Liddell (Alice's father), thought otherwise, and Dodgson reluctantly agreed. He arranged a compromise with Bishop Wilberforce that he would be ordained a deacon only, and never be ordained a priest. One reason for this was that he disliked preaching, having a slight stammer, and in fact he rarely preached. Also, the bishop had strong views on the theatre, believing that clergymen should not attend plays. Dodgson, on the other hand, loved the theatre, and was especially attached to the actress Ellen Terry; he frequently went to plays in London. Being somewhat independent on these matters, he interpreted the Bishop's ruling as applying to priests and not to deacons. The Bishop appears to have made no objection; perhaps he did not know.

Dodgson was thus in disagreement with his bishop on two matters, the theatre and the theory of evolution. There is an ironic aspect to Dodgson's acceptance of evolution. He was later convinced by arguments made by the eccentric biologist St George Jackson Mivart (1827–1900). As mentioned in the text, Mivart was a convert to the Roman Catholic Church, and for his writings on natural selection he was excommunicated and his books placed on the *Index*.

Several books by Richard Dawkins explain evolution with great clarity, but do not include much about molecular genetics:

*The selfish gene*, Oxford University Press, 1976, Oxford Paperbacks, 1978, 1989.

*The extended phenotype: the long reach of the gene*, Oxford University Press, 1982, Oxford Paperbacks, 1989,

*The blind watchmaker*, Longmans, London, 1986; Penguin Books, London, 1988; reprinted with an Appendix, 1991.

*River out of Eden*, Weidenfeld and Nicholson, London, 1995.

*Climbing Mount Improbable*, W W Norton, New York, 1996.

As discussed in the text, I find it regrettable that Dawkins, sometimes called 'Britain's leading atheist', seems to be reviving Huxley's attacks on religion.

The article by Einstein was entitled 'Science and religion', and was in *Nature*, **146**, 605 (1940), reproduced in *Bedside Nature*, p. 232.


## CHAPTER 8    PLANCK, EINSTEIN, THE QUANTUM THEORY, AND RELATIVITY

A number of books deal with the old quantum theory and quantum mechanics. A highly readable and entertaining account is G Gamow, *Thirty years that shook physics, the story of quantum theory*, Doubleday, New York, 1966; Dover Publications, New York, 1985. Also of great interest are B Hoffmann, *The strange story of the quantum*, Harper, New York, 1947; Dover Publications, New York, 1959; and B L Cline, *Men who made a new physics*, (1987). For an excellent biographical account of Max Planck and his work see J L Heilbron, *The dilemmas of an upright man; Max Planck as spokesman for German physics*, University of California Press, 1986.

At a more advanced level is W H Cropper, *The quantum physicists and an introduction to their physics*, Oxford University Press, 1970.

Several books by A Pais are authoritative and full of interest. One deals particularly with Einstein and his work: *Subtle is the Lord: the science and the life of Albert Einstein*, Clarendon Press, Oxford, 1982. A general account of modern physics is his *Inward bound: of matter and forces in the physical world*, Clarendon Press, Oxford, 1986.

An excellent account of Bohr's theory of the atom is to be found in Blanca L Haendler, 'Presenting the Bohr atom', *Journal of Chemical Education*, **59**, 372–376

(1982). For a more detailed account of Bohr's entire career see A Pais, *Niels Bohr's times, in physics, philosophy, and polity*, Clarendon Press, Oxford, 1991.

Some detailed biographies have been written of the pioneers of quantum mechanics. David C Cassidy's *Uncertainty: the life and science of Werner Heisenberg*, W H Freeman, New York, 1991, is of great interest; it analyses in some detail Heisenberg's ambiguous and controversial relationship with the Nazi regime.

There are many biographies of Einstein and accounts of his work, particularly on relativity. That theory, and its consequences, are clearly treated in Clifford M Will, *Was Einstein right? Putting relativity to the test* (paperback), Basic Books, New York, 1986. An interesting biography of Einstein, dealing more with personal aspects and including references to many previous publications, is Roger Highfield and Paul Carter, *The private lives of Albert Einstein*, St Martin's Press, New York, 1993. Other biographies that are interesting and explain the physics well include Michael White and John Gribbon, *Einstein: a life in science*, Simon and Schuster, London, 1993, paperback edition, 1994; and Denis Brian, *Einstein: a life*, Wiley, New York, 1996.

For an authoritative discussion of the differences of opinion between Einstein and Bohr on the significance of quantum mechanics and the uncertainty principle see Andrew Whittaker, *Einstein, Bohr and the quantum dilemma*, Cambridge University Press, 1996.

For general accounts of chemical bonding see C A Russell, *The history of valency*, Leicester University Press, 1971, and William H Brock's *History of chemistry*, Fontana Press, London, 1992 and W W Norton & Co, New York, 1993. For the work of G N Lewis and Linus Pauling on the chemical bond see the interesting account in W Servos, *Physical chemistry from Ostwald to Pauling: the making of a science in America*, Princeton University Press, 1990. Aspects of work on the chemical bond are also included in Mary Jo Nye, *From chemical philosophy to theoretical chemistry: dynamics of matter and dynamics of disciplines (1800–1950)*, University of California Press, 1993.

A readable and accurate account of nuclear physics and the development of the atomic bomb was given in a small book by E N da C Andrade, *The atom and its energy*, B Bell and Sons, London, 1947. This book is intended for the general reader. Interesting details regarding the construction of the atomic bombs are to be found in Pais' *Inward bound*. Information about German scientists during the Nazi regime are to be found in A D Beyerschen, *Scientists under Hitler: politics and the physics community in the Third Reich*, Yale University Press, 1977.

For fascinating information about the German work on the atomic bomb, and the reaction of German scientists to the Allied success in producing the bomb, see Jeremy Bernstein, *Hitler's uranium club: the secret recordings at Farm Hall*, Oxford University Press, 1996.

CHAPTER 9    SCIENTISTS, SCIENCE, AND SOCIETY

Some of the themes in this chapter are discussed in Martin Moskovits (Editor), *Science and society: the John C. Polanyi Nobel Laureate lectures*, House of Anansi Press, Toronto, 1995. For example, on pp. 31–42 C H Townes gives several examples in which great scientists ridiculed the possibility of what later turned out to be important practical consequences of their work. On pp. 97–106, H W Kendall deals with the important problem of future supplies of materials and energy in the light of the rapid increase in the world's population.

The early controversy regarding the temperature dependence of the rates of chemical reactions is covered in S R Logan, 'The origin and status of the Arrhenius equation', *Journal of Chemical Education*, **59**, 279–281 (1982), and K J Laidler, 'The development of the Arrhenius equation', *Journal of Chemical Education*, **61**, 494–498 (1984). The latter paper discusses in some detail the Harcourt-Esson equation, and why, although it was empirically the best equation, is was not the most useful.

The radiation hypothesis in chemical kinetics is discussed in detail in M Christine King, 'Chemical kinetics and the radiation hypothesis', *Archive for History of Exact Sciences*, **30**, 45–86 (1984) and in Mary Jo Nye, *From chemical philosophy to theoretical chemistry: dynamics of matter and dynamics of disciplines, 1800–1950*, University of California Press, 1993, especially pages 121–129.

For an interesting account of the ideas of Bernal and other left-wing scientists see Gary Werskey, *The visible college: a collective biography of British scientists and socialists of the 1930s*, Allan Lane, London, 1978.

J D Bernal's *The social function of science* (1939) is discussed by a number of persons, from several points of view in Maurice Goldsmith and Alan Mackay (Eds.), *The science of science*, Souvenir Press, London, 1964. Some of Bernal's ideas are clearly expressed in his memoir concerning Astbury: *Biographical Memoirs of Fellows of the Royal Society*, **9**, 1–35 (1963).

A useful and interesting account of some of the more unfortunate consequences of technology is given in Edward Tenner, *Why things bite back: technology and the revenge of unintended consequences*, Alfred A Knopf, New York, 1996. Sociological difficulties arising from the increasing use of computers and communications networks are well documented and discussed in Heather Menzies, *Whose brave new world? The information highway and the new economy*, Between the Lines, Toronto, 1996.

Some of the negative attitudes to science based on ignorance are well discussed in Eric S. Grace, *Biotechnology Unzipped; Promises and Realities*, Trifolium Books, Toronto, 1997.

The quotation at the end of this chapter is from the chorus in Act 1 of T S Eliot's pageant play *The Rock* (1934).

# *Index*